Parametrized Relativistic Quantum Theory

Fundamental Theories of Physics

An International Book Series on The Fundamental Theories of Physics: Their Clarification, Development and Application

Volume 56

Parametrized Relativistic Quantum Theory

by

John R. Fanchi
Littleton, Colorado, U.S.A.

KLUWER ACADEMIC PUBLISHERS
DORDRECHT / BOSTON / LONDON

Library of Congress Cataloging-in-Publication Data

Fanchi, John R.
Parametrized relativistic quantum theory / by John R. Fanchi.
p. cm. -- (Fundamental theories of physics ; v. 56)
Includes index.
ISBN 0-7923-2376-9 (alk. paper)
1. Relativistic quantum theory. I. Title. II. Series.
QC174.24.R4F36 1993
530.1'2--dc20 93-22797

ISBN 0-7923-2376-9

Published by Kluwer Academic Publishers,
P.O. Box 17, 3300 AA Dordrecht, The Netherlands.

Kluwer Academic Publishers incorporates
the publishing programmes of
D. Reidel, Martinus Nijhoff, Dr W. Junk and MTP Press.

Sold and distributed in the U.S.A. and Canada
by Kluwer Academic Publishers,
101 Philip Drive, Norwell, MA 02061, U.S.A.

In all other countries, sold and distributed
by Kluwer Academic Publishers Group,
P.O. Box 322, 3300 AH Dordrecht, The Netherlands.

Printed on acid-free paper

Printed in the Netherlands

To my beloved wife Kathy,

for encouraging me to remember

the important things in life.

TABLE OF CONTENTS

PART II: FOUNDATIONS OF PARAMETRIZED RELATIVISTIC QUANTUM MECHANICS

PART III: APPLICATIONS OF PARAMETRIZED RELATIVISTIC QUANTUM MECHANICS

PART IV: INTRODUCTION TO TACHYON PHYSICS

PREFACE

Over the past five decades researchers have sought to develop a new framework that would resolve the anomalies attributable to a patchwork formulation of relativistic quantum mechanics. This book chronicles the development of a new paradigm for describing relativistic quantum phenomena. What makes the new paradigm unique is its inclusion of a physically measurable, invariant evolution parameter. The resulting theory has been sufficiently well developed in the refereed literature that it is now possible to present a synthesis of its ideas and techniques. My synthesis is intended to encourage and enhance future research, and is presented in six parts.

The environment within which the conventional paradigm exists is described in the Introduction. Part I eases the mainstream reader into the ideas of the new paradigm by providing the reader with a discussion that should look very familiar, but contains subtle nuances. Indeed, I try to provide the mainstream reader with familiar "landmarks" throughout the text. This is possible because the new paradigm contains the conventional paradigm as a subset. The foundation of the new paradigm is presented in Part II, followed by numerous applications in the remaining three parts. The reader should notice that the new paradigm handles not only the broad class of problems typically dealt with in conventional relativistic quantum theory, but also contains fertile research areas for both experimentalists and theorists. To avoid developing a theoretical framework without physical validity, numerous comparisons between theory and experiment are provided, and several predictions are made.

I view this book as a snapshot of a work-in-progress. As an independent researcher, my work has been performed without the support of an academic or research institution. Although many would view this as a handicap, I believe my independence has one significant advantage: I have had virtually complete freedom to pursue my scientific interests. With this freedom came the opportunity to explore wide vistas in a search for equally broad "truths". These "truths" are presented as scientific principles and examined in the text.

I would be very pleased to find that the principles elucidated here are the "last word", yet I am vividly aware that history is replete with examples of scientific "truths" being upset by pesky anomalies. We must remember that nature is the ultimate judge of our attempts to understand the universe, and for this judgement we must turn to experiment. I learned great respect for the experimenters' craft when I spent five years as an experimentalist at the Universities of Denver and Mississippi. I have tried to convey this respect in the following pages with the hope that this book will encourage the design and completion of crucial experiments.

It is my pleasure to acknowledge the people who made this work possible. My wife Katherine F. Fanchi continued to support my work when everyone else turned away. Her willingness to listen and to assist whenever possible have been invaluable. Professor L.P. Horwitz's letters served to inspire me, and Professor Alwyn van der Merwe's willingness to gamble on a maverick shows great courage. Finally, I wish to thank all of you who take the time to consider what I have documented here.

John R. Fanchi
Littleton, Colorado, U.S.A.
April 5, 1993

CHAPTER 1
INTRODUCTION

Modern relativistic quantum theory recognizes two broad symmetry classes: external symmetries which preserve the invariance of physical properties under space-time transformations; and internal symmetries arising from transformations which change the wave function without changing the space-time coordinates [Aitchison and Hey, 1982; Chaichian and Nelipa, 1984]. External symmetries are usually characterized by the Poincaré group. Internal symmetries, which are most important for classifying members of the particle zoo, have been characterized by SU(2) ⊗ U(1) in the electroweak theory; SU(3) in QCD; SU(5) in grand unified theories; and various other groups.

Internal symmetries have received more attention than external symmetries in the modern physics community. Most physicists are content to accept the Poincaré group as the correct group for characterizing the external symmetries of a physical system. By contrast, parametrized relativistic quantum theories (PRQT) include an invariant evolution parameter of the type introduced by Fock [1937] and Stückelberg [1942]. They replace the Poincaré group, labelled by P, with the direct product group P ⊗ T where T is the translation group of a relativistic scalar evolution parameter [Fanchi, 1979].

PRQT was part of the physics mainstream in the late 1940's and early 1950's when quantum field theory was being formulated. It helped guide Feynman [1950], Schwinger [1951] and others in their treatment of propagators and divergences. By the end of the 1950's PRQT was diverging from the mainstream as researchers tried to develop a basis for parametrized theories that could be widely accepted as a paradigm. With a few notable exceptions, PRQT was neglected during most of the 1960's. Research in PRQT enjoyed a resurgence based on fundamentally new formulations appearing in the late 1960's and continuing through the 1970's. The 1980's saw an extension of these formulations as many seemingly disparate theories began to find a common ground. The 1990's promise to be a crucial period for establishing the credibility of PRQT.

In this chapter we briefly sketch the historical development of relativistic quantum mechanics, describe persistent anomalies of relativistic quantum theory, and discuss the role of competing theories. The scope and outline of the text are then provided.

§1.1: HISTORICAL SKETCH

Three of the most important relations of nonrelativistic quantum mechanics (NRQM) were established in the 1920s. They are Schrödinger's equation [Schrödinger, 1926a]:

$$i\hbar\frac{\partial\psi_S}{\partial t} = -\frac{\hbar^2}{2m}\nabla^2\psi_S + V\psi_S \tag{1.1.1}$$

Born's probability density [Born, 1926]:

$$\rho_S = \psi_S^*\psi_S \geq 0 \tag{1.1.2}$$

and the associated normalization condition

$$\int \rho_S d^3x = 1 \tag{1.1.3}$$

Shortly after Schrödinger's first paper on NRQM the following equation appeared [Schrödinger, 1926b; Gordon, 1926; and Klein, 1927]:

$$\left[\frac{\hbar}{i}\frac{\partial}{\partial x_\mu} - \frac{e}{c}A^\mu\right]\left[\frac{\hbar}{i}\frac{\partial}{\partial x^\mu} - \frac{e}{c}A_\mu\right]\Psi = m_0^2c^2\Psi \tag{1.1.4}$$

where m_0 is the particle rest mass and the metric has signature -2 (see §A.1). Originally, Eq. (1.1.4), now known as the Klein-Gordon equation, was thought to be the wave equation for relativistic particles. This belief was short lived.

Combining Eq. (1.1.4) and its complex conjugate yields the continuity equation

$$\frac{\partial j^{\mu}}{\partial x^{\mu}}=0 \tag{1.1.5}$$

where

$$j^{\mu} \equiv \frac{i\hbar}{2m_0}\left(\Psi^*\frac{\partial \Psi}{\partial x_{\mu}} - \Psi\frac{\partial \Psi^*}{\partial x_{\mu}}\right) - \frac{eA^{\mu}}{m_0 c}\Psi * \Psi \tag{1.1.6}$$

A continuity equation having the same form as Eq. (1.1.5) can also be derived from Schrödinger's equation. In an effort to mimic the successful NRQM, people assumed that j^0/c should be interpreted as the relativistic probability density. Problems of interpretation immediately arose because j^0 could have negative values. These problems cast a shadow of doubt on the adequacy of the Klein-Gordon equation as the wave equation for relativistic particles. An alternative was sought.

Dirac's relativistic theory [Dirac, 1928] was published a couple of years after the Klein-Gordon equation appeared. Negative probability densities occur in Dirac's theory also, but the ability of the Dirac equation to accurately describe the electron, including its magnetic moment, quickly gained Dirac's theory acceptance as the proper description of relativistic particles. Despite its successes, the Dirac equation has other problems besides negative probability densities. Some of them are discussed in more detail in §10.1. The Klein-Gordon equation was resurrected only after Pauli and Weisskoph [1934] reinterpreted Eq. (1.1.4) as the field equation for spinless particles in the second quantization formalism. The negative probability density problem was not resolved.

Wave equations for relativistic particles received attention in later years, but usually as background or preparatory work for second quantization, e.g. the Wigner [1939] and Bargmann [1947] studies of the Poincaré group, and the Bhabha study [1945] of relativistic wave equations. The failings of the wave mechanical theories purporting to describe relativistic particles were never eliminated.

Considerable interest in the theoretical description of relativistic particles

Considerable interest in the theoretical description of relativistic particles was generated when Lattes et al. discovered the pion in 1947. Eleven years later Feshbach and Villars [1958] published a review article that discussed the wave mechanical properties of relativistic particles as they were understood by the scientific community prior to 1958. In their two-component formalism, Feshbach and Villars separated the positive and negative energy components. They then associated a positive and negative probability norm with the positive and negative energy components respectively. An alternative formalism, as described by Schweber [1962], assumed that the only physically realizable energies of a free particle were positive. Most physicists view the failure of a single particle, wave mechanical description as justification for proceeding to a multiparticle field theory [e.g. Wu and Hwang, 1991]. These views represent the modern paradigm in the sense of Kuhn [1970] and correspond to his "normal science". To avoid the suggestion that alternative paradigms are "abnormal", I shall refer to widely accepted paradigms as "conventional theories". Alternative paradigms are then "nonconventional theories".

Many nonconventional theories have been proposed in the past three decades, but none has been adopted as a conventional theory. The effort continues, however, because anomalies persist in the conventional theory. A more comprehensive review of the anomalies will demonstrate the limitations of conventional theories and show why a new wave mechanical theory is needed.

§1.2 ANOMALIES OF CONVENTIONAL THEORIES

The first anomaly to be described was also the first to be recognized historically: negative probability densities. If the solution of the Klein-Gordon equation has an exponential time dependence given by exp(iωt), then j^0/c has the form

$$\frac{j^0}{c} = (\hbar\omega - eA^0)\,\Psi^*\Psi \tag{1.2.1}$$

The quantity j^0/c is negative whenever $\hbar\omega$-eA^0 is negative. Thus if j^0/c is identified as a probability density, the difficulty of negative probability densities must be confronted. The concept of a negative probability density was so unconventional when it was introduced that the entire theory was considered inadequate. Only when Pauli and Weisskoph resurrected the Klein-Gordon equation did people begin to reassess the meaning of j^0/c. The two-component formalism described by Feshbach and Villars [1958] is one such attempt.

In the theory of Feshbach and Villars, the second order Klein-Gordon equation is transformed into two first order equations. An attempt is then made to separate the positive and negative energy components of the solution to the Klein-Gordon equation. Once this is done, a positive and negative norm is associated with the positive and negative energy solutions respectively. This procedure only works for weak, slowly varying potentials; otherwise the Hamiltonian, obtained by applying the Foldy-Wouthuysen transformation [1950], will neither converge nor be Hermitian. Therefore the two-component formalism, for which a positive-definite probability density can be defined, is just an approximate formalism.

An alternative is to disregard mathematical custom and accept negative probability densities. When j^0/c is regarded as the probability density, as it is in conventional theories, the expectation value of a quantity such as radius r is defined as

$$\langle r\rangle = \frac{\int r(j^0/c)\,d^3x}{\int (j^0/c)\,d^3x} \tag{1.2.2}$$

Substituting Eq. (1.2.1) into Eq. (1.2.2) gives

$$\langle r\rangle = \frac{\int r\Psi^*\Psi(\hbar\omega - eA^0)\,d^3x}{\int \Psi^*\Psi(\hbar\omega - eA^0)\,d^3x} \tag{1.2.3}$$

Except in specific situations, such as A^0 being independent of (x^1,x^2,x^3), the value of $<r>$ does not agree in the nonrelativistic limit with the corresponding expectation value found using Schrödinger's nonrelativistic quantum mechanics:

$$\langle r \rangle_{NRQM} = \frac{\int r \Psi^* \Psi d^3x}{\int \Psi^* \Psi d^3x} \tag{1.2.4}$$

The problem with Eq. (1.2.3) is the appearance of A^0 in the integrand. For free particles, $A^0 = 0$ and this inconsistency is avoided. In general, however, j^0/c just does not work as a probability density.

Another approach to the problem of interpreting j^0/c is to assert that j^0/c must be multiplied by the particle charge e to get a meaningful quantity. The quantity ej^0/c is then interpreted as the charge density. Thus the continuity equation (1.1.5), upon multiplication by e, represents the conservation of charge. This is certainly a nontrivial approach to the problem for charged particles. For neutral particles such as the neutral pion, multiplying by the charge is nothing more than multiplying by zero. Equation (1.1.5) becomes the trivial identity $0 = 0$. Finding an interpretation of j^0/c that is nontrivial for both charged and neutral particles is a significant task.

The anomalies mentioned above are based on two facts: the existence of positive and negative energy solutions; and the existence of a continuity equation [Eqs. (1.1.5) and (1.1.6)]. These facts are responsible for even more difficulties.

Historically, the concept of negative energy solution was hard to accept. Dirac [1930] made the first attempt to understand the physical meaning of negative energy solutions with his now well-known hole theory. The hole theory, based on the Pauli exclusion principle, was quite suitable for fermions. Unfortunately, there are both conceptual and physical problems with the hole theory. The conceptual problem is that hole theory is a many-particle concept and is properly invoked in many-particle theories only. This complicates the mathematics and reduces conventional theories, which are single particle or at most few particle (such as the H atom problem), to theoretical approximations for describing particles. Field theories purportedly resolve this problem, but field theories work best in scattering problems with electroweak forces. Attempts to apply field theoretic techniques to bound state problems and strong interactions have been plagued by computational difficulties. Kim and Noz [1986] have pointed out that some modern attempts to develop alternative formulations of relativistic quantum theory are motivated in part by the difficulties associated with the application of field theory to bound state and strong interaction problems.

More important than the conceptual problem of hole theory is the physical problem: hole theory does not apply to bosons. In the case of bosons, the exclusion principle does not restrict the number of particles that may occupy a given state. Another physical interpretation of negative energy solutions was needed. It was provided by Stückelberg [1941b] and independently by Feynman [1949].

The Stückelberg-Feynman interpretation asserts that positive energy solutions represent particle propagation forward in time and negative energy solutions represent particle propagation backward in time. This interpretation can be applied to fermions and bosons, and it can refer to a single particle or to many particles. Thus the Stückelberg-Feynman interpretation resolves the difficulty in understanding negative energy solutions if we are willing to believe motion backward through time is physically realistic. The flaw with the interpretation is its ad hoc addition to conventional theories. A more complete theory will incorporate the Stückelberg-Feynman interpretation as an inherent part of its structure.

We previously demonstrated that j^0/c has not been adequately interpreted. This point is further illustrated by the Klein paradox. Klein [1929] used the interpretation of ej^0/c as the charge density to show that the reflection coefficient for an incident relativistic particle scattering from a step potential exceeds unity. Klein's prediction does not have a nonrelativistic counterpart and has not been tested experimentally. The question arises: is the Klein paradox physical, or is it a result of an erroneous theory? This question takes on added significance when it is realized that the Klein paradox does not arise in the parametrized theory presented here.

Another anomaly of conventional theories is the absence of proper time from their equations. Consequently, the equations of motion derived from conventional theories do not relate directly to the classical relativistic equations of motion. Hopes of resolving this inadequacy were partly responsible for the efforts of the early pioneers of parametrized quantum theories [Fanchi, 1993]. The lack of an invariant scalar parameter can be related to the inability of conventional theories, and the field theories built from them, to include mass operators as a fundamental part of the theoretical structure.

The final anomaly to be mentioned here is the inability of conventional single particle theories to describe particle decay. As shown by Eq. (1.1.3), these theories assume probability is conserved in space only. The description of particle decay requires probability densities in time that are not present in conventional theories. Any theory that purports to completely describe relativistic particles is deficient if it cannot model particle decay.

Role of Anomalies in Scientific Revolutions

Attempts to resolve the anomalies of conventional theories have led to the construction of parametrized relativistic quantum theories. These new theories attempt to replace an existing theoretical structure with a new theoretical structure. Such attempts lie at the root of scientific revolutions. Consequently, it is worthwhile to spend a few moments considering what is meant by a "scientific revolution".

Kuhn [1970] argued that science is characterized by belief systems he called "paradigms". A paradigm is an individual's basic set of beliefs regarding a particular aspect of nature. Paradigms have several functions: they determine the structure of knowledge; they help scientist's decide which problems are most important; and they provide a conceptual foundation upon which explanations of natural phenomena are based. The latter two functions play an important role in the day to day activities of what Kuhn calls "normal science".

Practitioners of normal science are involved in testing paradigm predictions. Another goal of normal science is to extend the applicability of a paradigm to new areas. Efforts to reformulate a paradigm into a more aesthetic form are also acceptable activities of normal science. The paradigm of normal science is the standard against which competing paradigms are measured. Normal science is the mainstream of scientific activity: its practitioners receive the largest share of research funding; they control the educational institutions; and their paradigm is the one taught in most textbooks. Scientific revolutions must dethrone this substantial infrastructure. Clearly scientific revolutions are the exception, but they have happened.

Scientific revolutions are characterized by the existence of competing paradigms. Periods of crisis in science do not exist without cause. They are usually due to a continuing failure on the part of the paradigm of normal science to explain important anomalies, such as those discussed above. New paradigms appear as structures which resolve existing anomalies of the mainstream paradigm. If the new paradigm is accepted, it will displace -- but not necessarily eliminate -- the old paradigm. Root-Bernstein [pg. 357, 1989] has pointed out that science does not forget its past. Old paradigms may be discarded, but some can be renewed. Indeed, several sections of this book show how to revise the conventional paradigm to resolve long-standing anomalies, e.g. the Klein paradox in Chapter 9 and particle stability in Chapter 13.

During periods of crisis, says Kuhn, it is not possible to rationally discuss conceptual differences or compare experimental results because competing paradigms may not overlap sufficiently for a common basis of communication to exist. Supporters of the old and new paradigms may not be speaking the same language. Although the form of the words they use are the same, the meanings have changed. The reader trained in the standard paradigm should keep this in mind. Indeed, in the pages to follow I will spend a good deal of time discussing the meaning of otherwise familiar concepts (such as time) because changes, though often subtle, are significant.

Kuhn's work was a pioneering effort to discover the mechanisms behind major changes in humanity's understanding of nature through science. His work contrasts with Karl Popper's notion of falsification. Popper argued that scientific knowledge must be falsifiable. An updated notion of falsification has been proposed by Lakatos [1970].

Lakatos says that scientists compare research programs instead of paradigms. According to Lakatos, a research program contains a hard core of basic assumptions. It includes perceptions of nature and its various constituents. The hard core of a research program also embodies unfalsified, widely accepted theories. In addition to a hard core, the research program has a heuristic part. The heuristic part consists of auxiliary hypotheses that help define the research to be done. From Lakatos' perspective, normal science exists as long as its research program can be extended to new areas, and testable predictions can be made. A research program becomes suspect if it must continually adopt new auxiliary hypotheses to explain persistent anomalies, or incorporate unexpected observations arising from competing research programs. The healthy research program is dubbed progressive, while the embattled research program is viewed as stagnating or regressive.

Many aspects of Lakatos' model have counterparts in Kuhn's model. Kuhn's paradigm, for example, encompasses much of what Lakatos calls the hard core part of the research program, as well as portions of the heuristic part. The feature which distinguishes the two models is Lakatos' assertion that scientists can reject one research program in favor of another as a result of rational considerations or decisive experimental tests.

Kuhn's model sees scientists converting to a new paradigm for irrational or scientifically irrelevant reasons (such as a loss of funding because he or she does not conform). Occasionally a scientist may be converted by rational means, as in Lakatos' model. In an ideal world Lakatos' model of a practicing scientist would be correct. In the real world, Kuhn's model acknowledges human behavior. Indeed, d'Abro [Chap. XIII, 1951] has noted how difficult it has been historically for radical new ideas to take hold. He points out that new discoveries or a crucial experiment are often needed to resolve disputes. Root-Bernstein [1989] has synthesized these views into a set of criteria that play a role in the conversion of a scientist from one paradigm to another. My personal experience has shown that Root-Bernstein's presentation is remarkably accurate.

Historians of science have not agreed on any particular model of a scientific revolution. The important point to note is that normal science is built upon a set of beliefs that is not permanent. Scientific revolutions are evidence that the cherished views of the scientific establishment can be replaced with a new set of equally cherished beliefs by succeeding generations. We need to remember this point when comparing two competing paradigms, especially when one of the paradigms has the authority of the scientific establishment behind it. The presence of persistent anomalies suggests that the paradigm of conventional relativistic quantum physics needs to be changed.

§1.3 SCOPE OF THE TEXT

The objectives of this work may be viewed in two contexts: philosophical and scientific. My scientific objectives are to present a new paradigm of relativistic quantum theory that resolves the persistent anomalies of conventional theories and makes experimentally testable predictions. But scientific objectives are not necessarily sufficient for creating a new paradigm. Werner Heisenberg has noted that "It is only a slight exaggeration to say that good physics has at times been spoiled by poor philosophy" [pg. 32, 1976]. My philosophical motivation for undertaking the work documented below was the desire to answer the question: can free will and determinism coexist?

Quantum physics of the early twentieth century provided the answer "yes" in the following sense. Nonrelativistic quantum mechanics, via the uncertainty principle and the Born interpretation, implied that the evolution of individual particles could not be predetermined on the microscopic level. Yet, on the average, the motion of an ensemble of particles could be accurately predicted. In other words, a 'free will' -- or lack of predestination -- exists in nature on the microscopic level for a single particle while, concurrently, determinism exists in nature for an aggregate of particles. This was a major break from the concept of Newtonian determinism and suggests the following hypothesis: 'free will' and determinism coexist if we regard determinism as the existence of a set of well-defined probabilistic laws that govern a statistically significant population, and 'free will' refers to an inherent uncertainty in predicting the evolution of a physical system. Our philosophical hypothesis may be translated into scientific terms if we make the assumption that the laws of physics stem from conventional probability and then pursue a careful investigation of the evolution of a physical system. The adoption of a philosophical position comes into play when decisions must be made in the construction of a scientific theory. These points are discussed throughout the book.

Relativistic quantum theory with an invariant evolution parameter is well suited for achieving the above objectives. My presentation is based on the view that the invariant evolution parameter is physically significant. Lines of research which view the parameter as a mathematical convenience with negligible physical content are reviewed elsewhere [Fanchi, 1993]. Justification for my perspective will be provided in several different contexts.

OUTLINE OF THE TEXT

A critique of conventional relativistic quantum theory (CRQT) was provided in §1.2. The historical evolution of three fundamental relations was sketched and some anomalies associated with CRQT were presented. The role of anomalies in indicating the need for a paradigm shift was discussed. Theories such as PRQT have been developed in an attempt to resolve the anomalies.

Part I presents parametrized relativistic classical theory. Parametrized special relativistic dynamics is introduced in Chapter 2 and alternative formulations such as parametrized Hamilton-Jacobi theory and path integral concepts appear in Chapter 3. Statistical PRCM is introduced in Chapter 4 with an eye towards a cosmological application in Chapter 5. A parametrized general relativistic basis is constructed so that the relativistic ideal gas equation of state derived in Chapter 4 can be applied to a modern astrophysical problem: calculation of the age of the universe.

Foundations of parametrized relativistic quantum theory are presented in Part II. A presentation of the symmetry principles upon which PRQT rests in Chapter 6 also displays fundamental differences between PRQT and CRQT. The symmetry principles of PRQT include the familiar symmetry principles of CRQT as a subset. One of the strengths of the formulation of PRQT advocated here is its ability to be constructed from a probabilistic basis. Field equations satisfying the symmetry principles and the form allowed by probability theory are constructed in Chapter 7. These field equations have historically appeared as heuristic extensions to the relativistic domain of non-relativistic quantum theory. By contrast, they appear naturally in PRQT. The correspondence between mathematics and physics is presented in Chapter 8. Introductory applications of the field equations to spin-0 and spin-½ particles are presented in Chapter 9 and Chapter 10. Finally the physical meaning of the evolution parameter and the construction of an evolution parameter clock are presented in Chapter 11.

Part III focuses on more sophisticated applications of PRQT with discussion of experimental results whenever possible. One of the strengths of PRQT -- practical applicability to bound state problems -- is introduced in Chapter 12. Models of the most basic bound state problems -- hydrogen-like atoms and the harmonic oscillator model of strong interactions -- are presented. The ability of PRQT to provide a consistent theoretical framework for describing particle stability is demonstrated in Chapter 13. Parameter independent perturbation theory is introduced in Chapter 14. Illustrations of the formalism include the interaction of a charged particle with a magnetic field and the treatment of a charged particle moving in the Coulomb field of an atom or molecule. The latter example illustrates the applicability of PRQT to relativistic chemical physics. Scattering theory and an illustrative application are presented in Chapter 15 and Chapter 16 respectively. A brief introduction to statistical PRQT is contained in Chapter 17.

Parameter dependent perturbation theory and its applicability to mass state transitions and tachyon kinematics are presented in Part IV. It is shown that quantum transitions across the light cone are theoretically possible. The ability of PRQT to make experimentally testable predictions is highlighted.

Parametrized relativistic quantum field theory (PRQFT) is introduced in Part V. The formalism is developed in Chapter 21 and a set of conservation laws is derived in Chapter 22. One new conservation law arises as a result of the appearance of a new operator identified as the mass operator. The mass operator is studied in the context of a simple quark model of mesons.

Standard notation, units, and frequently used acronyms are provided in the Appendix. Since much of the material presented in this book has been published in the refereed literature, my focus is on physical pedogogy with sufficient mathematical rigor to convince a skeptic that the theory is logically consistent. Appendix A.3 lists particularly significant primary references for each chapter, and an extensive bibliography is provided.

PART I:

PARAMETRIZED RELATIVISTIC CLASSICAL MECHANICS

CHAPTER 2
SPECIAL RELATIVISTIC DYNAMICS

Parametrized quantum theories have their origins in extensions of classical mechanics. As a point of reference, Part I is devoted to a presentation of classical relativistic dynamics from a perspective that is compatible with the parametrized quantum formulation introduced in Part II. Our interest in this chapter is to obtain special relativistic results within the context of a parametrized theory.

§2.1 HAMILTON'S VARIATIONAL PRINCIPLE

One of the most straightforward ways to develop a parametrized relativistic classical mechanics (PRCM) is to invoke Hamilton's variational principle [Goldstein, 1980; Barut, 1980]. We begin by defining a parameter-dependent action

$$A = \int_{s_1}^{s_2} L_s(q_1, \dots, q_n, \mathring{q}_1, \dots, \mathring{q}_n, s)\, ds \tag{2.1.1}$$

where the Lagrangian in the integrand is defined in terms of n generalized coordinates $(q_1, \dots, q_n)$ and an invariant parameter s with the unit of time. A dot over a generalized coordinate denotes differentiation with respect to the parameter s (Appendix A.1). The instantaneous configuration of a system is described by the values of the n generalized coordinates $q_1, \dots, q_n$ and corresponds to a point in an n-dimensional configuration space. The axes of the n-dimensional space are formed by the n generalized coordinates. The invariant parameter s traces the evolution of a system point in configuration space between the fixed parameter end points s_1 and s_2. Notice that the evolution parameter s is not a fifth coordinate as in a Kaluza-Klein theory. It is a relativistic scalar that is mathematically independent of the generalized coordinates.

The subscript s in the symbol L_s has been written to remind us that L_s is the Lagrangian of special relativity. As such, L_s should satisfy certain transformation properties. For example, the Lagrangian should be independent of the coordinate system used. Typical symmetry requirements for special relativity include invariance with respect to Lorentz transformations. A broader class of symmetry requirements would be invariance with respect to Poincaré transformations. Poincaré transformations include coordinate translations as well as coordinate rotations.

As the invariant parameter s monotonically increases, the system point moves in configuration space. Hamilton's principle for conservative systems states that the motion of a system point in configuration space from point s_1 to point s_2 must be such that the action is an extremum. A covariant formulation of Hamilton's principle requires that the variation of A must vanish:

$$\delta A = 0 = \int_{s_1}^{s_2} \delta L_s(q_1, \ldots, q_n, \ldots, \mathring{q}_1, \ldots, \mathring{q}_n, s)\, ds \tag{2.1.2}$$

For a single generalized coordinate q, the parametrized Lagrangian has the functional form

$$L_s = L_s(q, \mathring{q}, s) \tag{2.1.3}$$

where

$$\mathring{q} = \frac{dq}{ds}. \tag{2.1.4}$$

Variation of the Lagrangian gives

$$\delta L_s = \frac{\partial L_s}{\partial q} \delta q + \frac{\partial L_s}{\partial \mathring{q}} \delta \mathring{q}. \tag{2.1.5}$$

An integration by parts applied to the last term on the right hand side of Eq. (2.1.5) yields

$$\int_{s_1}^{s_2} \frac{\partial L_s}{\partial \mathring{q}} \delta \left[\frac{dq}{ds} \right] ds = \int_{s_1}^{s_2} \frac{\partial L_s}{\partial \mathring{q}} \frac{d}{ds} (\delta q) ds$$
$$= - \int_{s_1}^{s_2} \left[\frac{d}{ds} \frac{\partial L_s}{\partial \mathring{q}} \right] (\delta q) ds \tag{2.1.6}$$

where the condition that the variation of q vanish at the parameter end points has been invoked. Inserting Eq. (2.1.6) into Eq. (2.1.2) gives

$$\delta A = 0 = \int_{s_1}^{s_2} \left[\frac{\partial L_s}{\partial q} - \frac{d}{ds} \frac{\partial L_s}{\partial \mathring{q}} \right] (\delta q) ds \tag{2.1.7}$$

Equation (2.1.7) is valid for arbitrary variations only if the coefficients of the variations vanish. Consequently, the condition that Eq. (2.1.7) be valid requires that the coefficient of the variation of q vanish, thus

$$\frac{\partial L_s}{\partial q} - \frac{d}{ds} \frac{\partial L_s}{\partial \mathring{q}} = 0. \tag{2.1.8}$$

Equation (2.1.8) is the parametrized Euler-Lagrange equation for a dynamical system with a single generalized coordinate.

§2.2 HAMILTON'S EQUATIONS

A review of the literature shows that the Lagrangian formulation is just one of the approaches used to study parametrized relativistic theories. Another commonly used approach is the Hamiltonian formulation. The Lagrangian formulation is commonly used to study relativistic quantum field theory, while the Hamiltonian formulation is often the starting point for studying relativistic quantum mechanics. The two formulations are related.

A procedure for transforming between the classical Lagrangian and Hamiltonian formulations is demonstrated here for a single particle Lagrangian with the functional form

$$L_0 = L_0(q^\mu, \mathring{q}^\mu, s);\ \mu = 0,1,2,3. \tag{2.2.1}$$

The generalized coordinates are elements of the space-time four-vector. The Euler-Lagrange equations for Eq. (2.2.1) are

$$\frac{d}{ds}\frac{\partial L_0}{\partial \mathring{q}^\mu} - \frac{\partial L_0}{\partial q^\mu} = 0. \tag{2.2.2}$$

Legendre Transformation and Poisson Bracket

Canonical equations can be derived using a variety of methods such as Legendre transformations or differential forms. We first perform the more familiar Legendre transformation

$$K(q^\mu, p_\mu, s) = \mathring{q}^\mu p_\mu - L_0(q^\mu, \mathring{q}^\mu, s) \tag{2.2.3}$$

where K is the Hamiltonian function.

The differential of the Hamiltonian K is calculated directly from the left hand side of Eq. (2.2.3) as

$$dK = \frac{\partial K}{\partial q^\mu} dq^\mu + \frac{\partial K}{\partial p_\mu} dp_\mu + \frac{\partial K}{\partial s} ds. \tag{2.2.4}$$

The differential of the right hand side of Eq. (2.2.3) is

$$dK = \mathring{q}^\mu dp_\mu + p_\mu d\mathring{q}^\mu - \frac{\partial L_0}{\partial q^\mu} dq^\mu - \frac{\partial L_0}{\partial \mathring{q}^\mu} d\mathring{q}^\mu - \frac{\partial L_0}{\partial s} ds. \tag{2.2.5}$$

Equating Eqs. (2.2.4) and (2.2.5) leads to the relationships

$$\frac{\partial K}{\partial s} = -\frac{\partial L_0}{\partial s}, \tag{2.2.6}$$

$$p_\mu = \frac{\partial L_0}{\partial \mathring{q}^\mu}, \tag{2.2.7}$$

$$\mathring{q}^\mu = \frac{\partial K}{\partial p_\mu}, \tag{2.2.8}$$

$$\frac{\partial K}{\partial q^\mu} = -\frac{\partial L_0}{\partial q^\mu}. \tag{2.2.9}$$

Differentiating Eq. (2.2.7) with respect to s gives

$$\mathring{p}_\mu = \frac{d}{ds}\frac{\partial L_0}{\partial \mathring{q}^\mu} = \frac{\partial L_0}{\partial q^\mu} = -\frac{\partial K}{\partial q^\mu} \tag{2.2.10}$$

where Eq. (2.2.2) has been used. The resulting canonical equations are the parametrized Hamilton's equations

$$\mathring{p}_\mu = -\frac{\partial K}{\partial q^\mu}, \tag{2.2.11}$$

$$\mathring{q}^\mu = \frac{\partial K}{\partial p_\mu}. \tag{2.2.12}$$

The rate of change of an invariant function $F(q^\lambda, p^\lambda, s)$ with respect to s may be evaluated using Hamilton's equations. We begin with the expansion

$$\frac{d}{ds}F(q^\lambda, p^\lambda, s) = \frac{\partial F}{\partial s} + \frac{\partial F}{\partial q^\lambda}\frac{dq^\lambda}{ds} + \frac{\partial F}{\partial p^\lambda}\frac{dp^\lambda}{ds}. \tag{2.2.13}$$

Substituting Eqs. (2.2.11) and (2.2.12) into (2.2.13) gives

$$\frac{dF}{ds} = \frac{\partial F}{\partial s} + \{F,K\}, \tag{2.2.14}$$

where {F,K} is the Poisson bracket

$$\{F,K\} \equiv \frac{\partial F}{\partial q^{\lambda}}\frac{\partial K}{\partial p_{\lambda}} - \frac{\partial K}{\partial q_{\lambda}}\frac{\partial F}{\partial p^{\lambda}}. \tag{2.2.15}$$

Differential Forms

Horwitz and Piron [1973] formulated a classical dynamical principle that was a generalization of Hamilton's variational principle. Suppose an N-body system is characterized by the differential 1-form

$$\omega = \sum_{i=1}^{N} [p_{i\mu} dq_i^{\mu} - K(p_i, q_i) ds], \tag{2.2.16}$$

where K is a function of the 8N variables (p_i, q_i) of each particle i=1,...,N. The dynamical principle may be stated as follows [Reuse, 1979]: Given a closed curve C in

$$\Gamma = \{p_i, q_i, s\}, \tag{2.2.17}$$

the integral $\oint_C \omega$ is invariant for any (continuous) deformations of C generated by arbitrary displacements of its points along the trajectories corresponding to the evolution of the system. This principle is equivalent to the canonical equations

$$\frac{dq_i^{\mu}}{ds} = \frac{\partial K}{\partial p_{i\mu}}, \tag{2.2.18}$$

$$\frac{dp_{i\mu}}{ds} = -\frac{\partial K}{\partial q_i^{\mu}}, \tag{2.2.19}$$

where s is the evolution parameter. For more information about differential forms and their relation to mechanics, see Flanders [1989] or Burke [1985]. Horwitz and Piron [1973] called s historical time and interpreted s as a parameter analogous to time in nonrelativistic, Galilean invariant mechanics.

Equations (2.2.18) and (2.2.19) are covariant when the differential 1-form is invariant. Covariance is satisfied when K is a scalar with respect to the Lorentz group and the parameter s is invariant. Horwitz and Piron proposed expressions for K for a free particle, a particle in an electromagnetic field, and the relativistic 2-body problem. These examples are discussed later. Steeb and Miller [1982] presented a more detailed analysis of the classical theory in terms of differential forms, vector fields and the Lie derivative. Their work included a brief derivation of constants of motion.

§2.3 CLASSICAL FREE PARTICLE

As an example of transforming back and forth between the Lagrangian and Hamiltonian formulations, we construct a free particle Lagrangian from its Hamiltonian counterpart. The Hamiltonian for a free particle with rest mass M can be written as

$$K_f = \frac{g^{\mu\nu}}{2M} p_\mu p_\nu . \tag{2.3.1}$$

The state of the particle is characterized by eight independent numbers

$$\begin{aligned} q^\mu &= (q^0, q^1, q^2, q^3) = (ct, \vec{q}), \\ p_\mu &= (p_0, p_1, p_2, p_3) = (E/c, -\vec{p}) \end{aligned} \tag{2.3.2}$$

where c is the speed of light in vacuum. The q's are elements of the space-time four-vector and the p's are elements of the momentum-energy four-vector. Non-zero elements of the metric tensor are

$$g_{00} = 1 = -g_{11} = -g_{22} = -g_{33} . \tag{2.3.3}$$

The mass of the particle is a dynamical constant. Substituting Eq. (2.3.1) into the canonical equations (2.2.11) and (2.2.12) gives

$$\mathring{q}^{\mu} = p^{\mu}/M, \tag{2.3.4}$$

$$\mathring{p}_{\mu} = 0. \tag{2.3.5}$$

The Lagrangian corresponding to Eq. (2.3.1) follows immediately, thus

$$L_f = \frac{g^{\mu\nu}}{2M}(M\mathring{q}_{\mu})(M\mathring{q}_{\nu}) = \frac{M}{2}\mathring{q}^{\nu}\mathring{q}_{\nu}. \tag{2.3.6}$$

Some familiar results of special relativity follow from Eqs. (2.3.4) and (2.3.5). For example, from Eq. (2.3.5) we find that four-momentum is a constant:

$$p_{\mu} = p_{0\mu}. \tag{2.3.7}$$

Substituting Eq. (2.3.7) into (2.3.4) and integrating with respect to s gives

$$q^{\mu} = q_0^{\mu} + \frac{p_0^{\mu} s}{M} \tag{2.3.8}$$

where q_0^{μ} is an integration constant. Equation (2.3.8) shows that the trajectory of a free relativistic particle depends linearly on the evolution parameter s. Varying the space-time four-vector and forming its inner product gives

$$\delta q^{\mu}\delta q_{\mu} = \frac{p_0^{\mu} p_{0\mu}}{M^2}(\delta s)^2. \tag{2.3.9}$$

Taking the initial condition as

$$M^2 = \frac{p_0^{\mu} p_{0\mu}}{c^2} \tag{2.3.10}$$

yields the usual relation between classical invariant "proper time" and the space-time four-vector:

$$\delta q^{\mu}\delta q_{\mu} = c^2(\delta s)^2 \tag{2.3.11}$$

Remember that this "proper time" is the independent evolution parameter s associated with the action (2.1.1). An evaluation of Eq. (2.3.11) from an experimental point of view can convey a deeper understanding of the parameter s and its relationship to the space-time coordinates.

MEASURING THE WORLD-LINE OF A FREE PARTICLE

Suppose an observer at rest in a reference frame R wants to test the validity of Eq. (2.3.11). To do so, the observer must measure the world-line of a noninteracting particle P. A simple experiment is designed in which P is emitted from a source S and passes through two detectors D1 and D2 (Figure 2.1). The arrival times at D1 and D2 are measured by clocks C1 and C2, respectively. This apparatus provides the observer with two sets of measured data. The data sets M1 and M2 are obtained when P is observed first at D1 and then at D2.

FIGURE 2.1

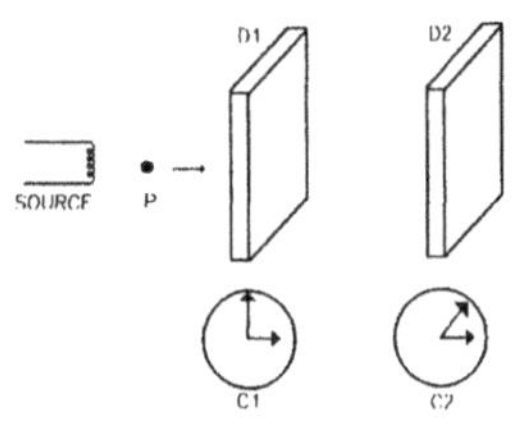

The following positions and arrival times are the result of the measurements:

$$M1: \; x_1 \pm \Delta x_1, \; t_1 \pm \Delta t_1,$$
$$M2: \; x_2 \pm \Delta x_2, \; t_2 \pm \Delta t_2.$$

The uncertainties in position and time (Δx_i and Δt_i), which may represent the width of the detectors or the limit of temporal resolution of the clocks, are explicitly shown because it is important to realize every physical measurement is imprecise to some extent. Aware of these uncertainties, the observer repeats the experiment until statistically reliable averages are obtained, namely $<x_1>$, $<x_2>$, $<t_1>$, $<t_2>$.

The observer now performs the same experiment except for one difference: the detectors and clocks are required to move at a velocity v parallel to the motion of particle P. Thus the new data sets M1' and M2' are what would be observed by an observer in a reference frame R' moving at a velocity v relative to R. The results for R' are

$$M1': x_1' \pm \Delta x_1',\ t_1' \pm \Delta t_1',$$
$$M2': x_2' \pm \Delta x_2',\ t_2' \pm \Delta t_2'.$$

Again the experiment with moving detectors and clocks is repeated until statistically reliable averages $<x'_1>$, $<x'_2>$, $<t'_1>$, $<t'_2>$ are obtained.

FIGURE 2.2

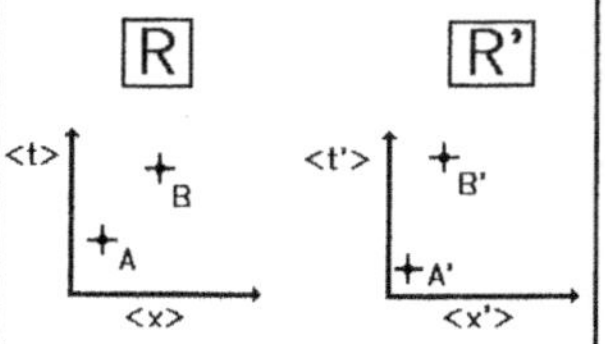

Equipped with the data sets {M1, M2, M1', M2'}, the observer constructs space-time diagrams for each reference frame (Figure 2.2). Notice the axes of these diagrams represent average values, and error bars corresponding to measurement uncertainties are also depicted to emphasize the imperfection of any physical measurement. Returning to the space-time diagrams, the special relativistic relationship Eq. (2.3.11) asserts that the length of the path from A to B should equal the length of the path from A' to B' for the Minkowski metric. This is readily checked with the result

$$c^2 \delta \langle t \rangle^2 - \delta \langle x \rangle^2 = c^2 \delta \langle t' \rangle^2 - \delta \langle x' \rangle^2 \tag{2.3.12}$$

where $\delta<t>=<t_2>-<t_1>$, $\delta<x>=<x_2>-<x_1>$, and similarly for $\delta<t'>$, $\delta<x'>$.

FIGURE 2.3

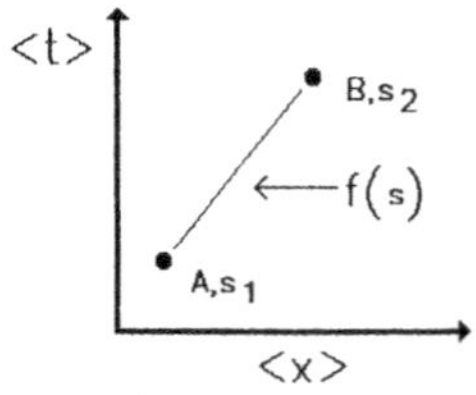

The world-line of P from A to B (or A' to B') can be parametrized by a continuously varying scalar (Figure 2.3). Observe the parameter depends on $<x>$ and $<t>$, not x and t. Therefore the parameter must be a conditioning parameter on the probability distribution used to determine the expectation values $<x>$ and $<t>$. The parameter is in fact just the scalar s, and the difference $\delta s = s_2 - s_1$ is given in the classical limit by

$$\delta s^2 = \delta \langle x^\mu \rangle \delta \langle x_\mu \rangle . \tag{2.3.13}$$

§2.4 CHARGED PARTICLE IN AN ELECTROMAGNETIC FIELD

The Hamiltonian for a particle with electric charge e in an external electromagnetic field described by the four-vector potential A_μ is obtained from Eq. (2.3.1) by introducing the concept of "minimal coupling", i.e.

$$K = \frac{g^{\mu\nu}}{2M}[p_\mu - \frac{e}{c}A_\mu][p_\nu - \frac{e}{c}A_\nu] \tag{2.4.1}$$

where the four-vector potential may depend on q^μ but not on s. Substituting Eq. (2.4.1) into (2.2.11) and (2.2.12) gives the canonical equations

$$\mathring{q}_\mu = \frac{\left[p_\mu - \frac{e}{c}A_\mu\right]}{M} = \frac{\pi_\mu}{M}, \tag{2.4.2}$$

$$\mathring{p}_\mu = \frac{e}{Mc}\left[p^\nu - \frac{e}{c}A^\nu\right]\partial_\mu A_\nu = \frac{e}{Mc}\pi^\nu \partial_\mu A_\nu \tag{2.4.3}$$

and we have defined kinetic momentum π^μ by

$$\pi^\mu = p^\mu - \frac{e}{c}A^\mu. \tag{2.4.4}$$

Pavsic [1991a] obtained the same equations by starting from a suitable Lagrangian. His procedure is outlined in §3.2.

LORENTZ FORCE

The canonical equations may be used to derive the Lorentz force. We begin by substituting Eq. (2.4.2) into (2.4.3) to get

$$\mathring{p}_\mu = \frac{e}{c}\mathring{q}^\nu \partial_\mu A_\nu . \tag{2.4.5}$$

Differentiating Eq. (2.4.4) with respect to s gives

$$\begin{aligned} \mathring{\pi}_\mu &= \mathring{p}_\mu - \frac{e}{c}\frac{dA_\mu}{ds} \\ &= \mathring{p}_\mu - \frac{e}{c}\frac{\partial A_\mu}{\partial q^\nu}\frac{dq^\nu}{ds} \\ &= \mathring{p}_\mu - \frac{e}{c}\,\partial_\nu A_\mu\,\mathring{q}^\nu . \end{aligned} \tag{2.4.6}$$

Rearranging Eq. (2.4.6) and combining with (2.4.5) lets us write

$$\mathring{p}_\mu = \mathring{\pi}_\mu + \frac{e}{c}\,\partial_\nu A_\mu\,\mathring{q}^\nu = \frac{e}{c}\mathring{q}^\nu\,\partial_\mu A_\nu . \tag{2.4.7}$$

Solving for the rate of change of kinetic momentum with respect to s yields the Lorentz force equation

$$\mathring{\pi}_\mu = \frac{e}{c}F_{\mu\nu}\mathring{q}^\nu \tag{2.4.8}$$

where $F_{\mu\nu}$ is the antisymmetric field strength tensor

$$F_{\mu\nu} = \partial_\mu A_\nu - \partial_\nu A_\mu . \tag{2.4.9}$$

Classical Mass

Classical mass is a constant of the motion associated with kinetic momentum [Pavsic, 1991a]. Let us form the inner product of the Lorentz force with π^μ:

$$\mathring{\pi}_\mu \pi^\mu = \frac{1}{2}\frac{d}{ds}(\pi_\mu \pi^\mu) = \frac{e}{c}F_{\mu\nu}\mathring{q}^\nu \pi^\mu . \tag{2.4.10}$$

Using Eq. (2.4.2) in (2.4.10) lets us write

$$\frac{1}{2}\frac{d}{ds}(\pi_\mu \pi^\mu) = \frac{e}{Mc} F_{\mu\nu} \mathring{q}^\nu \mathring{q}^\mu . \tag{2.4.11}$$

Summation over μ,ν in the latter term leads to

$$F_{\mu\nu} \mathring{q}^\nu \mathring{q}^\mu = 0 \tag{2.4.12}$$

since $F_{\mu\nu}$ is an antisymmetric tensor. Equation (2.4.11) becomes

$$\frac{d}{ds}(\pi_\mu \pi^\mu) = 0 \tag{2.4.13}$$

which implies

$$\pi_\mu \pi^\mu = \pi_{0\mu} \pi_0^\mu \equiv \kappa^2 \tag{2.4.14}$$

where κ is a constant of integration with respect to s. To be consistent with the initial condition specified in Eq. (2.3.10) for arbitrary A^μ, including $A^\mu = 0$ for all μ, we must have

$$\kappa^2 = M^2 c^2 . \tag{2.4.15}$$

The constant of the motion corresponds to classical mass.

It is well known [e.g. Barut, 1980] that taking energy and momentum as the primary concepts in Eq. (2.4.14) lets us include M=0 as a viable solution. It corresponds to lightlike particles such as photons. In this case, however, the Hamiltonian K in Eq. (2.4.1) would have to be changed. An acceptable free field Lagrangian is presented in §3.2. The important point to emphasize here is that energy and momentum -- not mass -- are the primary variables. The application of relativistic classical mechanics to physically realistic light like particles requires that mass be viewed as a dependent variable.

CHAPTER 3
ALTERNATIVE FORMULATIONS OF PRCM

There are many different methods for formulating parametrized relativistic classical mechanics. The different approaches are worth considering because they give us fresh perspectives on what a quantum formulation may look like. We are especially interested in Hamilton-Jacobi, variational and path integral treatments. These treatments have been used in the past to provide a better understanding of nonrelativistic quantum mechanics and have served to guide the development of relativistic quantum theory. They play a similar role here.

§3.1 PARAMETRIZED HAMILTON-JACOBI THEORY

Historically Hamilton-Jacobi theory has served as a classical stepping stone to wave mechanics. Borowitz [1967] gives a good discussion of the link between nonrelativistic Hamilton-Jacobi theory and Schrödinger wave mechanics. By surmising that the Hamilton-Jacobi equation was the short wavelength limit of a wave equation for matter, Schrödinger was led to his formulation of wave mechanics. A similar approach based on relativistic Hamilton-Jacobi theory gives a broader base for the parametrized relativistic quantum theory presented in Part II.

CANONICAL TRANSFORMATIONS

Before we can proceed to the Hamilton-Jacobi formulation, we must first introduce the concept of a canonical transformation. For ease of reference, let us recall Hamilton's equations of motion:

$$\mathring{q} = \frac{\partial K}{\partial p}, \ \mathring{p} = -\frac{\partial K}{\partial q}. \tag{3.1.1}$$

We are interested in transformations of coordinates and momenta from the set {q,p} to a new set {Q,P} which preserves the form of Hamilton's equations; thus

$$\mathring{Q} = \frac{\partial K}{\partial P}, \ \mathring{P} = -\frac{\partial K}{\partial Q}. \tag{3.1.2}$$

Transformations which preserve the form of Hamilton's equations are canonical transformations.

Another way to think about canonical transformations is to insist that the form of Hamilton's principle be preserved. Thus, if Hamilton's principle has the form

$$\delta \int_{s_1}^{s_2} [\sum_i p_i \mathring{q}_i - H(q,p,s)]\, ds = 0, \tag{3.1.3}$$

for the set {q,p} of generalized coordinates, then {Q,P} are related to {q,p} by a canonical transformation if Hamilton's principle has the form

$$\delta \int_{s_1}^{s_2} [\sum_i P_i \mathring{Q}_i - K(Q,P,s)]\, ds = 0. \tag{3.1.4}$$

The integrands must be equal to within the total derivative of a generating function F such that

$$\delta \int_{s_1}^{s_2} \frac{dF}{ds}\, ds = 0. \tag{3.1.5}$$

Possible forms of the generating function include $F_1(q,Q,s)$, $F_2(q,P,s)$, $F_3(p,Q,s)$ and $F_4(p,P,s)$.

As an illustration, let us consider the generating function F_1. The integrands are related by the expression

$$\sum_i p_i \mathring{q}_i - H = \sum_i P_i \mathring{Q}_i - K + \frac{d}{ds} F_1(q,Q,s). \tag{3.1.6}$$

Expanding dF_1/ds gives

$$\frac{dF_1}{ds} = \sum_i \frac{\partial F_1}{\partial q_i} \mathring{q}_i + \sum_i \frac{\partial F_1}{\partial Q_i} \mathring{Q}_i + \frac{\partial F_1}{\partial s}. \tag{3.1.7}$$

Recognizing that the coordinates q_i, Q_i are independent, the coefficients of dq_i/ds and dQ_i/ds must vanish. Combining terms yields

$$p_i = \frac{\partial F_1}{\partial q_i}, \; P_i = -\frac{\partial F_1}{\partial Q_i} \tag{3.1.8}$$

so that

$$K = H + \frac{\partial F_1}{\partial s}. \tag{3.1.9}$$

Equation (3.1.9) holds for all generating functions, i.e.

$$K = H + \frac{\partial F_j}{\partial s}, \; j = \{1,2,3,4\}. \tag{3.1.10}$$

If we write

$$F_2(q,P,s) = F_1(q,Q,s) + \sum_i P_i Q_i, \tag{3.1.11}$$

then we can use the above procedure to derive the transformation equations

$$p_i = \frac{\partial F_2}{\partial q_i}, \; Q_i = \frac{\partial F_2}{\partial P_i} \tag{3.1.12}$$

and Eq. (3.1.10) with $j=2$. We are especially interested in F_2 because of its relation to Hamilton's principal function, which we discuss in §3.2. First, let us consider a physically interesting example of a canonical transformation.

CANONICAL TRANSFORMATION FOR THE 2-BODY PROBLEM

The Hamiltonian for a potential interaction between two particles with masses M_1 and M_2 may be written as [Reuse, 1979]

$$K = \frac{1}{2M_1} g^{\mu\nu} p_{1\mu} p_{2\nu} + \frac{1}{2M_2} g^{\mu\nu} p_{2\mu} p_{2\nu} + \Phi(|q_1 - q_2|), \tag{3.1.13}$$

where we assume the scalar potential Φ depends on the four-coordinate length

$$|q_1 - q_2| \equiv [g_{\mu\nu}(q_1^\mu - q_2^\mu)(q_1^\nu - q_2^\nu)]^{1/2}. \tag{3.1.14}$$

Let us apply the transformation

$$\begin{aligned} P_\mu &= p_{1\mu} + p_{2\mu}, \quad Q^\mu = \frac{M_1 q_1^\mu + M_2 q_2^\mu}{M_1 + M_2} \\ p_\mu &= \frac{M_1 p_{2\mu} - M_2 p_{1\mu}}{M_1 + M_2}, \quad q^\mu = q_2^\mu - q_1^\mu \end{aligned} \tag{3.1.15}$$

to Eq. (3.1.13). The resulting Hamiltonian is

$$K = \frac{1}{2M_T} g^{\mu\nu} P_\mu P_\nu + \frac{1}{2M_R} g^{\mu\nu} p_\mu p_\nu + \Phi(|q|) \tag{3.1.16}$$

with total mass M_T and reduced mass M_R defined by

$$M_T \equiv M_1 + M_2, \quad M_R \equiv \frac{M_1 M_2}{M_1 + M_2}. \tag{3.1.17}$$

The form of the Hamiltonian has been preserved. Therefore Eq. (3.1.15) is a canonical transformation. Substituting the transformed Hamiltonian into Hamilton's equations gives

$$\mathring{Q}^{\mu} = \frac{\partial K}{\partial P_{\mu}} = \frac{P^{\mu}}{M_T}, \ \mathring{P}^{\mu} = -\frac{\partial K}{\partial Q_{\mu}} = 0,$$
$$\mathring{q}^{\mu} = \frac{\partial K}{\partial p_{\mu}} = \frac{p^{\mu}}{M_R}, \ \mathring{p}^{\mu} = -\frac{\partial K}{\partial q_{\mu}} = -\frac{\partial \Phi(|q|)}{\partial q_{\mu}}. \tag{3.1.18}$$

We see from the top two equations that the total energy-momentum P^{μ} is a constant of the motion, and the space-time coordinates Q^{μ} are proportional to s. For a more detailed discussion of this example, the reader should see Reuse [1979] or Horwitz and Piron [1973].

HAMILTON'S PRINCIPAL FUNCTION

Given the above mathematical tools, we want to find a canonical transformation to a coordinate system that has variables Q,P that are constant in s, i.e.

$$\frac{\partial K}{\partial P} = \mathring{Q} = 0,$$
$$-\frac{\partial K}{\partial Q} = \mathring{P} = 0. \tag{3.1.19}$$

Such a transformation can be found if the generating function F_2 is chosen to be the principal function S_{HJ} satisfying

$$K = H + \frac{\partial S_{HJ}}{\partial s} = 0 \tag{3.1.20}$$

or

$$H(q,p,s) + \frac{\partial S_{HJ}}{\partial s} = 0. \tag{3.1.21}$$

The principal function S_{HJ} is the solution of Eq. (3.1.10) with the additional constraint that the transformed Hamiltonian K=0. If the transformation exists, than K is the constant 0 and Eqs. (3.1.19) are satisfied.

We recast Eq. (3.1.21) by recalling the relation

$$p = \frac{\partial F_2}{\partial q} \tag{3.1.22}$$

from Eq. (3.1.12). The result is the Hamilton-Jacobi equation expressed in terms of the principal function:

$$H\left(q, \frac{\partial S_{HJ}}{\partial q}, s\right) + \frac{\partial S_{HJ}}{\partial s} = 0. \tag{3.1.23}$$

As an example, the Hamilton-Jacobi equation for a charged particle in an electromagnetic field becomes

$$\frac{g^{\mu\nu}}{2M}[\partial_\mu S_{HJ} - eA_\mu][\partial_\nu S_{HJ} - eA_\nu] + \frac{\partial S_{HJ}}{\partial s} = 0 \tag{3.1.24}$$

where the Hamiltonian is given by Eq. (2.4.1). Recalling that the nonrelativistic Hamilton-Jacobi description is a precursor to a wave mechanical description of matter in the short wavelength limit, Eq. (3.1.24) gives us an idea of what to expect from a wave equation with an invariant evolution parameter for the given Hamiltonian.

As a historical note, Fock [1937] used a Hamilton-Jacobi equation of the form of Eq. (3.1.24) as the basis for a quantum theory. He added the constraint that all of his final solutions must be stationary (independent of s). Although his work in this area was of limited utility, it was the first attempt to incorporate an evolution parameter into a relativistic formulation of quantum theory.

§3.2 VARIATIONAL PRINCIPLE WITHOUT THE MASS SHELL CONSTRAINT

The Hamilton-Jacobi treatment in §3.1 relies on the Hamiltonian description

description. We first present a Lagrangian formulation using a variational principle similar to that in Chapter 2, then proceed to a path integral formulation which incorporates the classical action as the phase of the quantum mechanical probability amplitude.

FIELD EQUATIONS

We begin by defining a parameter dependent action

$$A = \int_{s_1}^{s_2} \int_D \mathcal{L}\, dx\, ds \tag{3.2.1}$$

where $\mathcal{L}$ is the Lagrangian density. Much of the ensuing derivation will parallel the treatment of Hamilton's principle in Chapter 2. The primary difference is the use of a Lagrangian density rather than just a Lagrangian. The integral over the parameter s is for fixed end points.

Hamilton's principle requires that the variation of A vanish:

$$\delta A = 0 = \int_{s_1}^{s_2} \int_D \delta \mathcal{L}\, dx\, ds. \tag{3.2.2}$$

The Lagrangian density is allowed to have the functional form

$$\mathcal{L} = \mathcal{L}(q, \partial_\mu q, \mathring{q}, s). \tag{3.2.3}$$

The variation of the Lagrangian density is

$$\delta \mathcal{L} = \frac{\partial \mathcal{L}}{\partial q} \delta q + \frac{\partial \mathcal{L}}{\partial(\partial_\mu q)} \delta (\partial_\mu q) + \frac{\partial \mathcal{L}}{\partial \mathring{q}} \delta \mathring{q} + \frac{\partial \mathcal{L}}{\partial s} \delta s. \tag{3.2.4}$$

Assuming fixed parameter end points, an integration by parts on the δs term gives

$$\int_{s_1}^{s_2}\int_D \frac{\partial \mathcal{L}}{\partial s}\,\delta s\,dx\,ds = 0. \tag{3.2.5}$$

Another integration by parts applied to the δ(dq/ds) term yields

$$\begin{aligned}\int_{s_1}^{s_2}\int_D \frac{\partial \mathcal{L}}{\partial \mathring{q}}\,\delta\left[\frac{dq}{ds}\right]dx\,ds &= \int_{s_1}^{s_2}\int_D \frac{\partial \mathcal{L}}{\partial \mathring{q}}\left[\frac{d}{ds}(\delta q)\right]dx\,ds \\ &= -\int_{s_1}^{s_2}\int_D \left[\frac{d}{ds}\left(\frac{\partial \mathcal{L}}{\partial \mathring{q}}\right)\right]\delta q\,dx\,ds,\end{aligned} \tag{3.2.6}$$

and the condition that δq vanishes at the parameter end points has been used. A simplification of the $\partial_\mu q$ term is achieved by invoking the chain rule to find

$$\begin{aligned}\frac{\partial \mathcal{L}}{\partial(\partial_\mu q)}\,\delta(\partial_\mu q) &= \frac{\partial \mathcal{L}}{\partial(\partial_\mu q)}\,\frac{\partial(\delta q)}{\partial x^\mu} \\ &= \frac{\partial}{\partial x^\mu}[\pi_\mu\,\delta q] - \left[\frac{\partial \pi^\mu}{\partial x^\mu}\right]\delta q,\end{aligned} \tag{3.2.7}$$

where

$$\pi^\mu = \frac{\partial \mathcal{L}}{\partial(\partial_\mu q)}. \tag{3.2.8}$$

Collecting the results in Eqs. (3.2.5) through (3.2.8) gives

$$\begin{aligned}\int_{s_1}^{s_2}\int_D \delta\mathcal{L}\,dx\,ds &= \int_{s_1}^{s_2}\int_D \left\{\left[\frac{\partial \mathcal{L}}{\partial q} - \frac{\partial \pi^\mu}{\partial x^\mu} - \frac{d}{ds}\left(\frac{\partial \mathcal{L}}{\partial \mathring{q}}\right)\right]\delta q\right\}dx\,ds \\ &+ \int_{s_1}^{s_2}\int_D \left\{\frac{\partial}{\partial x^\mu}(\pi^\mu\,\delta q)\right\}dx\,ds.\end{aligned} \tag{3.2.9}$$

The divergence term

$$\int_{s_1}^{s_2}\int_D \frac{\partial}{\partial x^\mu}(\pi^\mu \delta q)\, dx\, ds$$

vanishes upon transformation by Gauss' theorem and integration over all space. Substituting the result into Eq. (3.2.9) yields

$$\delta A = 0$$
$$= \int_{s_1}^{s_2}\int_D \left\{ \left[\frac{\partial \mathcal{L}}{\partial q} - \frac{\partial \pi^\mu}{\partial x^\mu} - \frac{d}{ds}\left(\frac{\partial \mathcal{L}}{\partial \mathring{q}} \right) \right] \delta q \right\} dx\, ds. \tag{3.2.10}$$

Hamilton's principle, as embodied in Eq. (3.2.10), holds for arbitrary variations only if the coefficients of the variations vanish. Thus, the parametrized Euler-Lagrange field equation

$$\frac{\partial \mathcal{L}}{\partial q} - \frac{\partial \pi^\mu}{\partial x^\mu} - \frac{d}{ds}\left(\frac{\partial \mathcal{L}}{\partial \mathring{q}} \right) = 0 \tag{3.2.11}$$

results from the vanishing of the coefficient of δq.

Illustration: Particle In an Electromagnetic Field

Pavsic combined the action with a variational principle to derive parametrized equations of motion without the mass shell constraint in classical mechanics [Pavsic, 1991a], and both first and second quantized formulations [Pavsic, 1991b]. We are interested in the classical application here. In particular, Pavsic postulated the Lagrangian

$$L_P = p_\mu \mathring{x}^\mu - \frac{\lambda}{2}(\pi_\mu \pi^\mu - m^2), \quad \pi_\mu \equiv p_\mu - eA_\mu \tag{3.2.12}$$

with λ being a fixed constant (and not a Lagrange multiplier as in conventional approaches) so that all of the variables x^μ, p^μ are independent. To simplify the presentation, we are now working in natural units ($\hbar = c = 1$).

Varying x_μ and p_μ gives

$$\mathring{p}_\mu = \lambda e \pi_\nu \partial_\mu A^\nu \tag{3.2.13}$$

and

$$\pi^\mu = \frac{\mathring{x}^\mu}{\lambda} \tag{3.2.14}$$

respectively. The constant λ is obtained from Eq. (3.2.14) by forming the inner product $\pi_\mu \pi^\mu$ and solving for λ to get

$$\lambda^2 = \frac{\mathring{x}_\mu \mathring{x}^\mu}{\pi_\mu \pi^\mu}. \tag{3.2.15}$$

Equation (3.2.15) shows that the constant λ has units of mass. Mass appears as a constant of motion. The Lorentz force may be derived from Eqs. (3.2.13) and (3.2.14) using the procedure in §2.4.

As a final observation, let us recall from §2.4 that particles with vanishing mass are allowed if we treat energy and momentum as the primary variables. Physical examples of light like particles are photons. An acceptable Lagrangian L_A -- expressed in terms of the Lagrangian density $\mathscr{L}_A$ -- for the four-vector potential A^μ is

$$L_A = \int \mathscr{L}_A d^4x = -\frac{1}{4} \int F_{\mu\nu} F^{\mu\nu} d^4x \tag{3.2.16}$$

where A^μ is contained in the field strength tensor $F^{\mu\nu}$ (Appendix A.1). Applying the Euler-Lagrange equation yields the free field result

$$\partial_\mu F^{\mu\nu} = 0. \tag{3.2.17}$$

Notice that A^μ does not depend on s in Eq. (3.2.16). More discussion on this point is given in §9.3.

§3.3 PARAMETRIZED PATH INTEGRALS

Feynman's [1948] path integral formulation of nonrelativistic quantum mechanics included an outline for extending the formalism to relativistic systems. This extension included describing the space-time path of a particle with respect to an invariant evolution parameter s. We apply path integral concepts to the simple experiment described in Chapter 2. This application will help us evaluate the validity of the quantum theoretic formalism put forth in Part II.

PATH INTEGRAL POSTULATES

Feynman associated a probability amplitude with the motion of a particle as a function of s. In §2.3 we presented an experiment in which two space-time measurements were used to determine the space-time trajectory of a relativistic particle. Let us define the location of the apparatus on the left of Figure 2.1 as position A. The location of the apparatus on the right is defined as position C. We want to predict the result of a measurement at C after we have performed a measurement at A.

Measurements are made at successive parameter values that we call s_A and s_C. Following Feynman [1948], the difference between classical and quantum physics is clearly shown by supposing the existence of a third detector and clock at a position B between positions A and C. Let us denote the results of measurements at A,B,C as a,b,c respectively. We assume our measurement at position A has the outcome a. Our predictions of measurements at C are greatly influenced by how we treat the effect of a measuring apparatus at position B.

The set of measurements a,b,c is just one of a set of possible outcomes. We define P_{ab} as the probability that if measurement A gave the result a, then measurement B will give the result b. Similarly for P_{bc} and P_{ac}. The probability of all three measurements occurring is P_{abc}. Assuming that the events between a and b are independent of those between b and c, we know from probability theory [Feller, 1953; Kreyszig, 1979; Mayants, 1984] that

$$P_{abc} = P_{ab} P_{bc} \tag{3.3.1}$$

and

$$P_{ac} = \sum_b P_{abc}. \tag{3.3.2}$$

Equation (3.3.2) is true because we sum (or integrate in the case of a continuous variable) over all mutually exclusive outcomes for b.

Feynman states [pg. 369, 1948] that Eq. (3.3.2) is "the essential difference between classical and quantum physics". Equation (3.3.2) is always true in classical mechanics, and often false in quantum mechanics. Probabilities are replaced by complex numbers in quantum mechanics such that

$$P_{ab} = |\phi_{ab}|^2,\ P_{bc} = |\phi_{bc}|^2,\ P_{ac}^q = |\phi_{ac}|^2 \tag{3.3.3}$$

where P_{ac}^q is the quantum mechanical probability that a measurement at C has the outcome c if a measurement at A has the outcome a. The classical probability P_{ac} is obtained by combining Eqs. (3.3.1) and (3.3.2) to get a superposition of probabilities

$$P_{ac} = \sum_b P_{ab} P_{bc}. \tag{3.3.4}$$

The quantum probability P_{ac}^q is obtained from a superposition of probability amplitudes:

$$P_{ac}^q = |\phi_{ac}|^2 = |\sum_b \phi_{ab} \phi_{bc}|^2. \tag{3.3.5}$$

Equation (3.3.4) assumes that we made a measurement of b at B. If we do not attempt to verify the route of a particle in traveling from A to C, then Eq. (3.3.5) applies. Each available route of a particle is represented by a complex amplitude. This led to Feynman's first postulate [pg. 371, 1948]:

F-I. If an ideal measurement is performed to determine whether a particle has a path lying in a region of space-time, then the probability that the result will be affirmative is the absolute square of a sum of complex contributions, one from each path in the region.

Postulate F-I tells how to calculate probabilities from some complex function which we call the probability amplitude. Probability amplitudes represent the possible paths available to a particle. Feynman's second postulate [pg. 371, 1948] tells how to calculate probability amplitudes:

F-II. The paths contribute equally in magnitude, but the phase of their contribution is the classical action...

According to the path integral treatment, the classical path of the particle would be the path yielding the minimum action. We want to find an equation that describes the evolution of a relativistic free particle from the apparatus at position A to the apparatus at position C assuming no measurements are made at position B. Contributions from all possible paths must be considered.

ILLUSTRATION: RELATIVISTIC FREE PARTICLE

For simplicity, we work in one spacial dimension and adopt natural units ($\hbar=c=1$). The probability amplitude at space-time point x_{i+1},t_{i+1} and parameter $s+\varepsilon$ for an infinitesimal ε may be calculated from the probability amplitude at x_i,t_i,s using the relation

$$\phi(x_{i+1},\, t_{i+1},\, s+\varepsilon) = \frac{1}{A}\int e^{iS(x_i,\, t_i)}\phi(x_i,\, t_i,\, s)\,dx_i\, dt_i. \tag{3.3.6}$$

Equation (3.3.6) is a consequence of postulates F-I and F-II. The quantity A is a normalization constant and the action S is given by

$$S(x_i,t_i) = \int L(\mathring{x}_i,\mathring{t}_i)\,ds. \tag{3.3.7}$$

The Lagrangian for a free particle may be written as

$$L(\mathring{x},\mathring{t}) = \frac{m}{2}\left[\left(\frac{dt}{ds}\right)^2 - \left(\frac{dx}{ds}\right)^2\right]. \tag{3.3.8}$$

We are integrating over all possible paths. Assuming the trapezoidal rule is sufficiently accurate to solve the integral gives

$$S(x_i,t_i) = \varepsilon \frac{m}{2}\left[\left(\frac{t_{i+1}-t_i}{\varepsilon}\right)^2 - \left(\frac{x_{i+1}-x_i}{\varepsilon}\right)^2\right]. \tag{3.3.9}$$

Defining the variables

$$\begin{aligned} \delta_x &\equiv x_{i+1} - x_i \\ \delta_t &\equiv t_{i+1} - t_i \\ x &\equiv x_{i+1} \\ t &\equiv t_{i+1} \end{aligned} \tag{3.3.10}$$

lets us write Eq. (3.3.6) in the form

$$\begin{aligned} \phi(x,t,s+\varepsilon) = \frac{1}{A}\int e^{i\frac{m}{2\varepsilon}[\delta_t^2 - \delta_x^2]} d\delta_x d\delta_t \\ \cdot\phi(x-\delta_x, t-\delta_t, s). \end{aligned} \tag{3.3.11}$$

Performing Taylor series expansions of the probability amplitudes to first order in ε and second order in δ_x, δ_t gives

$$\begin{aligned} \phi(x,t,s) + \varepsilon\frac{\partial\phi}{\partial s} = \frac{1}{A}\int e^{i\frac{m}{2\varepsilon}[\delta_t^2 - \delta_x^2]} d\delta_x d\delta_t \\ \cdot[\phi(x,t,s) - \delta_t\frac{\partial\phi}{\partial t} - \delta_x\frac{\partial\phi}{\partial x} \\ - \delta_x\delta_t\frac{\partial^2\phi}{\partial x\partial t} + \frac{\delta_t^2}{2}\frac{\partial^2\phi}{\partial t^2} + \frac{\delta_x^2}{2}\frac{\partial^2\phi}{\partial x^2}]. \end{aligned} \tag{3.3.12}$$

Evaluating the integral on the right hand side lets us simplify Eq. (3.3.12) to the form

$$\begin{aligned} \phi(x,t,s) + \varepsilon\frac{\partial\phi}{\partial s} = -\frac{2\varepsilon\pi}{mA}\phi(x,t,s) \\ - \frac{i\varepsilon}{2m}\left(\frac{2\varepsilon\pi}{mA}\right)\left[\frac{\partial^2\phi}{\partial t^2} - \frac{\partial^2\phi}{\partial x^2}\right]. \end{aligned} \tag{3.3.13}$$

To assure agreement in the limit as ε goes to zero (the normalized case), we set

$$A = -2\varepsilon\pi/m$$

so that Eq. (3.3.13) becomes

$$\phi(x,t,s) + \varepsilon\frac{\partial\phi}{\partial s} = \phi(x,t,s) + \frac{i\varepsilon}{2m}\left[\frac{\partial^2\phi}{\partial t^2} - \frac{\partial^2\phi}{\partial x^2}\right]. \tag{3.3.14}$$

Equation (3.3.14) is satisfied when the coefficients of ε are equal, thus yielding the condition

$$\frac{\partial\phi}{\partial s} = \frac{i}{2m}\left[\frac{\partial^2\phi}{\partial t^2} - \frac{\partial^2\phi}{\partial x^2}\right]. \tag{3.3.15}$$

Multiplying Eq. (3.3.15) by i and introducing four-vector notation gives

$$i\frac{\partial\phi}{\partial s} = -\frac{1}{2m}\partial^\mu\partial_\mu\phi. \tag{3.3.16}$$

Equation (3.3.16) is a parametrized relativistic quantum mechanical equation derived from path integral concepts for a free particle. A different procedure is used in Part II to derive a more general field equation.

EXTENSION OF THE PATH INTEGRAL FORMALISM

Feynman outlined a procedure to extend his path integral formulation of nonrelativistic quantum mechanics to quantum electrodynamics (QED) in 1950. Morette [1951] succinctly presented the relativistic path integral formalism by writing the probability amplitude K for a particle to go from space-time point x^A to space-time point x^B as

$$K(x^B, x^A) = \int \exp(iS[x])\, d(paths) \tag{3.3.17}$$

The integral is over all parametrized paths x(s) from x^A to x^B (Figure 3.1). The action functional S[x] is given by

$$S[x] = \int_{s^A}^{s} L\{x(s), \dot{x}(s)\} ds \tag{3.3.18}$$

where L is the Lagrangian.

Following a suggestion made by Feynman [1948], Morette [1951] applied Feynman's path integral approach to a Dirac particle in a constant, external electromagnetic field. She used a Lagrangian of the form

FIGURE 3.1

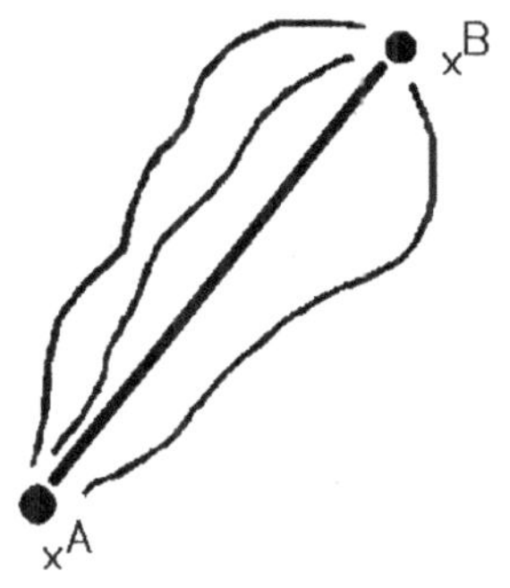

$$L_{Mor} = \frac{m}{2}\dot{x}_\mu \dot{x}^\mu + \frac{e}{2}(\gamma_\mu \dot{x}^\mu)(\gamma_\nu A^\nu) \tag{3.3.19}$$

to obtain the wave equation

$$2im\frac{\partial}{\partial s}\psi_{Mor}(x,s) = \{\gamma_\mu[\partial^\mu - ieA^\mu]\}^2 \psi_{Mor}(x,s) \tag{3.3.20}$$

where γ_μ are Dirac's gamma matrices.

Feynman [App. D, 1951] used a path integral formulation to derive Fock's parametrized spin-½ wave equation. His Lagrangian had the form

$$L_{FF} = -\left[\frac{1}{2}\mathring{x}_\mu \mathring{x}^\mu + e\mathring{x}_\mu A^\mu - \frac{e}{4}\sigma_{\mu\nu}F^{\mu\nu}\right] \tag{3.3.21}$$

where $\sigma_{\mu\nu}$ is the antisymmetric spin tensor (Appendix A.1). The interaction term is a result of "minimal coupling". The resulting spin dependent wave equations are essentially a parametrized version of the square of Dirac's equation.

We see from the Hamilton-Jacobi and path integral treatments that parametrization of the wave equation is a natural extension of classical relativistic mechanics. An interesting aspect of the path integral formalism -- from our perspective -- is the appearance of the first order derivative of the probability amplitude with respect to an invariant evolution parameter, e.g. $\partial\phi/\partial s$ in Eq. (3.3.16). We show in Chapter 6 that parametrized wave equations have the general form

$$i\frac{\partial\psi}{\partial s} = \boldsymbol{K}\psi \tag{3.3.22}$$

where the operator **K** is a generator of evolution parameter translations. Several forms have been proposed for **K** [Fanchi, 1993], but only a generator that is proportional to the inner product $\mathbf{P}^\mu\mathbf{P}_\mu$ of a four-momentum operator $\mathbf{P}^\mu$ can be derived from first principles, notably a probabilistic basis. The roles of the invariant evolution parameter s and the operator **K** are formalized in the parametrized relativistic quantum theory presented in Part II.

CHAPTER 4
STATISTICAL PRCM

The foundation of statistical mechanics in the context of parametrized relativistic classical mechanics was laid by Horwitz, Schieve, and Piron [1981]. Their work is of interest here because it is the basis for a calculation that leads to a test of the paradigm underlying the parametrized theory. To arrive at the test, we must first develop some of the basic ideas of classical statistical mechanics within the PRCM context. These ideas are presented below. The primary result is an expression for the free energy of a relativistic Gibbs ensemble. This expression is then used as part of a classical cosmological model in Chapter 5 to show that PRCM predictions can differ from predictions of the conventional nonparametrized paradigm.

§4.1 CLASSICAL N-BODY SYSTEM

The concepts of statistical mechanics may be applied to an N-body system of particles if N is sufficiently large to be statistically significant. Furthermore, it is sufficient for our purposes to assume that the number of particles N is conserved. This means that annihilation and creation processes do not play a significant role in the present analysis. Parametrized Hamilton's equations for this N-body system are

$$\frac{dq_i^\mu}{ds} = \frac{\partial K}{\partial p_{i\mu}}, \qquad \frac{dp_i^\mu}{ds} = -\frac{\partial K}{\partial q_{i\mu}}. \tag{4.1.1}$$

We have already considered the single free particle case in Chapter 2. Here we show that the parametrized formulation embodied by Eq. (4.1.1) may be used to derive expected results for a system of N noninteracting particles.

The Hamiltonian K_0 for N free particles may be written as

$$K_0 = \sum_{i=1}^{N} \frac{p_{i\mu} p_i^\mu}{2M_i}. \tag{4.1.2}$$

Substituting Eq. (4.1.2) into (4.1.1) gives Hamilton's equations for the free particle system:

$$\begin{aligned} \frac{dq_i^\mu}{ds} &= \frac{p_i^\mu}{M_i}, \\ \frac{dp_i^\mu}{ds} &= 0. \end{aligned} \tag{4.1.3}$$

Integrating the equations of motion gives

$$q_i^\mu = (q_0)_i^\mu + \frac{p_i^\mu}{M_i} s \tag{4.1.4}$$

where $\{q_0\}$ denotes the set of initial space-time positions. Calibrating our clocks so that $(q_0)_i^0 = 0$ lets us calculate the relationship between the temporal coordinate (q^μ with $\mu = 0$) and the parameter s as

$$q_i^0 \frac{M_i}{p_i^0} = s = t \frac{M_i}{\omega_i} \tag{4.1.5}$$

where we have replaced q_i^0 and p_i^0 with the coordinate time t and energy ω_i respectively. Notice that we have assumed all of the particles are calibrated and run at the same rate, thus all temporal coordinates $\{q_i^0\}$ are replaced with a single coordinate time.

For particle rest mass and energy fixed, as they are for the system under study, the differential of Eq. (4.1.5) is

$$ds = \frac{M_i}{\omega_i} dt. \tag{4.1.6}$$

Using Eq. (4.1.6) in Eq. (4.1.3) for the spacial components gives the expected result

$$\frac{d\vec{q}_i}{dt} = \frac{\vec{p}_i}{\omega_i}. \tag{4.1.7}$$

We now establish another desired result that will be of use in subsequent applications.

CENTER OF MASS MOTION

The center of mass motion for a parametrized system of particles with constant particle rest mass is obtained by making the coordinate transformation

$$\begin{aligned} P_i^\mu &= p_i^\mu - \frac{M_i}{M_{CM}} P_{CM}^\mu, \\ Q_i^\mu &= q_i^\mu - Q_{CM}^\mu \end{aligned} \tag{4.1.8}$$

where

$$\begin{aligned} P_{CM}^\mu &\equiv \sum_{i=1}^{N} p_i^\mu, \\ Q_{CM}^\mu &\equiv \sum_{i=1}^{N} \frac{M_i}{M_{CM}} q_i^\mu, \\ M_{CM} &\equiv \sum_{i=1}^{N} M_i. \end{aligned} \tag{4.1.9}$$

The new variables $\{P,Q\}$ are called relative coordinates because they represent the motion of the particles relative to the center of mass. The additional degrees of freedom represented by the coordinates $\{P_{CM},Q_{CM}\}$ require that a corresponding number of constraints must be satisfied for the mathematical problem to be well defined. These constraints are

$$\sum_{i=1}^{N} P_i^{\mu} = 0,$$
$$\sum_{i=1}^{N} M_i Q_i^{\mu} = 0. \tag{4.1.10}$$

In terms of the relative coordinates, the free particle Hamiltonian becomes

$$K_0 = \sum_{i=1}^{N} \frac{P_i^{\mu} P_{i\mu}}{2M_i} + \frac{P_{CM}^{\mu} P_{CM\mu}}{2M_{CM}}. \tag{4.1.11}$$

Equation (4.1.11) shows the separation of the center of mass and also the canonical character of the transformation since the form of K_0 has been preserved.

We may go a step further by allowing the system of particles to interact via a potential that depends only on the space-time coordinates. The resulting total Hamiltonian is

$$K = K_0 + K_I(q_1,\ldots,q_N) \tag{4.1.12}$$

which can be expressed in terms of the relative coordinates as

$$K = \sum_{i=1}^{N} \frac{P_i^{\mu} P_{i\mu}}{2M_i} + \frac{P_{CM}^{\mu} P_{CM\mu}}{2M_{CM}} + K_I(Q_1,\ldots,Q_N). \tag{4.1.13}$$

Hamilton's equations for the relative coordinates are

$$\frac{dQ_i^{\mu}}{ds} = \frac{P_i^{\mu}}{M_i},$$
$$\frac{dP_i^{\mu}}{ds} = -\frac{\partial K_I}{\partial Q_{i\mu}} \tag{4.1.14}$$

where the latter equation is a force equation. Specifying the initial condition

$$P_i^\mu P_{i\mu} = M_i^2 \tag{4.1.15}$$

lets us evaluate the scalar product

$$\frac{dQ_i^\mu}{ds}\frac{dQ_{i\mu}}{ds} = \frac{P_i^\mu P_{i\mu}}{M_i^2} = 1 \tag{4.1.16}$$

for the world-line of particle i relative to the center of mass. A similar calculation for the center of mass coordinates gives the equations of motion

$$\begin{aligned} \frac{dQ_{CM}^\mu}{ds} &= \frac{P_{CM}^\mu}{M_{CM}}, \\ \frac{dP_{CM}^\mu}{ds} &= 0 \end{aligned} \tag{4.1.17}$$

and the world-line

$$\frac{dQ_{CM}^\mu}{ds}\frac{dQ_{CM\mu}}{ds} = \frac{P_{CM}^\mu P_{CM\mu}}{M_{CM}^2} = 1. \tag{4.1.18}$$

§4.2 RELATIVISTIC IDEAL GAS

According to statistical mechanics, e.g. Landsberg [1978] and Chandler [1987], the state of a classical mechanical system is characterized by its trajectory in phase space. An ensemble is the collection of all states of the microscopic (or particle) system which are consistent with the constraints imposed on the macroscopic system. The primary assumption of statistical mechanics is that the observed value of a property should correspond to its ensemble average. Using these ideas, we derive the macroscopic thermodynamic relation governing the behavior of a relativistic ideal gas.

The system of N-particles described in §4.1 is a relativistic gas. The forces between particles in an ideal gas should be sufficiently weak that the relation

$$p_i^{\mu} p_{i\mu} \approx M_i^2 \tag{4.2.1}$$

holds for each particle. Particle rest masses do not change under this assumption, which is consistent with the set of assumptions discussed in §4.1. If we further assume that the system of N-particles is an isolated system, then it may be viewed from a statistical mechanical perspective as a microcanonical ensemble and should obey an ideal gas law, as we now show.

MICROCANONICAL ENSEMBLE

A microcanonical ensemble is characterized by specifying two macroscopic constraints, such as fixed total energy and size. We take the phase space for the microcanonical ensemble of PRCM to be

$$\Gamma(\kappa,\omega) = \int \delta(K-\kappa)\, \delta\left(\sum_{i=1}^{N} p_i^0 - \omega\right) d\mu_p\, d\mu_q \tag{4.2.2}$$

where the measures are defined as

$$\begin{aligned} d\mu_p &= d^4p_1 ... d^4p_N, \\ d\mu_q &= d^4q_1 ... d^4q_N. \end{aligned} \tag{4.2.3}$$

Both K and $\Sigma_i p_i^0$ are constants of the motion, i.e. their derivatives with respect to s vanish.

Given a phase space integral, we define entropy in the usual way as

$$S(\kappa,\omega) = k_B\, \ell n\, \Gamma(\kappa,\omega) \tag{4.2.4}$$

where k_B is Boltzmann's constant. The constraint

$$\sum_{i=1}^{N} p_i^0 = \omega \tag{4.2.5}$$

with ω a constant is needed so that temperature T can have the usual definition

$$T^{-1} \equiv \frac{\partial S(\kappa,\omega)}{\partial \omega}. \tag{4.2.6}$$

Elements of the microcanonical ensemble are uniformly distributed around

$$m_i \equiv \left[\left(p_i^0\right)^2 - \vec{p}_i \cdot \vec{p}_i\right]^{1/2} \tag{4.2.7}$$

where $m_i \approx M_i$ by Eq. (4.2.1). The space-time four-volume is limited to a finite volume of space V over a finite duration Υ. Recalling that $K \approx K_0$ since $K_0 \gg K_I$, we neglect the coordinate dependent interaction potential and evaluate the phase space integral over the space-time coordinates to get

$$\Gamma(\kappa,\omega) \approx V^N f(\kappa,\omega) \tag{4.2.8}$$

where

$$f(\kappa,\omega) \equiv \Upsilon^N \int \delta(K_0 - \kappa)\, \delta\left(\sum_{i=1}^{N} p_i^0 - \omega\right) d\mu_p. \tag{4.2.9}$$

IDEAL GAS LAW

The macroscopic thermodynamic quantities entropy and temperature are calculated from Eqs. (4.2.4) and (4.2.6). The results are

$$S = k_B \ell n\ V^N + k_B \ell n\ f(\kappa,\omega) \tag{4.2.10}$$

and

$$T^{-1} = \frac{k_B}{f(\kappa,\omega)} \frac{\partial f(\kappa,\omega)}{\partial \omega}. \tag{4.2.11}$$

The energy derivative in Eq. (4.2.11) is obtained by taking the implicit derivative of entropy S with respect to volume V for S constant. Taking the derivative gives

$$\frac{\partial S}{\partial V}\Big|_S = 0 = \frac{Nk_B}{V} + \frac{k_B}{f(\kappa,\omega)}\left[\frac{\partial f(\kappa,\omega)}{\partial \omega}\frac{\partial \omega}{\partial V}\right]_S. \tag{4.2.12}$$

Defining pressure P in the usual way as

$$P \equiv -\frac{\partial \omega}{\partial V}\Big|_S \tag{4.2.13}$$

lets us rewrite Eq. (4.2.12) in the form

$$0 = \frac{Nk_B}{V} - \frac{P}{T} \tag{4.2.14}$$

or

$$PV = Nk_B T. \tag{4.2.15}$$

Equation (4.2.15) is the expected result that the ideal gas law applies to an isolated, weakly interacting relativistic gas.

§4.3 FREE ENERGY OF THE IDEAL GAS

We showed in §4.2 that we could derive an expected result -- the ideal gas law -- by combining familiar thermodynamic and statistical mechanical concepts with PRCM. Our primary interest is in finding predictions of PRCM which may serve as tests of the new paradigm. The ideal gas law calculation showed that PRCM agrees with familiar experimental results. A calculation of the free energy of the ideal gas gives results which differ from the conventional paradigm in the ultrarelativistic regime. It is this calculation to which we now turn.

The system studied in §4.2 was an isolated system and was represented as a microcanonical ensemble with two macroscopic constraints. To best calculate the free energy of a relativistic ideal gas, we wish to reduce the number of macroscopic constraints to one and work with a canonical ensemble.

CANONICAL ENSEMBLE

In this case we are physically allowing one system to be in communication with another system. We continue to work with N particles, but the N particles are partitioned into two subsystems. We let the experimental subsystem with N_e particles exchange heat in the form of kinetic energy with the much larger "heat reservoir" subsystem containing $N_r = N - N_e \gg N_e$ particles. Assuming the interactions between particles are dominated by short-range forces, we can decompose the Hamiltonian into

$$K \approx K_e + K_r. \tag{4.3.1}$$

The equilibrium configuration of all particles is uniform in phase space, so the constraints of the experimental subsystem must be the same as the reservoir of N_r particles. The phase space integral for the microcanonical ensemble becomes

$$\begin{aligned}\Gamma(\omega,M) = \int &\delta(K_e - \kappa_e)\,\delta(K_r - \kappa_r)\,\delta(\omega_r + \omega_e - \omega)\\ &\cdot d^4p_1 ... d^4p_{N_e} d^4q_1 ... d^4q_{N_e}\\ &\cdot d^4p_{N_e+1} ... d^4p_N d^4q_{N_e+1} ... d^4q_N\end{aligned} \tag{4.3.2}$$

where

$$\omega_e = \sum_{i=1}^{N_e} \omega_i, \quad \omega_r = \sum_{i=N_e+1}^{N=N_e+N_r} \omega_i. \tag{4.3.3}$$

Writing

$$\begin{aligned}\Gamma_r(\omega - \omega_e, \kappa_r) = \int &\delta(K_r - \kappa_r)\,\delta(\omega_r - (\omega - \omega_e))\\ &\cdot d^4p_{N_e+1} ... d^4p_N d^4q_{N_e+1} ... d^4q_N\end{aligned} \tag{4.3.4}$$

for the phase space integral of the experimental subsystem lets us simplify Eq. (4.3.2). The result is

$$\Gamma(\omega,\kappa) = \int \delta(K_e - \kappa_e)\Gamma_r(\omega - \omega_e, \kappa_r) \cdot d^4p_1 ... d^4p_{N_e} d^4q_1 ... d^4q_{N_e}. \tag{4.3.5}$$

Introducing an integral over the Dirac delta function $\delta(\omega_e - \omega')$ gives

$$\Gamma(\omega,\kappa) = \int\int \delta(K_e - \kappa_e)\,\delta(\omega_e - \omega')\Gamma_r(\omega - \omega', \kappa_r) \cdot d^4p_1 ... d^4p_{N_e} d^4q_1 ... d^4q_{N_e} d\omega' \tag{4.3.6}$$

or

$$\Gamma(\omega,\kappa) = \int \Gamma_e(\omega', \kappa_e)\Gamma_r(\omega - \omega', \kappa_r) d\omega' \tag{4.3.7}$$

where

$$\Gamma_e(\omega', \kappa_e) = \int \delta(K_e - \kappa_e)\,\delta(\omega_e - \omega') \cdot d^4p_1 ... d^4p_{N_e} d^4q_1 ... d^4q_{N_e}. \tag{4.3.8}$$

Following the usual statistical arguments, we assume the existence of an energy ω^* that dominates the integral over ω' in a bimodal distribution such that

$$\begin{aligned} S &= k_B \ell n\ \Gamma(\omega,\kappa) \\ &\approx k_B \ell n\ \Gamma_e(\omega^*, \kappa_e) + k_B \ell n\ \Gamma_r(\omega - \omega^*, \kappa_r). \end{aligned} \tag{4.3.9}$$

Equation (4.3.9) implies the additivity of entropy

$$S = S_e + S_r, \tag{4.3.10}$$

and the existence of a maximum ω^* implies the equality of temperature for each subsystem:

$$\frac{1}{T} = \frac{\partial S_e}{\partial \omega'}\Big|_{\omega' = \omega^*} = \frac{\partial S_r}{\partial \omega'}\Big|_{\omega' = \omega - \omega^*}. \tag{4.3.11}$$

Expressing entropy in terms of the phase space integral lets us write

$$\frac{\partial \ln \Gamma_r}{\partial \omega'}\Big|_{\omega'=\omega-\omega^*} = \frac{1}{k_B T} \equiv \beta . \tag{4.3.12}$$

If temperature changes slowly with respect to ω' in the region near $\omega'=\omega-\omega^*$, we have

$$\Gamma_r(\omega - \omega_e, \kappa_r) \approx e^{\beta\omega'} = e^{\beta\omega} e^{-\beta\omega_e}. \tag{4.3.13}$$

Substituting Eq. (4.3.13) into (4.3.5) gives

$$\Gamma(\omega,\kappa) = e^{\beta\omega} \int \delta(K_e - \kappa_e) e^{-\beta\omega_e} \cdot d^4p_1 ... d^4p_{N_e} d^4q_1 ... d^4q_{N_e} . \tag{4.3.14}$$

FREE ENERGY CALCULATION

It is sufficient for our purposes to consider a gas of noninteracting, identical particles. The normalized distribution for the canonical ensemble is proportional to the integrand of Eq. (4.3.14) with the proportionality constant determined by probability conservation. The resulting distribution is

$$D(p,q) = \delta(K - \kappa) \frac{e^{-\beta\omega}}{Q_N(V^{(4)}, T, \kappa)} \tag{4.3.15}$$

where we have dropped the subscript e, $V^{(4)}$ is the space-time four-volume, and the partition function Q_N is

$$Q_N(V^{(4)}, T, \kappa) \equiv \int \delta(K - \kappa) e^{-\beta\omega} \frac{d\mu_p d\mu_q}{N!} . \tag{4.3.16}$$

The measures in Eq. (4.3.16) are given by (4.2.3), and N! accounts for the degeneracy of states associated with N identical particles. Equation (4.3.16) is the starting point for our calculation of the free energy of the relativistic ideal gas.

The partition function may be simplified by recognizing that $M_i = M_0$ where

M_0 is the particle rest mass for our system of identical particles. The resulting partition function is

$$Q_N(V^{(4)},T,\kappa) = \frac{1}{N!}\int \delta\left[\sum_{i=1}^{N}\left(\frac{p_{0i}^2}{2M_0} - \frac{\vec{p}_i\cdot\vec{p}_i}{2M_0}\right) - \kappa\right] \cdot e^{-\beta\sum_{i=1}^{N}\omega_i} d\mu_p d\mu_q . \tag{4.3.17}$$

Integrating over the space-time coordinates gives

$$Q_N(V^{(4)},T,\kappa) = \frac{(V^{(4)})^N}{N!}\int \frac{m_1 dm_1}{(\vec{p}_1\cdot\vec{p}_1 + m_1^2)^{1/2}} \cdots \frac{m_N dm_n}{(\vec{p}_N\cdot\vec{p}_N + m_N^2)^{1/2}} \cdot 2M_0\,\delta\left(\sum_{i=1}^{N} m_i^2 - NM_0^2\right) e^{-\beta\sum_{i=1}^{N}(\vec{p}_i\cdot\vec{p}_i + m_i^2)^{1/2}} d^3p_1 ... d^3p_N . \tag{4.3.18}$$

where the relations

$$p^0 = (\vec{p}\cdot\vec{p} + m^2)^{1/2},\ dp^0 = \frac{m\,dm}{(\vec{p}\cdot\vec{p} + m^2)^{1/2}} \tag{4.3.19}$$

and

$$\kappa = \frac{NM_0}{2} \tag{4.3.20}$$

have been used. To compare the parametrized expression with the result based on the conventional paradigm, we must restrict the range of masses close to M_0. The results of this restriction are the replacements

$$\frac{m_i}{(\vec{p}_i\cdot\vec{p}_i + m_i^2)^{1/2}} \approx \frac{M_0}{(\vec{p}_i\cdot\vec{p}_i + M_0^2)^{1/2}} \tag{4.3.21}$$

and

$$e^{-\beta\sum_{i=1}^{N}(\vec{p}_i\cdot\vec{p}_i + m_i^2)^{1/2}} \approx e^{-\beta\sum_{i=1}^{N}(\vec{p}_i\cdot\vec{p}_i + M_0^2)^{1/2}} \tag{4.3.22}$$

Substituting Eqs. (4.3.21) and (4.3.22) into (4.3.18) gives

$$Q_N(V^{(4)},T,\kappa) = \frac{(V^{(4)})^N}{N!} 2M_0 (M_0)^N \cdot \int dm_1 \ldots dm_n \, \delta\left(\sum_{i=1}^{N} m_i^2 - NM_0^2\right)(I_\beta)_i \tag{4.3.23}$$

where

$$(I_\beta)_i \equiv \prod_{i=1}^{N}\left[\int d^3p_i \frac{e^{-\beta(\vec{p}_i\cdot\vec{p}_i + M_0^2)^{1/2}}}{(\vec{p}_i\cdot\vec{p}_i + M_0^2)^{1/2}}\right]. \tag{4.3.24}$$

We are now in a position to calculate the free energy of the ensemble.

Free energy is calculated as the average

$$\begin{aligned}\langle\omega\rangle &= \frac{1}{N!}\int \omega\, D(p,q)\, d\mu_p d\mu_q \\ &= \frac{\frac{1}{N!}\int \omega\, \delta(K-\kappa) e^{-\beta\omega} d\mu_p d\mu_q}{Q_N(V^{(4)},T,\kappa)}.\end{aligned} \tag{4.3.25}$$

Equation (4.3.25) may be rewritten as

$$\langle\omega\rangle = \frac{\int \delta(K-\kappa)\left[-\frac{\partial e^{-\beta\omega}}{\partial\beta}\right]\frac{d\mu_p d\mu_q}{N!}}{Q_N(V^{(4)},T,\kappa)}. \tag{4.3.26}$$

Factoring out the derivative with respect to β gives

$$\begin{aligned}\langle\omega\rangle &= -\frac{1}{Q_N(V^{(4)},T,\kappa)}\frac{\partial}{\partial\beta}\int \delta(K-\kappa) e^{-\beta\omega}\frac{d\mu_p d\mu_q}{N!} \\ &= -\frac{1}{Q_N(V^{(4)},T,\kappa)}\frac{\partial}{\partial\beta} Q_N(V^{(4)},T,\kappa)\end{aligned} \tag{4.3.27}$$

or

$$\langle\omega\rangle \equiv \omega_H = -\frac{\partial \ln Q_N}{\partial\omega} \tag{4.3.28}$$

where we have simplified notation and written $<\omega>$ as ω_H. The subscript H reminds us that this calculation was first performed by Horwitz, et al [1981]. Evaluating the integral I_β in Eq. (4.3.24) yields

$$\begin{aligned} I_\beta &= \int d^3p \frac{e^{-\beta(\vec{p}\cdot\vec{p}+M_0^2)^{1/2}}}{(\vec{p}\cdot\vec{p}+M_0^2)^{1/2}} \\ &= -2\pi^2\frac{M_0}{\beta}H_1^{(1)}(i\sigma),\ \sigma \equiv M_0\beta. \end{aligned} \tag{4.3.29}$$

The functions $H_1^{(1)}$ above and $H_0^{(1)}$ below are Hankel functions [Abramowitz and Stegun, 1972; Arfken, 1970]. Substituting the result into Eq. (4.3.23) and carrying out the derivative in Eq. (4.3.28) gives

$$\omega_H = \frac{N}{\beta}\left[2 - i\sigma\frac{H_0^{(1)}(i\sigma)}{H_1^{(1)}(i\sigma)}\right]. \tag{4.3.30}$$

§4.4 FREE ENERGY COMPARISON

In the previous section, we fixed both the number of free particles and the mass of each particle. These restrictions are equivalent to specifying a system that is stationary, or independent of the relativistic scalar evolution parameter. They are needed to make possible a comparison of Eq. (4.3.30) with Jüttner's [1911] result:

$$\omega_J = \frac{N}{\beta}\left[1 - i\sigma\frac{H_2'^{(1)}(i\sigma)}{H_2^{(1)}(i\sigma)}\right]. \tag{4.4.1}$$

The prime in Eq. (4.4.1) denotes differentiation with respect to the argument of the Hankel function. A more modern discussion of Jüttner's calculation is given by Pauli [1958]. Differences between Eqs. (4.3.30) and (4.4.1) reflect differences in the dynamical laws of PRCM and the conventional paradigm. A more direct comparison can be made by manipulating the Hankel functions in Eq. (4.4.1) to arrive at the equivalent form

$$\omega_J = \frac{N}{\beta}\left[3 - i\sigma\frac{H_1^{(1)}(i\sigma)}{H_2^{(1)}(i\sigma)}\right]. \tag{4.4.2}$$

Equations (4.3.30) and (4.4.2) agree in the nonrelativistic limit ($\sigma \to \infty$):

$$(\omega_H)_{NR} = (\omega_J)_{NR} = NM_0 + \frac{1}{2}Nk_B T. \tag{4.4.3}$$

At the opposite extreme -- the ultrarelativistic limit ($\sigma \to 0$) -- the expressions have different values:

$$(\omega_H)_{UR} = 2Nk_B T \tag{4.4.4}$$

and

$$(\omega_J)_{UR} = 3Nk_B T. \tag{4.4.5}$$

The question arises: is this difference observable?

Free energy differences can effect the energy density of the stress-energy tensor used in stellar models (e.g. the Oppenheimer-Volkoff equation [Misner, Thorne and Wheeler, 1973]), relativistic gas dynamics [e.g. Konigl, 1980; Hiscock and Lindblom, 1985], and cosmology [e.g. Misner, et al, 1973; Brandenberger, 1985; Felten and Isaacman, 1986]. It was noted by Horwitz, et al [1981] that the difference between the calculated free energies in the UR limit was significant for temperatures in excess of 10^{12} K. This estimate assumes M_0 is the electron mass and represents a value of $\sigma = 0.001$. For hydrogen-like masses and the same value of σ, we have

$$\sigma_H \approx (1.1 \times 10^{12} K)/T. \tag{4.4.6}$$

At what temperature does the difference in calculated free energies become significant? Is the difference significant only in the UR limit, or does the difference manifest itself at a lower temperature? To answer these questions, we must evaluate the free energies as a function of temperature or, equivalently, σ.

Calculation of the free energies begins by first transforming the Hankel functions to confluent hypergeometric functions [Abramowitz and Stegun, 1972]:

$$\omega_H = \frac{N}{\beta}\left[2 + \frac{1}{2}\frac{U(0.5,1,Z)}{U(1.5,3,Z)}\right], \quad Z = 2\sigma \tag{4.4.7}$$

and

$$\omega_J = \frac{N}{\beta}\left[3 + \frac{1}{2}\frac{U(1.5,3,Z)}{U(2.5,5,Z)}\right], \quad Z = 2\sigma \tag{4.4.8}$$

where U(a,b,Z) is a confluent hypergeometric function. An advantage of expressing the free energies as in Eqs. (4.4.7) and (4.4.8) is that no complex numbers are needed. Integral representations of U(1.5,3,Z) and U(2.5,5,Z) were numerically evaluated using both four-point and six-point Gauss-Laguerre quadratures [Carnahan, et al, 1969]. The difference between each of the four-point and six-point quadrature values was less than one percent, which is sufficient accuracy for our purposes. Gauss-Laguerre does not yield asymptotically correct values for the remaining function U(0.5,1,Z). It is evaluated using Lebedev's [1965] series expansion for $Z < 1$, and by asymptotic approximation for $Z > 10$. Intermediate values are interpolated by nonlinear regression [Kuester and Mize, 1973]. Results of the calculations are shown in Figures 4.1 through 4.3. The functions are seen to be quite smooth on a log-log plot.

FIGURE 4.1

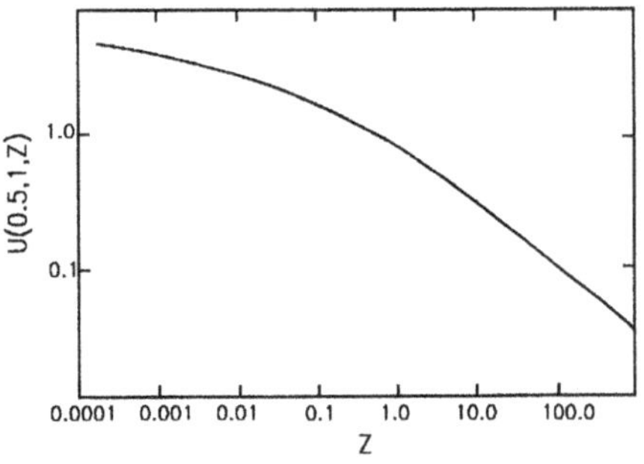

FIGURE 4.2

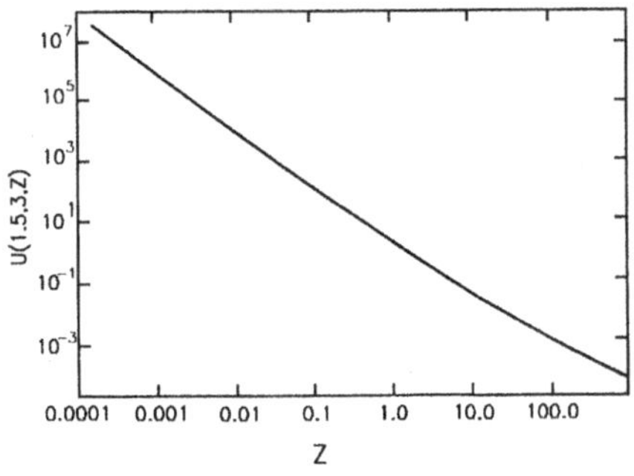

FIGURE 4.3

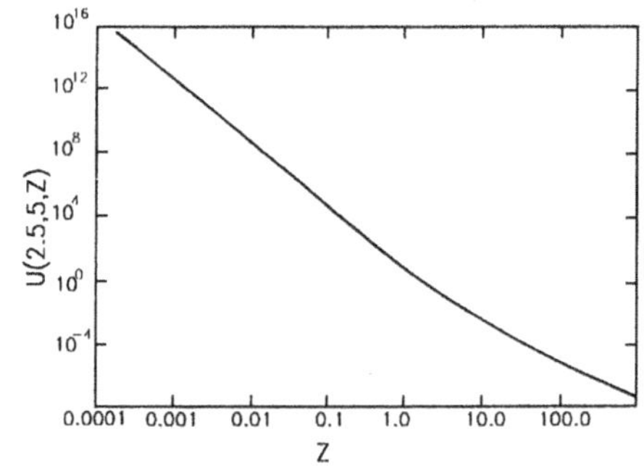

A physically more interesting plot is the semilogarithmic graph of ω_J/ω_H versus Z shown in Figure 4.4. According to Figure 4.4, the free energies are equivalent for values of Z greater than 10. The UR regime is reached for values of Z less than 0.1. The relatively narrow region from Z=0.1 to Z=10 is the region of transition from nonrelativistic to ultrarelativistic behavior. These values are consistent with those of Konigl [1980].

Using Eq. (4.4.6), the transition region occurs for temperatures ranging from $T=1.1\times10^{13}$ K to $T=1.1\times10^{11}$ K, and the UR limit is reached at $T=1.1\times10^{13}$ K. For comparison, the central temperature in the interior of the Sun is estimated [Kolb and Turner, 1990] to be about 1.6×10^{7} K, which is several orders of magnitude less than the temperature associated with the beginning of the transition region. It is unlikely that the different free energies shown in Eqs. (4.4.7) and (4.4.8) have observable consequences in either stellar astrophysics or earthbound relativistic gas dynamics. They can, however, affect cosmological models during epochs when temperatures are high enough to be in the UR regime. An illustration of this effect is presented in Chapter 5.

FIGURE 4.4

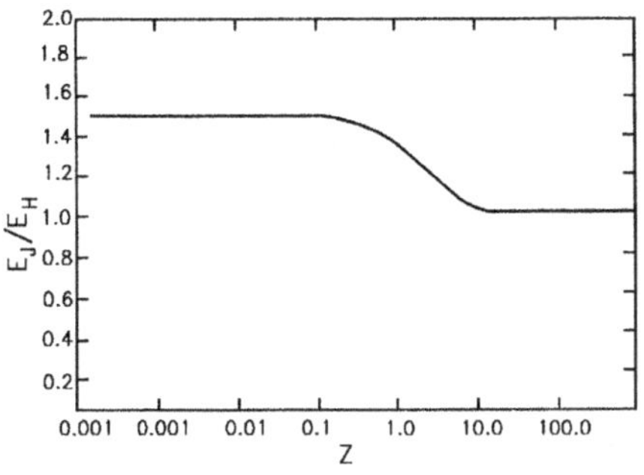

CHAPTER 5
GENERAL RELATIVITY IN PRCM

In all cases considered to date, PRQT has been compatible with observations while simultaneously removing some anomalies of the standard paradigm of relativistic quantum mechanics (see Chapter 1). Definitive tests of the validity or necessity of introducing a scalar parameter are needed. These tests must take advantage of differences between the parametrized and conventional formulations. One difference is the calculation of the free energy of an ensemble of independent, classical relativistic systems -- the Gibbs ensemble. We now turn consider some implications of this difference.

When the temperature of an ideal gas is sufficiently high, the mass-energy density of the stress-energy tensor becomes theory dependent and provides an opportunity for experimentally testing the validity of parametrized relativistic classical mechanics (PRCM). Horwitz, Schieve and Piron [1981] showed that the free energy of an ideal gas in the ultrarelativistic (UR) regime of PRCM is different from the free energy calculated by Jüttner [1911; Pauli, 1958] using the conventional theory of relativistic classical mechanics. The free energy difference noted by Horwitz, et al [1981] is an area of disagreement between PRCM and the conventional paradigm of relativistic classical mechanics (RCM) that may have measurable consequences.

The observation in Chapter 4 that the free energy difference is insignificant except at high temperatures leads us to consider only physical phenomena where the UR regime may be reached. Examples include the interior of stars, relativistic gas dynamics, and the first minute after the Big Bang. Of these, we showed in Chapter 4 that negligible differences are expected at realistic temperatures for relativistic gas dynamics and stellar models. This demonstration required an evaluation of the PRCM and RCM free energies over a wide range of temperatures extending from the non-relativistic limit to the UR limit. The greatest differences between PRCM and RCM are manifested in the density parameter of cosmological models. To study this effect, a general relativistic formalism is derived below from variational principles within the context of PRCM. Scale factors of Friedmann universes are then calculated. We show that the calculated age of the universe is theory dependent.

§5.1 VARIATIONAL FORMULATION

One of the most straightforward ways to develop a parametrized general relativistic formalism is to invoke a variational principle that is analogous to the Hilbert-Palatini variational principle [Misner, et al, 1973]. We begin by defining a parameter dependent action

$$A = \int_{s_1}^{s_2} \int_D \mathcal{L} \, dx \, ds \tag{5.1.1}$$

where $\mathcal{L}$ is the Lagrangian density. Equation (5.1.1) is the classical analog of the action used to develop a parametrized quantum field theory [Fanchi, 1979]. As in that case, the integral over the parameter is for fixed end points. The Lagrangian density $\mathcal{L}$ is a sum of geometric and particle field terms:

$$\mathcal{L} = \mathcal{L}_g + \mathcal{L}_f = \sqrt{-g}\,(L_g + L_f), \tag{5.1.2}$$

where L_g and L_f are Lagrangians. For simplicity, we assume the geometric factors L_g and the determinant g are formed from the elements of the fundamental metric tensor independent of s. The particle field term L_f may be parameter dependent.

Hamilton's principle requires that the variation of A vanish:

$$\delta A = 0 = \int_{s_1}^{s_2} \int_D [\delta(\sqrt{-g})\,L_g + \sqrt{-g}\,\delta L_g + \delta \mathcal{L}_f] \, dx \; ds. \tag{5.1.3}$$

The variations of the geometric terms may be treated in the usual manner [Misner, et al, 1973]. In particular, the geometric Lagrangian is expressed in terms of the Ricci tensor as

$$L_g = \frac{1}{16\pi} g^{\alpha\beta} R_{\alpha\beta}. \tag{5.1.4}$$

Given Eq. (5.1.4), the variation of the geometric Lagrangian is

$$\delta(\sqrt{-g} L_g) = \frac{\sqrt{-g}}{16\pi} \left[\left(R_{\alpha\beta} - \frac{1}{2} g_{\alpha\beta} R \right) \delta g^{\alpha\beta} + g^{\alpha\beta} \delta R_{\alpha\beta} \right] \tag{5.1.5}$$

where we have used the variation of the metric term in the form

$$\delta(\sqrt{-g}) = -\frac{1}{2} \sqrt{-g}\, g_{\alpha\beta} \delta g^{\alpha\beta}. \tag{5.1.6}$$

Except for straightforward generalizations to allow for a parameter dependence of the particle Lagrangian density, we proceed as in Landau and Lifshitz [1962] to evaluate the variation of the particle term in the action integral. The particle Lagrangian density is allowed to have the functional form

$$\mathcal{L}_f = \mathcal{L}_f(q, q_{,\mu}, \mathring{q}, g^{\alpha\beta}, s), \tag{5.1.7}$$

where

$$q_{,\mu} = \frac{\partial q}{\partial x^\mu}, \mathring{q} = \frac{dq}{ds}. \tag{5.1.8}$$

Notice that the evolution parameter s is not a fifth coordinate as in a Kaluza-Klein theory [Pauli, 1958; Iyer, et al, 1989]. It is a relativistic scalar as in Chapter 2 through Chapter 4. The variation of the particle Lagrangian density is

$$\begin{aligned} \delta \mathcal{L}_f = \frac{\partial \mathcal{L}_f}{\partial q} \delta q + \frac{\partial \mathcal{L}_f}{\partial q_{,\mu}} \delta q_{,\mu} + \frac{\partial \mathcal{L}_f}{\partial \mathring{q}} \delta \mathring{q} \\ + \frac{\partial \mathcal{L}_f}{\partial q^{\alpha\beta}} \delta q^{\alpha\beta} + \frac{\partial \mathcal{L}_f}{\partial s} \delta s. \end{aligned} \tag{5.1.9}$$

Assuming fixed parameter end points, we perform an integration by parts and obtain

$$\int_{s_1}^{s_2}\int_D \frac{\partial \mathcal{L}_f}{\partial s}\,\delta s\,dx\,ds = 0. \tag{5.1.10}$$

Another integration by parts applied to the δ(dq/ds) term yields

$$\begin{aligned}\int_{s_1}^{s_2}\int_D \frac{\partial \mathcal{L}_f}{\partial \mathring{q}}\,\delta\left[\frac{dq}{ds}\right] dx\,ds &= \int_{s_1}^{s_2}\int_D \frac{\partial \mathcal{L}_f}{\partial \mathring{q}}\frac{d}{ds}(\delta q)\,dx\,ds \\ &= -\int_{s_1}^{s_2}\int_D \left[\frac{d}{ds}\left(\frac{\partial \mathcal{L}_f}{\partial \mathring{q}}\right)\right]\delta q\,dx\,ds,\end{aligned} \tag{5.1.11}$$

where the condition that δq vanishes at the parameter end points has been used. A simplification of the $\delta q_{,\mu}$ term is achieved by invoking the chain rule to find

$$\begin{aligned}\frac{\partial \mathcal{L}_f}{\partial q_{,\mu}}\,\delta q_{,\mu} &= \frac{\partial \mathcal{L}_f}{\partial q_{,\mu}}\frac{\partial(\delta q)}{\partial x^{\mu}} \\ &= \frac{\partial}{\partial x^{\mu}}[\pi^{\mu}\,\delta q] - \left[\frac{\partial \pi^{\mu}}{\partial x^{\mu}}\right]\delta q,\end{aligned} \tag{5.1.12}$$

where

$$\pi^{\mu} = \frac{\partial \mathcal{L}_f}{\partial q_{,\mu}}. \tag{5.1.13}$$

Collecting the results in Eqs. (5.1.10) through (5.1.13) gives

$$\int_{s_1}^{s_2}\int_D \delta \mathcal{L}_f dx\, ds =$$

$$\int_{s_1}^{s_2}\int_D \left\{ \left[\frac{\partial \mathcal{L}_f}{\partial q} - \frac{\partial \pi^\mu}{\partial x^\mu} - \frac{d}{ds}\left(\frac{\partial \mathcal{L}_f}{\partial \mathring{q}} \right) \right] \delta q \right\} dx\, ds \tag{5.1.14}$$

$$+ \int_{s_1}^{s_2}\int_D \left\{ \frac{\partial}{\partial x^\mu} (\pi^\mu \delta q) + \frac{\partial \mathcal{L}_f}{\partial g^{\alpha\beta}} \delta g^{\alpha\beta} \right\} dx\, ds.$$

The divergence term

$$\int_{s_1}^{s_2}\int_D \frac{\partial}{\partial x^\mu} (\pi^\mu \delta q)\, dx\, ds$$

vanishes upon transformation by Gauss' theorem and integration over all space. Furthermore, the metric variation term can be written as

$$\frac{\partial \mathcal{L}_f}{\partial g^{\alpha\beta}} = \frac{\partial \sqrt{-g} L_f}{\partial g^{\alpha\beta}} = \frac{\partial \sqrt{-g}}{\partial g^{\alpha\beta}} L_f + \sqrt{-g} \frac{\partial L_f}{\partial g^{\alpha\beta}}$$

$$= \left[-\frac{1}{2} g_{\alpha\beta} L_f + \frac{\partial L_f}{\partial g_{\alpha\beta}} \right] \sqrt{-g}. \tag{5.1.15}$$

Substituting the results of Eqs. (5.1.4) through (5.1.15) in Eq. (5.1.3) lets us write

$$\delta A = 0$$

$$= \int_{s_1}^{s_2}\int_D \left[\left(\frac{1}{8\pi} G_{\alpha\beta} - T_{\mathrm{AB}} \right) \frac{\delta g^{\alpha\beta}}{2} + g^{\alpha\beta} \delta R_{\alpha\beta} \right] \sqrt{-g}\, dx\, ds \tag{5.1.16}$$

$$+ \int_{s_1}^{s_2}\int_D \left\{ \left[\frac{\partial \mathcal{L}_f}{\partial q} - \frac{\partial \pi^\mu}{\partial x^\mu} - \frac{d}{ds}\left(\frac{\partial \mathcal{L}_f}{\partial \mathring{q}} \right) \right] \delta q \right\} dx\, ds.$$

where the Einstein curvature tensor has been identified as

$$G_{\alpha\beta} \equiv R_{\alpha\beta} - \frac{1}{2} g_{\alpha\beta} R, \tag{5.1.17}$$

and the stress-energy tensor is

$$T_{\alpha\beta} \equiv g_{\alpha\beta} L_f - 2 \frac{\partial L_f}{\partial g^{\alpha\beta}} . \tag{5.1.18}$$

Hamilton's principle, as embodied in Eq. (5.1.16), can hold for arbitrary variations only if the coefficients of the variations vanish. The resulting equations are the general relativistic formalism we seek.

FIELD EQUATIONS

The vanishing of the coefficient of the variation of the Ricci tensor $R_{\alpha\beta}$ leads to the standard equation for the connection coefficients [Misner, et al, 1973]. Parametrized Euler-Lagrange field equations result from the vanishing of the coefficient of δq:

$$\frac{\partial \mathcal{L}_f}{\partial q} - \frac{\partial \pi^\mu}{\partial x^\mu} - \frac{d}{ds}\left(\frac{\partial \mathcal{L}_f}{\partial \mathring{q}} \right) = 0. \tag{5.1.19}$$

Finally, the requirement that the coefficient of the variation of the metric tensor vanish yields

$$G_{\alpha\beta} = 8\pi T_{\alpha\beta} \tag{5.1.20}$$

Equation (5.1.20) is Einstein's second-order equation for space-time geometry.

Each of the above equations is expressed in terms of Lagrangian functions. Most applications of the parametrized formulation have been in terms of Hamiltonian functions. All parametrized Hamiltonians used to date are based on a flat space-time metric. The classical Lagrangian and Hamiltonian formulations can be related in the usual way for flat space-time. The presence of a curved background metric complicates the connection and goes beyond the scope of this work.

§5.2 COSMOLOGICAL APPLICATION

Einstein's curvature tensor $G_{\alpha\beta}$ is parameter dependent if the stress-energy tensor $T_{\alpha\beta}$, through the Lagrangian L_f, is parametrized. If L_f is stationary with respect to the parameter s, i.e. L_f is not parametrized, then Eq. (5.1.20) is the usual relativistic classical mechanical expression. It was pointed out in §5.1 that this is the case for the comparison we are considering. We can therefore use the standard model of cosmology [Brandenberger, 1985; Felten and Isaacman, 1986; Particle Data Group, 1992] as our test model for evaluating the impact of Eqs. (4.4.7) and (4.4.8) on cosmology. In particular, we are interested in studying the effect of Eqs. (4.4.7) and (4.4.8) on the Friedmann scale factor. We recognize that the classical treatment, especially at the high temperatures needed to reach the ultrarelativistic (UR) regime, is only approximate [Brandenberger, 1985]. Nevertheless, the comparison given below indicates that a more rigorous treatment could lead to significant differences between the cosmologies of parametrized and nonparametrized formulations of relativistic physics.

STANDARD COSMOLOGICAL MODEL

We begin our comparison by summarizing the derivation of the standard cosmological model. Invoking the cosmological principle, i.e. three-space is homogeneous and isotropic, and assuming the metric is Riemannian leads to the Robertson-Walker metric:

$$ds^2 = -dt^2 + R^2(t)\left[\frac{dr^2}{1-K_0 r^2} + r^2(d\theta^2 + \sin^2\theta\, d\phi^2)\right]. \tag{5.2.1}$$

R(t) is a time dependent scale factor with the initial condition

$$R_0 \equiv R(0) = 1 \tag{5.2.2}$$

where K_0 is an arbitrary constant. The model universe is either closed, open, or spacially flat if K_0 is +1, -1, or 0 respectively. The Gaussian curvature of three-space is given in terms of R(t) and K_0 as

$$K(t) = \frac{K_0}{R^2},\ K(0) \equiv K_0, \tag{5.2.3}$$

where the initial condition of K(t) is derived from Eq. (5.2.2).

Substitution of Eq. (5.2.1) in the Einstein field equations leads to the Lemaître equations:

$$\left[\frac{1}{R}\frac{dR}{dt}\right]^2 = \frac{8}{3}\pi G\rho(t) - c^2\frac{K_0}{R^2} + \frac{\Lambda}{3}, \tag{5.2.4}$$

and

$$\frac{d\rho(t)}{dt} = -3\left[\rho(t) + \frac{p(t)}{c^2}\right]\left(\frac{1}{R}\frac{dR}{dt}\right). \tag{5.2.5}$$

The speed of light c has been made explicit, $\rho(t)$ is the mass density including the mass equivalent of all energy present, p(t) is pressure, G is the gravitational constant, and Λ is the cosmological constant. The Lemaître equations together with an equation of state relating p and ρ, determine the behavior of R(t).

The Friedmann scale factor is the solution of the Friedmann equation, which is derived from Eqs. (5.2.4) and (5.2.5) by making the assumption $p \ll \rho c^2$. In this case Eq. (5.2.5) can be solved to give

$$\rho(t) = \frac{\rho_0}{R^3},\ \rho(0) \equiv \rho_0, \tag{5.2.6}$$

and Eq. (5.2.4) takes on the simplified form

$$\left[\frac{1}{R}\frac{dR}{dt}\right]^2 = \frac{8}{3}\pi G\frac{\rho_0}{R^3} - c^2\frac{K_0}{R^2} + \frac{\Lambda}{3}. \tag{5.2.7}$$

Following Felten and Isaacman [1986], we introduce the Hubble parameter and dimensionless density (or closure) parameter with the definitions

$$H(t) = \frac{1}{R}\frac{dR}{dt}, H_0 \equiv H(0) \tag{5.2.8}$$

and

$$\Omega(t) = \frac{8}{3}\pi G \frac{\rho(t)}{H^2(t)}, \Omega_0 \equiv \Omega(0) = \frac{8}{3}\pi G \frac{\rho_0}{R_0^3 H_0^2}. \tag{5.2.9}$$

The Friedmann equation can be rewritten into terms of these parameters as

$$c^2 K(t) = \frac{8}{3}\pi G \rho(t) + \frac{\Lambda}{3} - H^2(t). \tag{5.2.10}$$

Introducing the dimensionless time

$$t_D \equiv H_0 t, \tag{5.2.11}$$

and substituting Eqs. (5.2.8) through (5.2.11) in Eq. (5.2.7) gives

$$\left[\frac{1}{R}\frac{dR}{dt_D}\right]^2 = \Omega_0 \frac{(1-R)}{R^3} + \frac{1}{R^2} + \frac{1}{3}\left[\frac{\Lambda}{H_0^2}\right]\frac{(R^2-1)}{R^2}. \tag{5.2.12}$$

Age of the Universe

A comparison of the parametrized and nonparametrized results in the UR case is made by first writing down distinct density parameters for Jüttner and Horwitz, et al: they are

$$\Omega_{0J} = \frac{8}{3}\pi G \frac{\rho_{0J}}{R_{0J}^3 H_{0J}^2} \tag{5.2.13}$$

and

$$\Omega_{0H} = \frac{8}{3}\pi G \frac{\rho_{0H}}{R_{0H}^3 H_{0H}^2}. \tag{5.2.14}$$

Their ratio is

$$\frac{\Omega_{0J}}{\Omega_{0H}} = \frac{\rho_{0J}}{\rho_{0H}} \frac{R_{0H}^3 H_{0H}^2}{R_{0J}^3 H_{0J}^2}. \tag{5.2.15}$$

Neither the scale factor nor the Hubble constant depend on the parameter s. It is reasonable to treat the factors as theory independent. In this case Eq. (5.2.15) simplifies to the form

$$\frac{\Omega_{0J}}{\Omega_{0H}} = \frac{\rho_{0J}}{\rho_{0H}} = \frac{\omega_J}{\omega_H}, \tag{5.2.16}$$

where we have included the relation between the ratios of free energy and mass density. Equation (5.2.16) shows that the density parameters are significantly different during the UR regime. The difference in theoretical density parameters leads to substantial differences in such calculated quantities as the age of Friedmann universes.

Several model ages have been calculated from Eq. (5.2.12) using the numerical procedure described by Felten and Isaacman [1986]. They are presented in Table 5.1. Each model age is shown as a range extending over the numerical step size (0.0025) of the dimensionless time t_D. Accompanying each pair of density parameter values for a given free parameter (Λ/H_0^2) value is the ratio of model ages corresponding to Eq. (5.2.16).

TABLE 5.1 AGE OF FRIEDMANN UNIVERSES

FREE PARAMETER Λ/H_0^2	DENSITY PARAMETER Ω_0	HUBBLE TIME RANGE UNITS OF $1/H_0$	HUBBLE TIME RATIO
-10	0.15	0.5575-0.5600	
	0.10	0.5650-0.5675	0.987
0	0.15	0.8700-0.8725	
	0.10	0.8975-0.9000	0.969
3	0.15	1.2625-1.2650	
	0.10	1.3750-1.3775	0.918
-10	1.50	0.4700-0.4725	
	1.00	0.4925-0.4950	0.954
0	1.50	0.6100-0.6125	
	1.00	0.6650-0.6675	0.917
3	1.50	0.6950-0.6975	
	1.00	0.7850-0.7875	0.885

Equation (5.2.16) does not lead to a definitive observational test because of uncertainties associated with important cosmological parameters, e.g. the Hubble constant (50-100 km sec/Mpc) and the cosmological constant. For example, the value of a Hubble time unit ($1/H_0$) varies from $(10\text{-}20)\times 10^9$ yr depending on the value of Hubble's constant [Felten and Isaacman, 1986; Particle Data Group, 1992]. Thus, even though theoretical differences are calculable, the free energy difference first noted by Horwitz et al [1981] cannot be used at present as the basis for an observational test of PRCM. Our results do show, however, that cosmology may eventually become a fruitful testing ground for parametrized relativistic theories.

PART II:

FOUNDATIONS OF PARAMETRIZED RELATIVISTIC QUANTUM MECHANICS

CHAPTER 6
SYMMETRY PRINCIPLES

We are concerned with determining the correct partial differential equation(s) for describing a relativistic particle. Several forms were suggested in Part I. All of them included an invariant evolution parameter. In this chapter we present a physical foundation for using a parametrized wave equation with the general form

$$i\frac{\partial \psi}{\partial s} = \boldsymbol{K}\psi$$

where s is the invariant evolution parameter and **K** is the generator of translations in s. By insisting that the theory retain a probabilistic basis, we obtain mathematical constraints on the form of the generator **K** in Chapter 7. A link between the mathematical formalism and physics is postulated in Chapter 8. The remaining chapters test the fundamental assumptions of Chapter 6 through Chapter 8 by examining applications of the formalism.

§6.1 FUNDAMENTAL PHYSICAL ASSUMPTIONS

Scientific theories consist of a mathematical formalism with a set of rules for relating mathematical quantities and operations to physical quantities and measurements. Relationships between data measurements are described by equations. One approach to constructing the mathematical formalism is to specify how physical variables should behave when observed from different reference frames. We adopt the fundamental physical assumption -- the principle of relativity -- that says relationships between physically significant quantities should remain invariant with respect to transformations between different reference frames.

Invariance requirements may be represented mathematically as transformations between coordinate systems. Coordinate transformations are represented by continuous symmetry groups, notably Lie groups. An r-parameter Lie group is a set of transformations with the properties of a group and labeled by r continuously varying parameters. Infinitesimal transformations must exist arbitrarily close to the identity element of the group. Infinitesimal generators may be constructed from the infinitesimal transformations to linearize the law of group multiplication. The generators form a vector space called the Lie algebra.

We present the symmetry properties of PRQT by first describing the invariance requirements. From the invariance requirements we obtain a set of transformations which lead naturally to the Lie algebra of PRQT.

INVARIANCE REQUIREMENTS

One expression of the principle of relativity is to assert that two reference frames are equivalent only if they are at rest with respect to each other. Such transformations include three-dimensional rotations and translations in Euclidean space, and are described by the Euclidean group. The Euclidean group requires that the equations of physics be the same for two observers at rest with respect to each other, but the equations may change if the observers are moving relative to one another. If we want the equations of physics to be valid whether we are standing on land, traveling in sea, or moving through space, then the Euclidean group is too restrictive.

We extend the principle of relativity by requiring that the equations of physics be invariant for observers in relative motion. Suppose reference frame F2 is moving relative to frame F1. If we are at rest in frame F1, we can measure the motion of frame F2 as the change in the spacial location of its origin divided by the change in a monotonically increasing evolution parameter. We require that the evolution parameter increase at the same rate in both reference frames, i.e. it is independent of the frame in which it is measured. We say the two reference frames are in uniform motion relative to one another, and we obtain the Galilei group associated with Newtonian mechanics. The evolution parameter is just the absolute time of Galileo and Newton.

The equations of Newtonian mechanics are invariant with respect to transformations associated with the Galilei group, but Maxwell's equations are not. The Michelson-Morley experiment showed that the invariance of Maxwell's equations should be preserved, even at the expense of Newtonian symmetry. New transformation requirements were needed. They were provided by Einstein's special relativity.

Special relativity requires that inertial (non-accelerating) observers in equivalent reference frames measure the same speed for the propagation of light in vacuum. The invariant evolution parameter of Newtonian mechanics is replaced by an additional coordinate on a four-dimensional Minkowski space. The symmetry requirements of special relativity are embodied in the inhomogeneous Lorentz group, which is also known as the Poincaré group.

The Poincaré group presumes that the mass of a particle is a constant parameter. Mass is not calculated by the theory, but is put into the theory. One of the outstanding experimental findings of the past half century is the proliferation of particles of varying mass. The concepts of mass state and transitions between mass states are precluded by the assumption of Poincaré invariance. If we hope to improve our understanding of mass states, we must go beyond the transformations associated with the Poincaré group.

In this chapter we present a set of transformations that form the basis for quantizing the classical theory discussed in Part I. A basis for interpreting the formalism, particularly the wave function, is given in Chapter 7 and Chapter 8. Spin-0 and spin-½ wave equations are considered in Chapter 9 and Chapter 10 respectively. A procedure for measuring the evolution parameter is presented in Chapter 11.

§6.2 THE JARS GROUP

POSTULATES

We begin by specifying four postulates for quantizing the classical theory:

P1. The states of a physical system are characterized by unit rays in a normed, linear vector space (Hilbert space). A transition probability is defined as the modulus of the inner product between any two states.

P2. We assume the principle of linear superposition: if two states of the system are represented by ψ_1 and ψ_2, then

$$\Psi = a_1 \psi_1 + a_2 \psi_2 \tag{6.2.1}$$

is also a state of the system.

P3. We associate a Lie algebra with a set of transformations. The transformations map states of a system into themselves and must preserve transition probabilities. Generators of the transformations will be associated with physical measurements. Representations of the algebra are linear operators on a linear vector space. Each physical observable is represented by a linear Hermitian operator that has real eigenvalues under suitable mathematical restrictions.

P4. We assume the dynamical development of the system is given by a unitary operator that is a function of an invariant evolution parameter.

Postulates P1 through P3 are familiar from the conventional formulation of quantum theory. Postulate P4 introduces the idea of an invariant evolution parameter. We show below that it leads to a parametrized wave equation.

ALGEBRAIC PROPERTIES

The Lie algebra of PRQT is constructed by assuming that relationships between physically significant quantities are invariant with respect to linear transformations of the form

$$x_\mu{}' = \Lambda_\mu^\nu x_\nu + a_\mu + b_\mu s \tag{6.2.2}$$

where $\{\Lambda_\mu{}^\nu\}$ represent a homogeneous Lorentz transformation, $\{a_\mu\}$ represent translations along the $\{x_\mu\}$ axes, and the four-velocity transformation ("relativistic Galilean boost" of Aghassi, Roman and Santilli [1970]) is characterized by the four parameters $\{b_\mu\}$. In the parametrized theory with an independent variable s, the theory is assumed to be invariant with respect to the linear transformation

$$s' = s + \Delta s \tag{6.2.3}$$

where Δs represents translations along the s axis. We refer to the set of transformations represented by Eqs. (6.2.2) and (6.2.3) as the JARS transformation. The acronym comes from the names of the people who first studied the properties of the set, i.e. Johnson [1969], and Aghassi, Roman and Santilli [1970].

A direct comparison of the JARS transformation with the nonrelativistic Galilei transformation was presented by Aghassi, et al [1970] and is illustrated in Table 6.1. The physical quantities in the Galilei transformation are the rotation matrix **R**, the nonrelativistic boost v_G, and the spacial translation a_G. Several analogies are obvious. Perhaps the most interesting from our perspective is the relationship between the invariant evolution parameter s and Galilean time. Both may be viewed as monotonically increasing ordering parameters for characterizing the evolution of a physical system. The relationship between the relativistic and nonrelativistic ordering parameters is explored in more detail in Chapter 11.

TABLE 6.1 TRANSFORMATIONS

- Homogenous Lorentz Transformation

$$x'^{\mu} = \Lambda^{\mu}_{\nu} x^{\nu}$$

- Inhomogenous Lorentz Transformation

$$x'^{\mu} = \Lambda^{\mu}_{\nu} x^{\nu} + a^{\mu}$$

- JARS Transformation

$$x'^{\mu} = \Lambda^{\mu}_{\nu} x^{\nu} + a^{\mu} + b^{\mu} s$$
$$s' = s + \Delta s$$

- Galilei Transformation

$$x_G'^{j} = R^{j}_{i} (x_G)^{i} + (a_G)^{j} + (v_G)^{j} t_G$$
$$t_G' = t_G + \Delta t_G$$

The JARS transformations may be denoted by $\{a,b,\Lambda,\Delta s\}$, which is the group element of a 15-parameter continuous group. The identity element is $\{0,0,1,0\}$ and the inverse element is $\{-\Lambda^{-1}(a-b\Delta s),-\Lambda^{-1}b,\Lambda^{-1},-\Delta s\}$. The composition law of two group elements is

$$\{a_2, b_2, \Lambda_2, \Delta s_2\}\{a_1, b_1, \Lambda_1, \Delta s_1\} = \{a_2 + \Lambda_2 a_1 + (\Delta s_1) b_2, b_2 + \Lambda_2 b_1, \Lambda_2 \Lambda_1, \Delta s_1 + \Delta s_2\} \tag{6.2.4}$$

which expresses the product of two transformations. We call the group associated with the JARS transformation the JARS group.

Johnson [1969] and Aghassi, et al [1970] presented an algebraic structure with 15 generators. For an infinitesimal transformation the set $\{a_\mu,b_\mu,\Lambda_\mu{}^\nu,\Delta s\}$ becomes $\{\epsilon_{p\mu},\epsilon_{x\mu},g_\mu{}^\nu+\epsilon_\mu{}^\nu,\epsilon\}$ where $\{\epsilon_{p\mu},\epsilon_{x\mu},\epsilon_\mu{}^\nu,\epsilon\}$ are first order infinitesimals. The infinitesimal generator for dynamical development of the system is

$$U = 1 + iF_J \tag{6.2.5}$$

where

$$F_J = \frac{1}{2} J^{\mu\nu} \epsilon_{\mu\nu} + P^{\mu} (\epsilon_p)_{\mu} + X^{\mu} (\epsilon_x)_{\mu} + K\epsilon . \tag{6.2.6}$$

Ten generators are the six independent elements of the antisymmetric angular momentum tensor $\mathbf{J}^{\mu\nu}$ and the four elements of the energy-momentum four-vector $\mathbf{P}^{\mu}$. The generators of the Poincaré group satisfy the 45 usual commutation relations

$$[P_{\mu}, P^{\nu}] = 0, \tag{6.2.7}$$

$$[J_{\mu\nu}, P_{\lambda}] = i(P_{\nu} g_{\mu\lambda} - P_{\mu} g_{\nu\lambda}), \tag{6.2.8}$$

$$[J_{\mu\nu}, J_{\lambda\sigma}] = i(g_{\mu\sigma} J_{\nu\lambda} + g_{\mu\lambda} J_{\nu\sigma} + g_{\nu\sigma} J_{\mu\lambda} + g_{\lambda\nu} J_{\sigma\mu}). \tag{6.2.9}$$

The generator **K** generates infinitesimal translations along s. Ten commutators are determined by taking the commutator of the generator **K** with each of the ten generators of the Poincaré group. The generator **K** commutes with all of the generators of the Poincaré group:

$$[J_{\mu\nu}, K] = 0, \tag{6.2.10}$$

$$[P_{\mu}, K] = 0. \tag{6.2.11}$$

The additional four generators are the space-time position operators that satisfy the commutators

$$[X^{\mu}, X^{\nu}] = 0, \tag{6.2.12}$$

$$[P^{\mu}, X^{\nu}] = i g^{\mu\nu}, \tag{6.2.13}$$

$$[J^{\mu\nu}, X^{\lambda}] = i[g^{\lambda\nu} X^{\mu} - g^{\lambda\mu} X^{\nu}]. \tag{6.2.14}$$

In the classical theory, the commutator $[\mathbf{P}^\mu, \mathbf{X}^\mu]$ vanishes. Horwitz and Piron [1973; also see Horwitz, 1983] have shown that the position operator in Eq. (6.2.13) is the covariant relativistic generalization of the Newton-Wigner [1949] position operator.

The choice of the form of the generator **K** of s translations is not unique. Several forms have been proposed [Fanchi, 1993]. The one that is suitable for our purposes is introduced in Chapter 7. First, however, we show that the combination of symmetry principles and canonical transformations leads to a wave equation. We begin by reviewing the concept of canonical transformation and show that the JARS transformation is a canonical transformation. Commutation relations for the operators associated with the JARS transformation are then used to derive a parametrized wave equation in §6.4.

§6.3 CANONICAL TRANSFORMATIONS

The canonical transformation of a field $\psi(x,s)$ to $\psi'(x,s)$ by a unitary operator **U** is defined by

$$\psi'(x,s) = U\,\psi(x,s)\,U^{-1}. \tag{6.3.1}$$

Since **U** is unitary, its adjoint (conjugate transpose) $\mathbf{U}^+$ satisfies the relation

$$U^+ = U^{-1} \tag{6.3.2}$$

so that

$$U\,U^+ = U^+\,U = 1. \tag{6.3.3}$$

All physical properties of the system described by the original field $\psi(x,s)$ are unchanged in the new system described by the transformed field $\psi'(x,s)$. A quantity which is invariant with respect to a canonical transformation is considered physically significant.

An infinitesimal canonical transformation can be obtained by writing **U** as

$$U = 1 + iF \tag{6.3.4}$$

where **F** is a first order infinitesimal. If we apply the condition for unitarity in Eq. (6.3.3) to Eq. (6.3.4) and neglect second order infinitesimals we find

$$U^{+}U = (1 - iF^{+})(1 + iF) \approx 1 - i(F^{+} - F). \tag{6.3.5}$$

The right hand side of Eq. (6.3.5) must equal one by Eq. (6.3.3). This is satisfied for all **F** only if **F** is Hermitian, thus

$$F^{+} = F\,. \tag{6.3.6}$$

The operator **F** is the generator of the infinitesimal canonical transformation **U**. The finite canonical transformation is obtained by iterating Eq. (6.3.4), thus

$$U = \lim_{n\to\infty}\left(1 + \frac{i}{n}F\right)^{n} = e^{iF} \tag{6.3.7}$$

and

$$U^{+} = \lim_{n\to\infty}\left(1 - \frac{i}{n}F\right)^{n} = e^{-iF}, \tag{6.3.8}$$

where **F** is now considered a finite operator.

Substituting Eq. (6.3.4) into (6.3.1) gives

$$\psi'(x,s) = (1 + iF)\,\psi(x,s)\,(1 - iF) \tag{6.3.9}$$

or, to first order,

$$\psi'(x,s) = \psi(x,s) + i[F, \psi(x,s)]. \tag{6.3.10}$$

If the total variation or change of the field ψ(x,s) by an infinitesimal canonical transformation is defined by

$$\delta_0\psi(x,s) = \psi'(x,s) - \psi(x,s), \tag{6.3.11}$$

then substituting Eq. (6.3.11) into (6.3.10) expresses the variation of the field in terms of the commutator of **F** with ψ(x,s); thus

$$\delta_0\psi(x,s) = i[F, \psi(x,s)]\,. \tag{6.3.12}$$

A similar expression can be obtained for any operator **Ω** satisfying Eq. (6.3.1), hence

$$\delta_0 \boldsymbol{\Omega} = i[\boldsymbol{F}, \boldsymbol{\Omega}]. \tag{6.3.13}$$

The corresponding finite transformation is found using Eqs. (6.3.1), (6.3.7) and (6.3.8):

$$\psi(x,s) = e^{iF}\,\psi(x,s)\,e^{-iF}. \tag{6.3.14}$$

A similar result is obtained for $\boldsymbol{\Omega}$:

$$\boldsymbol{\Omega}' = e^{iF}\,\boldsymbol{\Omega}\,e^{-iF}. \tag{6.3.15}$$

§6.4 THE JARS TRANSFORMATION AS A CANONICAL TRANSFORMATION

An infinitesimal JARS transformation can be written as

$$\begin{aligned} \Lambda_\mu^\nu &= g_\mu^\nu + \epsilon_\mu^\nu, \\ a_\mu &= \epsilon_{x\mu}, \\ b_\mu &= \epsilon_{p\mu}, \\ \delta s &= \epsilon \end{aligned} \tag{6.4.1}$$

where $(\epsilon, \epsilon_{p\mu}, \epsilon_{x\mu}, \epsilon_\mu^{\ \nu})$ are first order infinitesimals. The infinitesimal generator for dynamical development of the system has the unitary operator representation

$$\boldsymbol{U}_F = 1 + i\boldsymbol{F} \tag{6.4.2}$$

where

$$\boldsymbol{F} = \frac{1}{2}\boldsymbol{J}^{\mu\nu}\epsilon_{\mu\nu} + \boldsymbol{P}^\mu(\epsilon_p)_\mu + \boldsymbol{X}^\mu(\epsilon_x)_\mu + \boldsymbol{K}\epsilon. \tag{6.4.3}$$

It is easy to prove that $\mathbf{U}_F$ is unitary to first order in infinitesimals by showing that

$$\boldsymbol{U}_F\boldsymbol{U}_F^+ = \boldsymbol{U}_F^+\boldsymbol{U}_F = 1 \tag{6.4.4}$$

as in §6.3. A finite JARS transformation is obtained by iterating Eq. (6.4.2). Since $\mathbf{U}_F$ is unitary, the JARS transformation can be interpreted as a canonical transformation.

A set of commutation relations of the operators $\mathbf{J}^{\mu\nu}$, $\mathbf{P}^\mu$, $\mathbf{X}^\mu$ and $\mathbf{K}$ with the field ψ is obtained by developing each side of Eq. (6.3.12) separately in terms of the infinitesimals $(\epsilon, \epsilon_{p\mu}, \epsilon_{x\mu}, \epsilon_\mu{}^\nu)$. The coefficients of the linear terms on both sides are then equated.

The term $\delta_0\psi$ may be expressed in an alternate form by introducing the local variation

$$\delta\psi \equiv \psi'(x', s') - \psi(x,s) \tag{6.4.5}$$

where x,s and x',s' refer to the same physical point as viewed from canonically transformed reference frames. Performing a Taylor series expansion of ψ'(x',s') and keeping only first order terms gives

$$\delta\psi = \psi'(x,s) + \partial^\mu \psi'(x,s)\,\delta x_\nu + \partial_s \psi'(x,s)\,\delta s - \psi(x,s). \tag{6.4.6}$$

Rearranging and using Eq. (6.3.11) yields

$$\delta_0\psi = \delta\psi - \partial^\nu\psi\,\delta x_\nu - \partial_s\psi\,\delta s. \tag{6.4.7}$$

The increments $\delta\psi$, δx_ν, δs are specified in terms of the infinitesimal JARS transformation:

$$\delta x_\nu = \epsilon_{\nu\mu} x^\mu + \epsilon_{p\nu} + \epsilon_{x\nu} s, \tag{6.4.8}$$

$$\delta s = \epsilon, \tag{6.4.9}$$

$$\delta\psi = \frac{1}{2}\Sigma^{\nu\mu}\psi(x,s)\,\epsilon_{\nu\mu} \tag{6.4.10}$$

where $\Sigma^{\nu\mu}$ are the infinitesimal generators of the proper homogeneous Lorentz group [Wu and Hwang, 1991; Bjorken and Drell, 1965]. Note that translations along x,s do not cause the variation $\delta\psi$, but do cause the variation $\delta_0\psi$.

Combining Eqs. (6.4.7) through (6.4.10) gives

$$\delta_0\psi = \frac{1}{2}[(\Sigma^{\nu\mu}\psi + (x^\nu\partial^\mu - x^\mu\partial^\nu)\psi]\epsilon_{\nu\mu} - (\partial^\nu\psi)\epsilon_{p\nu} - (\partial^\nu\psi)\epsilon_{x\nu}s - \dot{\psi}\epsilon \tag{6.4.11}$$

where the $\epsilon_{\nu\mu}$ term has been rewritten in the form

$$(\partial^\nu\psi)\epsilon_{\nu\mu}x^\mu = \frac{1}{2}[(x^\nu\partial^\mu - x^\mu\partial^\nu)\psi]\epsilon_{\nu\mu} \tag{6.4.12}$$

and the antisymmetry relation

$$\epsilon_{\nu\mu} = -\epsilon_{\mu\nu} \tag{6.4.13}$$

has been used. The commutator $i[\mathbf{F},\psi]$ is obtained by substituting Eq. (6.4.3) into Eq. (6.3.12), thus

$$i[\boldsymbol{F},\psi] = \frac{1}{2}i[\boldsymbol{J}^{\nu\mu},\psi]\epsilon_{\nu\mu} + i[\boldsymbol{P}^\nu,\psi]\epsilon_{p\nu} + i[\boldsymbol{X}^\nu,\psi]\epsilon_{x\nu} + i[\boldsymbol{K},\psi]\epsilon. \tag{6.4.14}$$

Comparing Eqs. (6.4.11) with (6.4.14) yields the commutation relations

$$i[\boldsymbol{J}^{\nu\mu},\psi] = \Sigma^{\nu\mu}\psi + (x^\nu\partial^\mu - x^\mu\partial^\nu)\psi, \tag{6.4.15}$$

$$i[\boldsymbol{P}^\nu,\psi] = -\partial^\nu\psi, \tag{6.4.16}$$

$$i[\boldsymbol{X}^\nu,\psi] = -s(\partial^\nu\psi), \tag{6.4.17}$$

and

$$i[\boldsymbol{K},\psi] = -\partial_s\psi. \tag{6.4.18}$$

Equations (6.4.15) through (6.4.18) can be considered the defining relations for the operators $\mathbf{J}^{\nu\mu}$, $\mathbf{P}^\nu$, $\mathbf{X}^\nu$, $\mathbf{K}$.

WAVE EQUATION

We may derive a wave equation from the results presented above. In particular, rearrange Eq. (6.4.18) so that it appears as an operator acting on an arbitrary field ϕ:

$$\{i(\partial_s\psi) - [\boldsymbol{K},\psi]\}\phi = 0. \tag{6.4.19}$$

The parentheses around the first term mean that the operator ∂_s operates only on the field ψ. Expanding the commutator bracket gives

$$\begin{aligned}[\boldsymbol{K},\psi]\phi &= [\boldsymbol{K}\psi - \psi\boldsymbol{K}]\phi \\ &= \psi(\boldsymbol{K}\phi) + (\boldsymbol{K}\psi)\phi - \psi(\boldsymbol{K}\phi) \\ &= (\boldsymbol{K}\psi)\phi .\end{aligned} \tag{6.4.20}$$

Combining Eq. (6.4.20) with Eq. (6.4.19) yields

$$\{i(\partial_s\psi) - (\boldsymbol{K}\psi)\}\phi = 0. \tag{6.4.21}$$

A wave equation is obtained by observing that Eq. (6.4.21) is satisfied for any field ϕ when the bracketed term vanishes, thus

$$i\frac{\partial\psi}{\partial s} = \boldsymbol{K}\psi. \tag{6.4.22}$$

Several forms for **K** have been proposed [Fanchi, 1993]. Each of the forms can be constructed from more basic principles, but in every formulation except one the order of $\mathbf{P}^\mu$ relative to the order of s is determined by a model of the system -- such as an assumed Lagrangian -- rather than fundamental assumptions such as probability conservation. Only a probabilistic formulation has resulted in a relationship between the order of $\mathbf{P}^\mu$ and the order of s that arises from first principles. Before presenting the probabilistic formulation in Chapter 7, we review alternative constructions of parametrized wave equations. This review should be viewed as representative of the constructions that have been made in the past. A more comprehensive review is given in Fanchi [1993].

Proposed Forms of the Parameter Translation Generator

The most common forms of **K** are the Dirac-like form

$$K_1 = \alpha_1 \gamma_\mu P^\mu , \tag{6.4.23}$$

the Schrödinger-like form

$$K_2 = \alpha_2 P_\mu P^\mu , \tag{6.4.24}$$

and a third form suggested by Johnson [1969]:

$$K_3 = \alpha_3 \sqrt{P_\mu P^\mu} . \tag{6.4.25}$$

Each generator applies to the dynamics of a free particle. The quantities α_j are constants with respect to the operators. For example, Johnson [1969] defined $\alpha_1 = i/m$ and $\alpha_3 = 1$; Aghassi, et al [1970] chose $\alpha_2 = 1$, while Pearle [1968] and Johnson [1969] considered $\alpha_2 = 1/2m$ where m is the mass of the particle.

Johnson [1969] argued that Eq. (6.4.25) could be used to eliminate certain unphysical representations by requiring that the Hamiltonian be Hermitian and positive-definite. These "unphysical" representations correspond to spacelike (tachyonic) states that may be physical but their existence has not been established [Fanchi, 1990], although experimentalists have measured spacelike masses for neutrinos [Particle Data Group, 1992; Chodos, et al, 1992; Chapter 20]. By presenting a physical rationale for eliminating these mathematically admissable states, Johnson had to explain why exact localization was not possible for a physical state. Furthermore, the positive-definite requirement applied to the Hamiltonian conflicts with the allowed existence of negative energy states in the rest frame of a particle.

Johnson interpreted Eq. (6.4.25) as applying to massive, free particles. Johnson's attempts to describe massive, spin-½ particles led to an ambiguity in the definition of the Hamiltonian. In particular, he found that Eq. (6.4.23) has the same commutator as Eq. (6.4.25) when taken with the Poincaré group operators

$J^{\mu\nu}$ and P^{μ}. He argued that Eq. (6.4.25) worked better with other physical observables such as position and spin than did Eq. (6.4.23), but he was unable to directly connect Eq. (6.4.25) with the Dirac theory of spin-½ particles.

Another problem with Eq. (6.4.25) arises when interactions are considered. Johnson and Chang [1974] used a transformation like the Foldy-Wouthuysen transformation [Bjorken and Drell, 1964] to solve the problem of a Dirac particle in an external electromagnetic field for a Hamiltonian with the form of Eq. (6.4.23). They noted that Eq. (6.4.25) was difficult to transform because of the asymmetrical treatment of its zeroth component. These problems discourage us from considering Eq. (6.4.25) any further.

We have already noted in Chapter 3 that parametrized Hamilton-Jacobi theory, parametrized variational principles, and parametrized path integrals can lead to equations that are formally the same as Eq. (6.4.22) with generators $\mathbf{K}_1$ and $\mathbf{K}_2$. Horwitz and Piron [1973] introduced a parametrized quantum theory with a generator like $\mathbf{K}_2$ as a quantum mechanical analog to a Hamiltonian formulation of relativistic classical mechanics. More recently, Evans [1991] tried to justify $\mathbf{K}_1$, but his development was based on an extension of a classical Hamiltonian and his choice of probability density was not positive definite. In every case, parametrized wave equations developed as a quantum mechanical extension of relativistic classical physics invariably depend on an assumed Lagrangian or Hamiltonian model. We are looking for a formulation of a wave equation that does not depend on a particular physical model, but is the result of fundamental physical insight. It is possible to construct such a formalism with the attributes of quantum theory from fundamental concepts of probability theory. This approach is presented in Chapter 7 and Chapter 8.

CHAPTER 7
PROBABILISTIC BASIS

One of the distinguishing characteristics of our approach to developing a new quantum theory is the insistence that the concepts of probability be retained. Collins has shown [1977a, 1977b, 1979a] that it is possible to construct a formalism that has the attributes of nonrelativistic quantum mechanics from fundamental concepts of probability theory. His approach may be readily adapted to describe a parametrized, relativistic quantum system. Of the many possible interpretations proposed for the wave function [e.g. Ballentine, 1970 and 1990; Mayants, 1984; Barrow and Tipler, 1986; Cramer, 1986], we find the statistical interpretation to be most complementary with our probabilistic approach. Several other attempts have been made to base quantum theory on a probabilistic or stochastic basis. For a review of those attempts that include an evolution parameter, see Fanchi [1993].

The probabilistic concepts of immediate interest to us are outlined in §7.1. In §7.2 through §7.4, we use the probabilistic basis to derive a formalism for describing the physical behavior of a many-body system characterized by internal variables. The formalism is a relativistic extension of Collins' [1977a] single particle, nonrelativistic formalism. A physical correspondence between the formalism and physical variables is presented in Chapter 8.

§7.1 CONSERVATION OF PROBABILITY

An understanding of the relationship between the mathematical concept of probability and the empirical aspects of statistics is important as a prerequisite to understanding subsequent results. A good discussion of this relationship has been provided by Mayants [1973; Ch. 10, 1984; 1993].

Mayants views any natural science as containing two parts: experimental and theoretical. Each part deals with distinct subjects. The subjects of the experimental part of natural science are the individual objects that are observed or experimented upon. Mayants calls them "concrete objects". The subjects of the theoretical part of natural science are idealizations which Mayants has named "abstract objects". Abstract objects are devoid of individual features which distinguish concrete objects from one another but are unessential in theoretical considerations. To obtain a meaningful interpretation of the link between theory and experiment, Mayants argues that a strict distinction between concrete and abstract objects must be maintained. Given this key methodological principle, Mayants created "probabilistics", or the science of probability. The theoretical and experimental parts of probabilistics are, respectively, probability theory and experimental statistics. The connection between the two parts is contained in the approximate equality of probabilities with relevant experimental statistical frequencies obtained after a sufficiently large number of random tests. The *gedanken* experiment in §2.3 illustrates Mayants' ideas.

From our point of view, Mayants' approach is philosophically sound but its technical implementation lacks an invariant evolution parameter. Historically, relativistic quantum theories were constructed by analogy with a nonrelativistic counterpart. It was the custom to retain spacial normalization {Eq. (1.1.3)} while discarding Born's probability density {Eq. (1.1.2)}. Many of the anomalies discussed in Chapter 1 are a consequence of efforts which use time as both an evolution parameter and an explicit dynamical variable. To avoid overworking "time", a new evolution parameter must be introduced.

We take the approach that a Born-like, positive-definite probability density should be retained while the domain for normalization should be extended to include coordinate time. The probability density ρ is treated as a joint distribution in the space and time coordinates, though ρ may be conditioned by some invariant parameters. If we assume that ρ is conditioned by a single invariant scalar s, the normalization condition is

$$\int_D \rho(x|s)\, d^4x = 1 \tag{7.1.1}$$

where D is the hypervolume of interest.

The probability of locating the particle between x and x+dx when the evolution parameter has the value s is $\rho(x|s)d^4x$. According to probability theory, the joint probability density $\rho(x|s)$ can always be written as the product of a marginal probability density $\rho(x^0|s)$ and a conditional probability density $\rho(x^1,x^2,x^3|x^0,s)$, thus

$$\rho(x|s) = \rho(x^0|s)\,\rho(\vec{x}|x^0,s). \tag{7.1.2}$$

If $\rho(x|s)d^4x$ is integrated over a three-space hypersurface R encompassing the spacial domain only, the remaining probability $\rho(x^0|s)dx^0$ is the probability of finding the particle anywhere in R during the time interval x^0 to x^0+dx^0 at s. When $\rho(x^0|s)dx^0$ is zero, the particle cannot be found anywhere in R, i.e. the particle does not exist in the space-time domain Rdx^0. This unique feature of parametrized theories is relevant to the description of particles which may decay or participate in creation/annihilation processes. The existence of the particle anywhere in the three-space R during an interval dx^0 depends on $\rho(x^0|s)$.

It is significant that the region of integration in the above integrals is the entire space-time hypervolume D. Contributions to the integrals from space-time regions outside of the special relativistic light cone are allowed in PRQT. This is justified by realizing that a probability $\rho(x|s)d^4x$ may be assigned to every point $x \in D$. A nonzero value of $\rho(x|s)d^4x$ simply means that in a finite fraction of the experiments performed a particle was observed at x corresponding to the parameter value s. Nothing at all is said about the location of that particle, or any other particle, before or after the measurement. We show in Chapter 9 that the light cone constraint enters PRQT as a constraint on the most probable trajectory of a free particle; it does not apply to the allowed region of integration.

Differentiation of Eq. (7.1.1) with respect to s gives

$$\int_D \frac{\partial \rho(x|s)}{\partial s} d^4x = 0 \tag{7.1.3}$$

which asserts that probability is conserved in the hypervolume D. Equation (7.1.3) is an integral expression of probability conservation. A differential expression of probability conservation is derived by requiring that the probability entering an element of the hypervolume D across a closed boundary ∂D (Figure 7.1) minus the probability leaving the element equals the probability accumulating in the element during a duration ds.

FIGURE 7.1

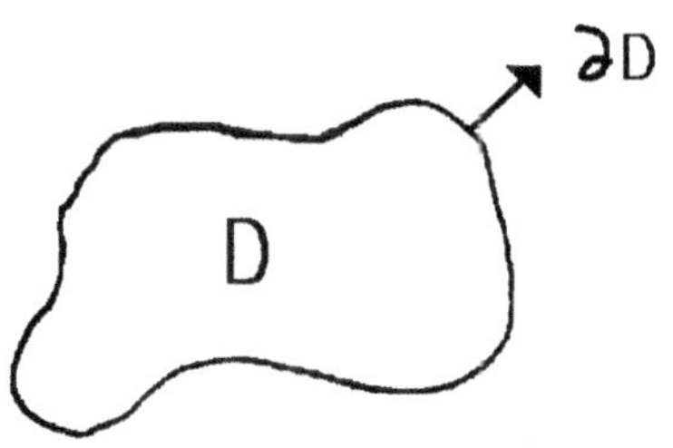

Define ρV^μ as the flux of probability density in space-time. The four-vector V^μ represents the movement of probability density in space-time during an interval s to s+ds as measured by the invariant evolution parameter. Invoking conservation of probability thus implies the existence of a continuity equation

$$\frac{\partial \rho}{\partial s} + \frac{\partial}{\partial x_\mu}(\rho V^\mu) = 0 \qquad (7.1.4)$$

where ρ vanishes at the hypersurface of a hypervolume D (typically at spacial and temporal infinity). Equation (7.1.4) plays a key role in the development of a relativistic quantum formalism, as we show in §7.2.

§7.2 FIELD EQUATIONS

Construction of field equations that are consistent with probability theory begins with the assumption that a conditional probability density $\rho(y|s)$ exists. The symbol y denotes a set of 4N coordinates with N being the number of particles in the hypervolume D^N for which $\rho(y|s)$ has nonzero values. The μth component of the position four-vector of particle a is written as y_a^μ where $\mu=0$, 1, 2, 3. Indices 1, 2, 3 signify space components, and 0 signifies the geometric time component.

Notice that only one evolution (or conditioning) parameter s has been assumed. The corresponding physical assumption is that the rest frame clock of one particle proceeds at the same rate as the rest frame clock of any other particle. A more detailed discussion of evolution parameter clocks is given in Chapter 11.

We restrict our discussion to the formalism for a single evolution parameter. The following derivation is based on a format suggested by Collins [1979b] and developed by Fanchi [1981b]. It may also be extended to include an evolution parameter for each particle [Fanchi and Wilson, 1983].

According to probability theory, $\rho(y|s)$ must be positive definite and normalizable, thus

$$\rho(y|s) \geq 0 \tag{7.2.1}$$

and

$$\int_{D^N} \rho(y|s)\, dy = 1 \tag{7.2.2}$$

where

$$dy = \prod_{a=1}^{N} d^4 y_a, \quad d^4 y_a \equiv dy_a^0 dy_a^1 dy_a^2 dy_a^3. \tag{7.2.3}$$

Conservation of probability implies

$$\frac{\partial \rho}{\partial s} + \sum_{a=1}^{N} \frac{\partial \rho V_a^\mu}{\partial y_a^\mu} = 0 \tag{7.2.4}$$

where Einstein's summation convention for Greek indices is assumed and the term ρV_a^μ represents the μth component of probability flux of particle a.

Implicit within the form of Eq. (7.2.4) is the assumption that the metric elements are constant. We shall work with a Minkowski metric with signature -2 (Appendix A.1). The following derivation would apply to any metric for which the covariant derivative may be written as in Eq. (7.2.4), i.e. the Christoffel symbols vanish. Relaxing this assumption may yield a probabilistic basis for a quantum mechanical formalism which is valid in the general relativistic sense. This approach is not pursued here.

Suppose ρ depends on L internal variables, e.g. spin or isospin, having discrete parameter values $\{l_i\}$ such that

$$\rho(y|s)\,dy = \sum_{l_1}\sum_{l_2}\cdots\sum_{l_L} \rho(y,l_1,l_2,\ldots,l_L|s)\,dy \tag{7.2.5}$$

where the ith sum extends over the entire range of allowed values of the ith parameter l_i. The term $\rho(y,l_1,\ldots,l_L|s)dy$ is the probability of observing a system having parameter values $l_1,\ldots,l_L$ in the configuration y within an infinitesimal 4N-volume dy at the historical time s.

Denote the allowed values of the ith discrete parameter l_i as L_i. Then the number of terms Λ in the L sums of Eq. (7.2.5) is just the product of L_i, namely

$$\Lambda = \prod_{i=1}^{L} L_i. \tag{7.2.6}$$

The L sums can be replaced by a single sum over the range $1 \leq \lambda \leq \Lambda$. This requires assigning a one-to-one correspondence between λ and each term of the L sums. Doing so lets us write Eq. (7.2.5) as

$$\begin{gathered} \rho(y|s)\,dy = \sum_{\lambda=1}^{\Lambda} \rho_\lambda(y|s)\,dy, \\ \rho_\lambda(y|s) \equiv \rho(y,l_1,\ldots,l_L|s) \end{gathered} \tag{7.2.7}$$

where the subscript λ of $\rho_\lambda(y|s)$ represents a set of values of the discrete parameters $\{l_i\}$.

For ρ to be differentiable and non-negative [Collins, 1977a, 1977b, 1979a], its derivative must satisfy $\partial\rho/\partial s = 0$ if $\rho = 0$. The positive-definite requirement for $\rho_\lambda(y|s)$ is satisfied by writing $\rho_\lambda(y|s)$ in the Born representation:

$$\rho_\lambda(y|s) = \psi_\lambda^*(y,s)\,\psi_\lambda(y,s). \tag{7.2.8}$$

Equation (7.2.8) implies the scalar eigenfunctions ψ_λ are specified only to within a gauge transformation of the first kind. Consequently ψ_λ can be written as

$$\psi_\lambda(y,s) = [\rho_\lambda(y|s)]^{1/2} \exp[i\,\xi(y,s)] \tag{7.2.9}$$

where the real scalar function $\xi(y,s)$ is assumed independent of the discrete parameters. This restriction is realistic if the probability density of the physical system can be decomposed such that the parameters $\{l_i\}$ characterize mutually independent events. Systems characterized by an eigenfunction with the product form ψ(space-time)$\otimes$ ψ(spin)$\otimes$ ψ(isospin)... fulfill the independence requirement.

THE STÜCKELBERG EQUATION

The Born representation and conservation of probability are at the heart of a formalism which has the attributes of quantum theory for an N-body system. One of the most important features of the formalism is the derivation of a field equation which satisfies the assumptions of conventional probability theory.

Our derivation [Fanchi, 1981b] begins by expressing the four-velocity of the ath particle V_a^μ as

$$V_a^\mu(y,s) = \varepsilon_a \left[\varepsilon' \frac{\partial \xi(y,s)}{\partial y_{a\mu}} + \varepsilon'' A_a^\mu(y,s) \right] \tag{7.2.10}$$

where ε', ε'', and ε_a are real c-numbers whose values are not yet specified. We assume A_a^μ and, consequently, V_a^μ are independent of λ. To make a physically meaningful interpretation of the four-velocity V_a^μ, we assume A_a^μ is a real, scalar function. From Eq. (7.2.9) we derive the relation

$$\frac{\partial \xi}{\partial y_{a\mu}} = \frac{-i}{2\rho_\lambda} \left[\psi_\lambda^* \frac{\partial \psi_\lambda}{\partial y_{a\mu}} - \psi_\lambda \frac{\partial \psi_\lambda^*}{\partial y_{a\mu}} \right] \tag{7.2.11}$$

where there is no sum over the repeated index λ. Combining Eqs. (7.2.7) through (7.2.11) gives

$$\rho V_a^\mu = \sum_{\lambda=1}^{\Lambda} \varepsilon_a \left\{ \psi_\lambda^* \left[-i\varepsilon' \frac{\partial}{\partial y_{a\mu}} + \varepsilon'' A_a^\mu \right] \psi_\lambda + \frac{i\varepsilon'}{2} \frac{\partial \psi_\lambda^* \psi_\lambda}{\partial y_{a\mu}} \right\}. \tag{7.2.12}$$

Differentiating Eq. (7.2.12) with respect to y_a^μ gives

$$\frac{\partial \rho V_a^\mu}{\partial y_a^\mu} = \sum_{\lambda=1}^{\Lambda} \varepsilon_a \{ \varepsilon'' \psi_\lambda^* [\frac{1}{2} A_a^\mu \frac{\partial \psi_\lambda}{\partial y_a^\mu} + \frac{1}{2} \frac{\partial}{\partial y_a^\mu} (A_a^\mu \psi_\lambda)]$$

$$+ \varepsilon'' \psi_\lambda [\frac{1}{2} A_a^\mu \frac{\partial \psi_\lambda^*}{\partial y_a^\mu} + \frac{1}{2} \frac{\partial}{\partial y_a^\mu} (A_a^\mu \psi_\lambda^*)] \tag{7.2.13}$$

$$- \frac{i\varepsilon'}{2} \psi_\lambda^* \frac{\partial^2 \psi_\lambda}{\partial y_a^\mu \partial y_{a\mu}} + \frac{i\varepsilon'}{2} \psi_\lambda \frac{\partial^2 \psi_\lambda^*}{\partial y_a^\mu \partial y_{a\mu}} \}.$$

Substituting Eq. (7.2.13) into the continuity equation {Eq. (7.2.4)}, multiplying by iε' and rearranging yields

$$\sum_{\lambda=1}^{\Lambda} \psi_\lambda^* F_\lambda = \sum_{\lambda=1}^{\Lambda} F_\lambda^* \psi_\lambda \tag{7.2.14}$$

where

$$F_\lambda = i\varepsilon' \frac{\partial \psi_\lambda}{\partial s} + \sum_{a=1}^{N} \varepsilon_a \{ \frac{\varepsilon'^2}{2} \frac{\partial^2 \psi_\lambda}{\partial y_a^\mu \partial y_{a\mu}}$$

$$+ \frac{i\varepsilon' \varepsilon''}{2} [A_a^\mu \frac{\partial \psi_\lambda}{\partial y_a^\mu} + \frac{\partial}{\partial y_a^\mu} (A_a^\mu \psi_\lambda)] \}. \tag{7.2.15}$$

In matrix notation Eq. (7.2.14) becomes

$$\underline{\Psi}^+ \underline{F} = \underline{F}^+ \underline{\Psi} \tag{7.2.16}$$

where $^+$ denotes conjugate transpose and we have used the assumption that A_a^μ is a real, scalar function. The λth element of the column vector $\underline{F}$ is F_λ, and $\underline{\Psi}$ is the column vector

$$\underline{\Psi} = \begin{bmatrix} \psi_1 \\ \vdots \\ \psi_\Lambda \end{bmatrix}. \tag{7.2.17}$$

Equation (7.2.15) is decoupled by observing that $\underline{\Psi}^{+}\underline{F}$ must be real since

$$\underline{\Psi}^{+}\underline{F} = \underline{F}^{+}\underline{\Psi} = (\underline{\Psi}^{+}\underline{F})^{+}. \tag{7.2.18}$$

This constraint is generally satisfied by writing

$$\underline{F} = \underline{\underline{V}}'\underline{\Psi} \tag{7.2.19}$$

where $\underline{\underline{V}}'$ must be a ΛxΛ Hermitian matrix. The general form of the subsequent field equation is

$$i\varepsilon'\frac{\partial\underline{\Psi}}{\partial s} + \sum_{a=1}^{N}\varepsilon_a\Big[\frac{\varepsilon'^2}{2}\frac{\partial^2\underline{\Psi}}{\partial y_a^{\mu}\partial y_{a\mu}} + i\frac{\varepsilon'\varepsilon''}{2}\Big(A_a^{\mu}\frac{\partial\underline{\Psi}}{\partial y_a^{\mu}} + \frac{\partial A_a^{\mu}\underline{\Psi}}{\partial y_a^{\mu}}\Big)\Big] - \underline{\underline{V}}'\underline{\Psi} = 0. \tag{7.2.20}$$

Equation (7.2.19) may be put in a more suggestive form by replacing the Hermitian matrix $\underline{\underline{V}}'$ with

$$\underline{\underline{V}}' = \sum_{a=1}^{N}\frac{\varepsilon_a\varepsilon''^2}{2}A_a^{\mu}A_{a\mu}\underline{\underline{I}} + \underline{\underline{V}} \tag{7.2.21}$$

where $\underline{\underline{I}}$ is the identity matrix. The first term on the right hand side of Eq. (7.2.21) is real and therefore Hermitian. To retain the Hermiticity of $\underline{\underline{V}}'$, we require $\underline{\underline{V}}$ to be Hermitian. The resulting field equation is

$$i\varepsilon'\frac{\partial\underline{\Psi}}{\partial s} = \sum_{a=1}^{N}\frac{\varepsilon_a}{2}\pi_a^{\mu}\pi_{a\mu}\underline{\Psi} + \underline{\underline{V}}\underline{\Psi} \tag{7.2.22}$$

where we define the operators $\boldsymbol{\pi}_a^{\mu}$ and $\mathbf{p}_a^{\mu}$ by

$$\pi_a^{\mu} \equiv p_a^{\mu} + \varepsilon'' A_a^{\mu} \tag{7.2.23}$$

and

$$p_a^{\mu} \equiv \frac{\varepsilon'}{i} \frac{\partial}{\partial y_{a\mu}} . \tag{7.2.24}$$

The single particle form of Eq. (7.2.22) resembles an equation first studied by Fock [1937] and Stückelberg [1942]. For ease of reference, it is worthwhile to associate a name with Eq. (7.2.22). A case could be made for naming Eq. (7.2.22) after Fock because of his chronological priority. I prefer to acknowledge Stückelberg's contribution, however, because his overall theory was much more compatible with the theory being advanced here. For example, the solutions to Fock's equation had to be stationary states with zero eigenvalues. Fock, and later Nambu [1950], viewed the invariant evolution parameter as a mathematical convenience, i.e. the parameter had no physical significance. By contrast, Stückelberg was the first to develop a parametrized quantum theory that included a new normalization condition and a new interpretation of the probability function in addition to a parametrized wave equation. Consequently, we refer to Eq. (7.2.22) as the Stückelberg equation.

A nonrelativistic many-body formalism can be easily derived as a special case of the above formulation by replacing the space-time Minkowski metric by the three-space Euclidean metric. In such a formulation, the nonrelativistic evolution parameter corresponds to "time" in the sense of Galileo and Newton. The physical meaning of "time" is discussed in more detail in Chapter 11.

BOUNDARY CONDITIONS

Solution of the partial differential field equations requires that we specify boundary conditions. We recall that ρ and V^{μ} have been assumed single-valued and continuous. Two familiar boundary conditions are obtained by assuming that $\underline{\Psi}$ and A^{μ} are also single-valued and continuous. Integrating the continuity equation {Eq. (7.2.4)} over a "pillbox" in space-time, letting its length normal to the boundary go to zero, and employing the divergence theorem yields

$$(\rho V_a^\mu)_I = (\rho V_a^\mu)_{II} \tag{7.2.25}$$

which shows the continuity of flux across the boundary between regions I and II.

Combining Eqs. (7.2.9) through (7.2.11) lets us rewrite the probability flux as

$$\rho V_a^\mu = -\frac{i\varepsilon_a \varepsilon'}{2}\left(\psi_\lambda^* \frac{\partial \psi_\lambda}{\partial y_{a\mu}} - \psi_\lambda \frac{\partial \psi_\lambda^*}{\partial y_{a\mu}}\right) + \varepsilon_a \varepsilon'' \psi_\lambda^* \psi_\lambda A_a^\mu \tag{7.2.26}$$

for each element of $\underline{\Psi}$. Substituting Eq. (7.2.26) into (7.2.25) and letting $\partial/\partial n$ represent the normal derivative we obtain

$$\begin{aligned} -\frac{i\varepsilon'}{2}\left(\psi_{\lambda I}^* \frac{\partial \psi_{\lambda I}}{\partial n} - \psi_{\lambda I} \frac{\partial \psi_{\lambda I}^*}{\partial n}\right) + \varepsilon'' \psi_{\lambda I}^* \psi_{\lambda I} A_{aI}^\mu n_\mu &= \\ -\frac{i\varepsilon'}{2}\left(\psi_{\lambda II}^* \frac{\partial \psi_{\lambda II}}{\partial n} - \psi_{\lambda II} \frac{\partial \psi_{\lambda II}^*}{\partial n}\right) + \varepsilon'' \psi_{\lambda II}^* \psi_{\lambda II} A_{aII}^\mu n_\mu & \end{aligned} \tag{7.2.27}$$

where n_μ is the unit normal vector and all quantities are evaluated at the boundary between regions I and II.

Since $\underline{\Psi}$ is single-valued we have

$$\underline{\Psi}_I = \underline{\Psi}_{II} \tag{7.2.28}$$

on the boundary. Combining Eqs. (7.2.27) and (7.2.28) with the assumption that A^μ is single-valued implies the other boundary condition, namely

$$\frac{\partial \psi_{\lambda I}}{\partial n} = \frac{\partial \psi_{\lambda II}}{\partial n}. \tag{7.2.29}$$

Equations (7.2.28) and (7.2.29) state that the eigenvector $\underline{\Psi}$ and its normal derivative are continuous at the boundary.

§7.3 EXPECTATION VALUES

The last item needed to connect the probabilistic formalism to observable quantities is a definition of expectation values. Expectation value is a well-defined concept in probability theory. By applying the probabilistic concept of expectation value to the calculation of two common physical quantities -- average four-position and four-velocity -- we justify a form of the definition of expectation value that is well suited to the operator concepts introduced in Chapter 6.

MOTIVATION

The average value of the μth component of the four-position of particle a is

$$\langle y_a^\mu \rangle = \int_{D^N} \rho\, y_a^\mu \, dy \tag{7.3.1}$$

or

$$\begin{aligned} \langle y_a^\mu \rangle &= \sum_{\lambda=1}^{\Lambda} \int_{D^N} \psi_\lambda^* y_a^\mu \psi_\lambda \, dy \\ &= \int_{D^N} \underline{\Psi}^+ y_a^\mu \underline{\Psi} \, dy . \end{aligned} \tag{7.3.2}$$

Calculation of the rate of change of $<y_a^\mu>$ with respect to the evolution parameter is readily accomplished by differentiating Eq. (7.3.2) with respect to s:

$$\frac{d\langle y_a^\mu \rangle}{ds} = \int_{D^N} y_a^\mu \frac{\partial \rho}{\partial s} \, dy . \tag{7.3.3}$$

Using the continuity equation {Eq. (7.3.4)} lets us write

$$\frac{d\langle y_a^\mu \rangle}{ds} = - \sum_{b=1}^{N} \int_{D^N} y_a^\mu \frac{\partial \rho V_b^\nu}{\partial y_b^\nu} \, dy . \tag{7.3.4}$$

This can be rewritten as

$$\frac{d\langle y_a^\mu\rangle}{ds} = \sum_{b=1}^{N}\left\{\int_{D^N} \rho V_b^\nu \frac{\partial y_a^\mu}{\partial y_b^\nu}\, dy - \int_{D^N} \frac{\partial \rho V_b^\nu y_a^\mu}{\partial y_b^\nu}\, dy\right\}. \tag{7.3.5}$$

The latter term on the right hand side of Eq. (7.3.5) vanishes because we can use the divergence theorem to convert the hypervolume integral to a hypersurface integral and then recall that ρ goes to zero on the hypersurface of the hypervolume D^N. Equation (7.3.5) simplifies to the form

$$\frac{d\langle y_a^\mu\rangle}{ds} = \int_{D^N} \rho V_a^\mu\, dy = \langle V_a^\mu\rangle \tag{7.3.6}$$

where we have used the relation

$$\frac{\partial y_a^\mu}{\partial y_b^\nu} = \delta_\nu^\mu \delta_{ab}, \tag{7.3.7}$$

and δ_ν^μ, δ_{ab} are Kronecker deltas.

A quantum mechanical expression for $<V_a^\mu>$ is obtained by rewriting Eq. (7.2.12) in terms of π_a^μ, thus

$$\rho_\lambda V_a^\mu = \psi_\lambda^* \varepsilon_a \pi_a^\mu \psi_\lambda + \frac{i\varepsilon_a \varepsilon'}{2} \frac{\partial \psi_\lambda^* \psi_\lambda}{\partial y_{a\mu}}. \tag{7.3.8}$$

Substituting Eq. (7.3.8) into Eq. (7.3.6) and using the divergence theorem with the requirement that ρ vanishes on the hypersurface of the hypervolume gives

$$\langle V_a^\mu\rangle = \int_{D^N} \psi_\lambda^* \varepsilon_a \pi_a^\mu \psi_\lambda\, dy. \tag{7.3.9}$$

From Eq. (7.3.9) we see the operator corresponding to V_a^μ is $\varepsilon_a \pi_a^\mu$.

DEFINITION

The expressions for the expectation values of x_a^μ and V_a^μ suggest that we should adopt the usual definition of the expectation value of an operator $\mathbf{\Omega}$, thus

$$\langle \Omega \rangle \equiv \int_{D^N} \underline{\Psi}^+ \Omega \underline{\Psi} dy \tag{7.3.10}$$

in the space-time coordinate representation. Expectation values defined by Eq. (7.3.10) are equivalent to Lorentz invariant scalar products of square integrable functions $\underline{\Phi}, \underline{\Psi} \in L^2(y)$, i.e.

$$\langle \Omega \rangle = \langle \underline{\Psi} | \underline{\Phi} \rangle = \int_{D^N} \underline{\Psi}^+ \underline{\Phi} dy \tag{7.3.11}$$

where

$$\underline{\Phi} = \Omega \underline{\Psi} . \tag{7.3.12}$$

The inner product defined by Eq. (7.3.10) is compared with the inner product of conventional quantum theory in §8.2 after a physical interpretation of the unspecified constants ε_a, ε' and ε'' is adopted. Given Eq. (7.3.10), we are in a position to formally solve Eq. (7.2.22).

§7.4 SOLUTION OF THE STÜCKELBERG EQUATION

The Stückelberg equation -- Eq. (7.2.22) -- may be written in the form

$$i\varepsilon' \frac{\partial \underline{\Psi}}{\partial s} = \boldsymbol{K} \underline{\Psi} \tag{7.4.1}$$

where

$$K = \sum_{a=1}^{N} \frac{\varepsilon_a}{2} \pi_a^{\mu} \pi_{a\mu} \underline{\underline{I}} + \underline{\underline{V}}. \tag{7.4.2}$$

The operator **K** is Hermitian because it is the sum of two Hermitian operators. Since **K** is Hermitian, there exists a set of solutions $\{\underline{\Phi}_\kappa\}$ of an eigenvalue equation

$$K \underline{\Phi}_\kappa = \kappa \underline{\Phi}_\kappa \tag{7.4.3}$$

which constitute an orthogonal basis of eigenvectors that span the Hilbert space $L^2(y)$ and obey the orthonormality condition

$$\int_{D^N} \underline{\Phi}_{\kappa'}^{+}(y)\ \underline{\Phi}_\kappa(y)\ dy = \delta_{\kappa'\kappa}. \tag{7.4.4}$$

The general solution of Eq. (7.4.1) may be written as a superposition of eigenfunctions $\{\underline{\Phi}_\kappa(y)\}$, namely

$$\underline{\Psi}(y,s) = \uplus\, \eta_\kappa(s) \underline{\Phi}_\kappa(y) \tag{7.4.5}$$

with $\eta_\kappa(s)$ denoting expansion coefficients and $\uplus$ implying a sum over a discrete eigenvalue range, an integral over a continuous eigenvalue range, or a combination as described in Appendix A.1. Substituting Eq. (7.4.5) into (7.4.1) gives

$$\begin{aligned} i\varepsilon' \uplus \frac{\partial \eta_\kappa}{\partial s} \underline{\Phi}_\kappa &= K \underline{\Psi} \\ &= \uplus\, \eta_\kappa K \underline{\Phi}_\kappa \\ &= \uplus\, \eta_\kappa \kappa \underline{\Phi}_\kappa \end{aligned} \tag{7.4.6}$$

since **K** does not operate on the expansion coefficients $\eta_\kappa(s)$. Combining coefficients of $\underline{\Phi}_\kappa$ yields

$$i\varepsilon' \frac{\partial \eta_\kappa}{\partial s} = \kappa \eta_\kappa. \tag{7.4.7}$$

Integrating Eq. (7.4.7) gives

$$\eta_\kappa(s) = \eta_\kappa(s_0)\ \exp[-i\kappa(s-s_0)/\varepsilon'] \tag{7.4.8}$$

where s_0 and $\eta_\kappa(s_0)$ are integration constants and s_0 denotes an initial value of s. The general solution becomes

$$\underline{\Psi}(y,s) = \biguplus \eta_\kappa(s_0)\ \underline{\Phi}_\kappa(y)\ \exp[-i\kappa(s-s_0)/\varepsilon'] . \tag{7.4.9}$$

The expansion coefficients are found by specifying the initial condition $\underline{\Psi}(y,s_0)$ so that Eq. (7.4.9) simplifies to the form

$$\underline{\Psi}(y,s_0) = \biguplus\ \eta_\kappa(s_0)\ \underline{\Phi}_\kappa(y) . \tag{7.4.10}$$

Multiplying Eq. (7.4.10) by $\underline{\Phi}^+_{\kappa'}(y)$ and integrating over dy gives

$$\eta_{\kappa'}(s_0) = \int_{D^N} \underline{\Phi}^+_{\kappa'}(y)\ \underline{\Psi}(y,s_0)\,dy \tag{7.4.11}$$

where we have used the orthonormality condition Eq. (7.4.4). Equations (7.4.9) and (7.4.11) comprise the formal expression for the general solution of Eq. (7.4.1) To use these equations, we must have the solutions of Eq. (7.4.3) and an initial condition $\underline{\Psi}(y,s_0)$. An example of an initial condition is the minimum wave packet presented in §8.3. Examples of solutions of the Stückelberg equation are given in subsequent chapters.

CHAPTER 8
PHYSICAL CORRESPONDENCE

By focusing on the mathematics in Chapter 7, we found that many attributes of quantum theory may be viewed as a consequence of our probabilistic basis. To achieve our goal of creating a new paradigm of modern physics, it is not enough to establish a formal relationship between mathematical structures. We need to demonstrate that the mathematics can be related to measurable quantities in a meaningful way. Our interest here is to establish a correspondence between physical concepts and the mathematical formalism of Chapter 7. Most of the rest of the book examines the validity and predictive power of the parametrized relativistic quantum theory (PRQT) outlined in Chapters 6 through 8.

We begin by associating the constants ε_a, ε' and ε'' with physical variables. Given these identifications, we are in a position to compare the PRQT inner product defined in §7.4 with the conventional inner product. This is followed by a discussion of dispersion relations and the derivation of Heisenberg's uncertainty principle. Correspondence between quantum theory and classical physics is then established by deriving a matrix representation of the quantum theoretic equations of motion.

§8.1 PHYSICAL INTERPRETATION

A physical interpretation of the mathematical formalism developed in Chapter 6 and Chapter 7 is obtained by considering a particle located in the space-time hypervolume D. The particle, which may be interacting, has a world-line that is specified by successive measurements of its four-space position $x \in D$. Repetition of this series of measurements yields a distribution of measured values of x. From a theoretical perspective, the number of repetitions should be unbounded. From a practical point of view, the number of repetitions needs to be statistically significant. The resulting distribution of measured values is used to construct a probability distribution $\rho(x|s)d^4x$ which is conditioned by a smoothly varying, monotonically increasing, invariant scalar s. The scalar s parametrizes the world-line of the particle in a manner analogous to that outlined in §2.3 for the classical free particle. We argued in Chapter 7 that the Stückelberg equation -- Eq. (7.2.22) -- is the appropriate equation for describing this system.

The physical correspondence between the constants ε_a, ε' and ε'' introduced in the derivation of the Stückelberg equation and physically meaningful quantities is obtained by comparing Eq. (7.2.22) with the Klein-Gordon equation, i.e. Eq. (1.1.4). Field equations for a many-body system which explicitly exhibits the "minimal coupling" electromagnetic interaction are obtained by making the identifications

$$\begin{aligned} \varepsilon' &= \hbar \\ \varepsilon'' &= -e/c \\ \varepsilon_a &= 1/m_a \end{aligned} \tag{8.1.1}$$

where m_a is a c-number with mass units, and $\hbar$,c are made explicit. We identify the four-vector A^μ as the external four-vector potential acting on the ath particle. Additional interactions may be incorporated via the presently unspecified Hermitian matrix $\underline{\underline{V}}$ of Eq. (7.2.22). Applications in subsequent chapters will further justify the above identifications. For now we view Eq. (8.1.1) as a set of plausible postulates. With these identifications, π_a^μ becomes the four-momentum operator with "minimal coupling" for the ath particle.

Before leaving this section, let us note that Eq. (7.2.22) may be written in the form

$$i\hbar\frac{\partial \underline{\Psi}}{\partial s} = \boldsymbol{K}\,\underline{\Psi} \tag{8.1.2}$$

where **K** corresponds to the generator of s translations in Eq. (6.4.22). The generator **K** has the explicit form

$$\boldsymbol{K} \equiv \Sigma_{a=1}^{N}\left(\frac{\pi_a^{\mu}\pi_{a\mu}}{2m_a}\right)\underline{\underline{I}} + \underline{\underline{V}}. \tag{8.1.3}$$

If we consider a non-interacting particle in which A^{μ} and $\underline{\underline{V}}$ are zero, **K** in Eq. (8.1.3) corresponds to the Schrödinger-like operator $\mathbf{K}_2$ discussed in §6.4. Unlike theories which postulate a Hamiltonian or Lagrangian model, we have obtained **K** from the probabilistic basis described in Chapter 7. An approach which leads to a Dirac-like representation of the Stückelberg equation is presented in §10.1.

§8.2 INNER PRODUCT REPRESENTATION

The expectation value defined in §7.3 was motivated by our reliance on conventional probability theory and the derivation of plausible results by calculating the expected value of position and velocity. Once an expectation value was defined, it was a simple step to generalize the definition to encompass the inner product of the probability amplitudes. The resulting PRQT inner product is a special case of the general definition of an inner product presented by Wigner [pg. 185, 1939]. For ease of reference, we write Wigner's inner product definition in four-momentum space in the form

$$\langle\psi|\phi\rangle = \sum_{\zeta}\int \psi^{*}(k,\zeta)\,\phi(k,\zeta)\,df(k,\zeta) \tag{8.2.1}$$

where k denotes four-momentum and the index ζ represents discrete parameters characterizing such physical quantities as spin. The probability amplitudes ψ, ϕ and the measure df(k, ζ) may depend on k, ζ. Before discussing the PRQT inner product, we first recall the conventional inner product.

CONVENTIONAL INNER PRODUCT

It is instructive to review the conventional normalization constraint before presenting the conventional inner product. The conventional normalization is [Aitchison and Hey, pg. 33, 1983]

$$\begin{Bmatrix} 1 \\ 2\omega_0 \end{Bmatrix} = \sum_{\zeta} \iiint_{-\infty}^{\infty} \psi^*(k,\zeta)\,\psi(k,\zeta)\,dk^1 dk^2 dk^3 \tag{8.2.2}$$

where

$$\omega_0 = (K + \vec{k}\cdot\vec{k})^{1/2} \tag{8.2.3}$$

and K is a nonzero, positive constant. The unit value in Eq. (8.2.2) is the conventional norm of the nonrelativistic theory. It is replaced in the relativistic theory by $2\omega_0$ where ω_0 is the energy of the system. Notice that these norms differ from a normalization using j^0 of Eq. (1.1.6) as a "probability density". The conventional inner product corresponding to the $2\omega_0$ norm is

$$\langle\psi|\phi\rangle_{con} = \sum_{\zeta} \iiint_{-\infty}^{\infty} \psi^*(k,\zeta)\,\phi(k,\zeta)\,\frac{dk^1 dk^2 dk^3}{2\omega_0} \tag{8.2.4}$$

with the normalization condition

$$\langle\psi|\psi\rangle_{con} = 1 = \sum_{\zeta} \iiint_{-\infty}^{\infty} \psi^*(k,\zeta)\,\psi(k,\zeta)\,\frac{dk^1 dk^2 dk^3}{2\omega_0}. \tag{8.2.5}$$

The measure in Eq. (8.2.4) may be written as the Lorentz invariant volume element [Bjorken and Drell, pg. 112, 1964]

$$\frac{dk^1 dk^2 dk^3}{2\omega_0} = \int_{-\infty}^{\infty} \delta(k^\mu k_\mu - \omega_0^2)\,\theta(\omega_0)\,d^4k,$$
$$\theta(\omega_0) \equiv \begin{Bmatrix} 1 \text{ for } \omega_0 > 0 \\ 0 \text{ for } \omega_0 < 0 \end{Bmatrix}. \tag{8.2.6}$$

The step function $\theta(\omega_0)$ selects only positive values of ω_0. Consequently, the integration only includes timelike contributions. Rewriting the conventional inner product with the Lorentz invariant volume element lets us express Eq. (8.2.4) as

$$\langle \psi | \phi \rangle_{con} = \sum_{\zeta} \iiiint_{-\infty}^{\infty} \psi^*(k,\zeta)\, \phi(k,\zeta)\, \delta(k^{\mu}k_{\mu} - \omega_0^2)\, \theta(\omega_0)\, d^4k. \tag{8.2.7}$$

PRQT Inner Product

Following §7.3, we may write the PRQT normalization condition and the corresponding inner product in the momentum representation as

$$\langle \psi | \psi \rangle = 1 = \sum_{\zeta} \iiiint_{-\infty}^{\infty} \psi^*(k,\zeta)\, \psi(k,\zeta)\, d^4k \tag{8.2.8}$$

and

$$\langle \psi | \phi \rangle = \sum_{\zeta} \iiiint_{-\infty}^{\infty} \psi^*(k,\zeta)\, \phi(k,\zeta)\, d^4k. \tag{8.2.9}$$

Equation (8.2.9) is the PRQT replacement for Eq. (8.2.7). In the PRQT case, only the unit normalization constant is needed.

Another perspective is obtained by reducing the PRQT inner product to the conventional inner product. The reduction is achieved by first assuming that the dispersion relation in Eq. (8.2.3) applies. Physically this assumption corresponds to assuming the particle has a fixed mass. The resulting inner product is

$$\langle \psi | \phi \rangle_{fix} = \sum_{\zeta} \iiiint_{-\infty}^{\infty} \psi^*(k,\zeta)\, \phi(k,\zeta) \cdot \delta((k^0)^2 - \vec{k}\cdot\vec{k} - K)\, d^4k \tag{8.2.10}$$

where the subscript "fix" reminds us that we have imposed an on mass shell assumption. Equation (8.2.10) may be rearranged by using the delta function expansion [Dennery and Krzywicki, 1967]

$$\delta(\omega^2 - \vec{k}\cdot\vec{k} - K) = \frac{\delta(\omega - \omega_0) + \delta(\omega + \omega_0)}{2\omega_0} \tag{8.2.11}$$

to obtain

$$\begin{aligned}\langle\psi|\phi\rangle_{fix} &= \sum_{\zeta} \int\!\!\int\!\!\int\!\!\int_{-\infty}^{\infty} \psi^*(k,\zeta)\,\phi(k,\zeta)\,\frac{\delta(\omega - \omega_0)}{2\omega_0}\,d^4k \\ &+ \sum_{\zeta} \int\!\!\int\!\!\int\!\!\int_{-\infty}^{\infty} \psi^*(k,\zeta)\,\phi(k,\zeta)\,\frac{\delta(\omega + \omega_0)}{2\omega_0}\,d^4k.\end{aligned} \tag{8.2.12}$$

Integrating over ω yields

$$\begin{aligned}\langle\psi|\phi\rangle_{fix} &= \sum_{\zeta} \int\!\!\int\!\!\int_{-\infty}^{\infty} \psi^*(\omega_0,\vec{k},\zeta)\ \phi(\omega_0,\vec{k},\zeta)\,\frac{d^3k}{2\omega_0} \\ &+ \sum_{\zeta} \int\!\!\int\!\!\int_{-\infty}^{\infty} \psi^*(-\omega_0,\vec{k},\zeta)\ \phi(-\omega_0,\vec{k},\zeta)\,\frac{d^3k}{2\omega_0}.\end{aligned} \tag{8.2.13}$$

Comparing Eq. (8.2.13) with Eq. (8.2.4) shows that the conventional inner product is a special case of the PRQT inner product. Physically, the PRQT inner product allows contributions from more than one mass state and from spacelike and timelike domains. Further justification for this interpretation is provided in subsequent chapters. These additional contributions to the inner product represent opportunities for developing experimental tests of PRQT. Some specific applications -- notably mass state transitions and tachyon kinematics -- are considered in Part IV.

§8.3 DISPERSION RELATIONS

The existence of uncertainty relations is a mathematical consequence of the probabilistic basis, as we now show. We begin by noting that operators of interest here are Hermitian, i.e. an operator Ω must satisfy

$$\int \underline{u}^{+}\mathbf{\Omega}\underline{v}dy = \int (\mathbf{\Omega}^{+}\underline{u}^{+})\underline{v}dy = \left[\int \underline{u}^{+}\mathbf{\Omega}\underline{v}dy\right]^{+} \tag{8.3.1}$$

where $\underline{u}$, $\underline{v}$ are elements of $L^2(y)$ and we have simplified the notation. We use Eq. (8.3.1) and Schwartz's inequality to construct dispersion relations. The dispersion relations will provide a mathematical basis for the uncertainty principle presented below.

SCHWARTZ'S INEQUALITY

The form of Schwartz's inequality of interest to us is

$$\int \underline{u}^{+}\underline{u}dy \int \underline{v}^{+}\underline{v}dy \geq \frac{1}{4}\left[\int (\underline{u}^{+}\underline{v} + \underline{v}^{+}\underline{u})dy\right]^2. \tag{8.3.2}$$

The equality in Eq. (8.3.2) is true only when $\underline{u}$, $\underline{v}$ are proportional, i.e.

$$\underline{v} = \alpha \underline{u} \tag{8.3.3}$$

where α is a complex proportionality constant. We show the equality by substituting Eq. (8.3.3) into (8.3.2) to get:

$$\begin{aligned} \alpha^{*}\alpha\left[\int \underline{u}^{+}\underline{u}dy\right]^2 &\geq \frac{1}{4}\left[\int (\underline{u}^{+}\underline{u}\alpha + \alpha^{*}\underline{u}^{+}\underline{u})dy\right]^2 \\ &\geq \frac{1}{4}(\alpha + \alpha^{*})^2 \left[\int \underline{u}^{+}\underline{u}dy\right]^2. \end{aligned} \tag{8.3.4}$$

Invoking normalization of the eigenfunctions, the bracketed term on the right hand side equals one and Eq. (8.3.4) becomes

$$\alpha^{*}\alpha \geq \frac{1}{4}(\alpha + \alpha^{*}) = \frac{[2\ Re(\alpha)]^2}{4} = \alpha^{*}\alpha \tag{8.3.5}$$

as asserted.

To obtain dispersion relations, we use Eq. (8.3.2) with $\underline{u}$, $\underline{v}$ defined as

$$\underline{u} \equiv (\mathbf{\Omega}_1 - \langle\omega_1\rangle)\underline{\mathbf{X}}$$
$$\underline{v} \equiv i(\mathbf{\Omega}_2 - \langle\omega_2\rangle)\underline{\mathbf{X}} \tag{8.3.6}$$

where $\underline{\mathbf{X}} \in L^2$ and $<\omega_j>$ is the expectation value of the operator $\mathbf{\Omega}_j$. We see the form $\mathbf{\Omega}_j - <\omega_j>$ is Hermitian because we have merely subtracted a real number $<\omega_j>$ from the Hermitian operator $\mathbf{\Omega}_j$. Thus $\mathbf{\Omega}_j - <\omega_j>$ satisfies Eq. (8.3.1). Substituting Eq. (8.3.6) into (8.3.2) and using the fact that $\mathbf{\Omega}_j - <\omega_j>$ is Hermitian we find

$$\begin{aligned} &\int \underline{\mathbf{X}}^+(\mathbf{\Omega}_1 - \langle\omega_1\rangle)^2 \underline{\mathbf{X}}\,dy \int \underline{\mathbf{X}}^+(\mathbf{\Omega}_2 - \langle\omega_2\rangle)^2 \underline{\mathbf{X}}\,dy \\ &\geq -\frac{1}{4}\Big[\int \underline{\mathbf{X}}^+(\mathbf{\Omega}_1 - \langle\omega_1\rangle)(\mathbf{\Omega}_2 - \langle\omega_2\rangle)\underline{\mathbf{X}}\,dy \\ &- \int \underline{\mathbf{X}}^+(\mathbf{\Omega}_2 - \langle\omega_2\rangle)(\mathbf{\Omega}_1 - \langle\omega_1\rangle)\underline{\mathbf{X}}\,dy\Big]^2. \end{aligned} \tag{8.3.7}$$

Expanding the right hand side of Eq. (8.3.7) and simplifying yields

$$\begin{aligned} &\int \underline{\mathbf{X}}^+(\mathbf{\Omega}_1 - \langle\omega_1\rangle)^2 \underline{\mathbf{X}}\,dy \int \underline{\mathbf{X}}^+(\mathbf{\Omega}_2 - \langle\omega_2\rangle)^2 \underline{\mathbf{X}}\,dy \\ &\geq -\frac{1}{4}\left[\int \underline{\mathbf{X}}^+[\mathbf{\Omega}_1,\mathbf{\Omega}_2]\underline{\mathbf{X}}\,dy\right]^2 \end{aligned} \tag{8.3.8}$$

and the commutator bracket has been used. Defining dispersion in the usual way as

$$(\Delta\omega_j)^2 \equiv \int \underline{\mathbf{X}}^+(\mathbf{\Omega}_j - \langle\omega_j\rangle)^2 \underline{\mathbf{X}}\,dy \tag{8.3.9}$$

lets us write Eq. (8.3.8) in the form

$$(\Delta\omega_1)^2(\Delta\omega_2)^2 \geq -\frac{1}{4}\left[\int \chi^+[\mathbf{\Omega}_1,\mathbf{\Omega}_2]\chi\,dy\right]^2. \tag{8.3.10}$$

Equation (8.3.10) expresses the dispersion relation between two Hermitian operators.

Relativistic Minimum Eigenfunction

An eigenfunction analogous to a nonrelativistic Gaussian wave packet may be constructed in PRQT by the appropriate choice of initial condition. We look for an initial condition that satisfies Eq. (8.3.3) where $\underline{u}$, $\underline{v}$ are given by Eqs. (8.3.6), (7.2.24) and

$$\mathbf{\Omega}_1 \equiv \boldsymbol{x}_\mu,\ \mathbf{\Omega}_2 \equiv \boldsymbol{p}_\mu = \frac{\hbar}{i}\frac{\partial}{\partial x^\mu},\ \underline{\mathbf{X}} = \underline{\mathbf{\Phi}}_{\min}(x). \tag{8.3.11}$$

Rather than complicating the mathematics by considering a multi-particle eigenfunction with internal variables, we assume the eigenfunction $\underline{\Phi}_{\min}$ represents a single scalar particle. This is sufficient for our application of the results to the description of a free spinless particle in Chapter 9.

Separating variables yields the product

$$\mathbf{\Phi}_{\min}(x) = \prod_{\mu=0}^{3} \phi_{\min}(x^\mu). \tag{8.3.12}$$

To avoid confusion, we explicitly write summations for the rest of this section. Substituting Eqs. (8.3.11) and (8.3.12) into Eq. (8.3.3) and rearranging gives

$$\frac{\partial \phi_{\min}}{\partial x^\mu} = -\frac{g_{\mu\mu}}{\hbar}\left[\alpha\left(x^\mu - \langle x^\mu\rangle\right) + i\langle p^\mu\rangle\right]\phi_{\min}. \tag{8.3.13}$$

where we have used the relation

$$x_\mu = \sum_{\nu=0}^{3} g_{\mu\nu} x^\nu = g_{\mu\mu} x^\mu. \tag{8.3.14}$$

Notice that Eq. (8.3.14) is possible because all off-diagonal elements of the metric tensor $g_{\mu\nu}$ are zero. The solution of Eq. (8.3.13) must satisfy the normalization condition

$$\int_{-\infty}^{\infty} \phi_{min}^* \phi_{min} dx^\mu = 1 \tag{8.3.15}$$

and the dispersion relation

$$\int_{-\infty}^{\infty} \phi_{min}^* [x - \langle x^\mu \rangle]^2 \phi_{min} dx^\mu \equiv (\Delta x^\mu)^2 . \tag{8.3.16}$$

We have defined the domain of integration to range from $-\infty$ to $+\infty$ because of our planned application to the description of an unbounded, noninteracting spinless particle in Chapter 9. The resulting solution is

$$\phi_{min}(x^\mu) = \eta_{min} \exp\left[-i\frac{g_{\mu\mu}\langle p_0^\mu \rangle}{\hbar}(x^\mu - \langle x_0^\mu \rangle) - \frac{(x^\mu - \langle x_0^\mu \rangle)^2}{4(\Delta x_0^\mu)^2}\right] \tag{8.3.17}$$

where

$$\eta_{min} = [2\pi(\Delta x_0^\mu)^2]^{-1/4} . \tag{8.3.18}$$

Equations (8.3.12) and (8.3.17) are used to construct a Gaussian wave packet in Chapter 9.

THE UNCERTAINTY PRINCIPLE

A problem in conventional theories is their inadequate explanation of the origin of the time-energy uncertainty relation. For a discussion of the problem, see Mayants [Sec. 14.7, 1984], and Kim and Noz [Sec. 6.1, 1986]. This problem disappears in PRQT because of Eq. (8.3.10). The dispersion relation in Eq. (8.3.10) is a general result which we use to derive uncertainty relations by evaluating the product $(\Delta x_\mu)(\Delta p_\mu)$. Notice that the summation convention for repeated indices has been suspended for the rest of this section.

The operators $\boldsymbol{\Omega}_j$ are

$$\boldsymbol{\Omega}_1 = \boldsymbol{x}_\mu, \ \boldsymbol{\Omega}_2 = \boldsymbol{p}_\mu = \frac{\hbar}{i}\frac{\partial}{\partial x^\mu}. \tag{8.3.19}$$

Evaluating the commutator bracket gives

$$[\boldsymbol{x}_\mu, \boldsymbol{p}_\mu] = -\frac{\hbar}{i} g_{\mu\mu}. \tag{8.3.20}$$

The important result in Eq. (8.3.20) is the appearance of $g_{\mu\mu}$. It assures the proper sign of the usual commutation relations, but it is based on a derivation which treats space and time in the same manner. Using Eq. (8.3.20) in Eq. (8.3.10) gives

$$(\Delta x_\mu)^2 (\Delta p_\mu)^2 \geq \frac{\hbar^2}{4}(g_{\mu\mu})^2 = \frac{\hbar^2}{4}. \tag{8.3.21}$$

Simplifying Eq. (8.3.21) yields

$$|\Delta x_\mu| \ |\Delta p_\mu| \geq \frac{\hbar}{2}. \tag{8.3.22}$$

Equation (8.3.22) is the set of uncertainty relations for space-momentum and energy-time.

Applications of the uncertainty relations are presented in subsequent chapters. For example, a Gaussian wave packet is constructed in §9.2 using the relativistic minimum wave packet derived above, and a mass-lifetime relationship is obtained in §13.4.

§8.4 CLASSICAL CORRESPONDENCE

We wish to establish a correspondence between the classical and quantum equations of motion. To do so, we first construct a matrix representation.

Equation (8.1.2) can be written as

$$i\hbar \frac{\partial \underline{\Psi}_{\kappa}(y,s)}{\partial s} = \boldsymbol{K}\, \underline{\Psi}_{\kappa}(y,s) = \kappa \underline{\Psi}_{\kappa}(y,s) \tag{8.4.1}$$

where the set of eigenfunctions $\{\underline{\Psi}_{\kappa}(y,s)\}$ satisfy the orthonormality condition

$$\langle \underline{\Psi}_{\kappa} | \underline{\Psi}_{\kappa'} \rangle \equiv \int_{D^N} \underline{\Psi}_{\kappa}^{+}(y,s)\, \underline{\Psi}_{\kappa'}(y,s)\, dy = \delta_{\kappa\kappa'} . \tag{8.4.2}$$

The functions $\underline{\Psi}$ and $\{\underline{\Psi}_{\kappa}\}$ are state vectors in an infinite-dimensional Hilbert space. As such, the state vectors can be expressed in an infinite number of alternative representations which are related by unitary transformations. These transformations correspond to rotations of the axes in Hilbert space. The norm of the respective state vectors remains unchanged because the transformations are unitary. We work in the coordinate representation in this section.

MATRIX REPRESENTATION

Given a Hermitian operator $\boldsymbol{\Omega}$, the function $\boldsymbol{\Omega}\underline{\Psi}_{\kappa}$ can be expanded in terms of the complete set of eigenfunctions $\{\underline{\Psi}_{\kappa}(y,s)\}$ such that

$$\boldsymbol{\Omega} \underline{\Psi}_{\kappa} = \sum_{\iota} \underline{\Psi}_{\iota}\, \omega_{\iota\kappa} \tag{8.4.3}$$

where $\omega_{\iota\kappa}$ are the matrix elements of the operator $\boldsymbol{\Omega}$. Premultiplying Eq. (8.4.3) by $\underline{\Psi}_{\gamma}^{+}$ and integrating over dy yields

$$\omega_{\gamma\kappa} = \langle \underline{\Psi}_{\gamma} | \boldsymbol{\Omega} \underline{\Psi}_{\kappa} \rangle \tag{8.4.4}$$

where Eq. (8.4.2) has been used. These results show that every operator $\boldsymbol{\Omega}$ can be associated with an array $\{\omega_{\gamma\kappa}\}$. Since $\boldsymbol{\Omega}$ is Hermitian, it has the property

$$\langle \boldsymbol{\Omega} \underline{\Psi}_{\gamma} | \underline{\Psi}_{\kappa} \rangle = \langle \underline{\Psi}_{\gamma} | \boldsymbol{\Omega} \underline{\Psi}_{\kappa} \rangle \tag{8.4.5}$$

which implies that

$$\omega_{\gamma\kappa} = \omega_{\kappa\gamma}^{*} . \tag{8.4.6}$$

EQUATION OF MOTION

The matrix formulation is particularly useful for examining how the matrix elements of an operator change with respect to s. Differentiating Eq. (8.4.4) with respect to s gives

$$\frac{d\omega_{\gamma\kappa}}{ds} = \frac{d}{ds}\langle \boldsymbol{\Psi}_\gamma | \boldsymbol{\Omega} \boldsymbol{\Psi}_\kappa \rangle$$
$$= \langle \frac{\partial \boldsymbol{\Psi}_\gamma}{\partial s} | \boldsymbol{\Omega} \boldsymbol{\Psi}_\kappa \rangle + \langle \boldsymbol{\Psi}_\gamma | \frac{\partial \boldsymbol{\Omega}}{\partial s} \boldsymbol{\Psi}_\kappa \rangle + \langle \boldsymbol{\Psi}_\gamma | \boldsymbol{\Omega} \frac{\partial \boldsymbol{\Psi}_\kappa}{\partial s} \rangle . \tag{8.4.7}$$

Using Eq. (8.4.1) and the Hermitian properties of the operators lets us write Eq. (8.4.7) in the form

$$\frac{d\omega_{\gamma\kappa}}{ds} = \frac{i}{\hbar}\langle \boldsymbol{\Psi}_\gamma | (\boldsymbol{K}\boldsymbol{\Omega} - \boldsymbol{\Omega}\boldsymbol{K}) \boldsymbol{\Psi}_\kappa \rangle + \langle \boldsymbol{\Psi}_\gamma | \frac{\partial \boldsymbol{\Omega}}{\partial s} \boldsymbol{\Psi}_\kappa \rangle . \tag{8.4.8}$$

Defining $(\mathbf{A})_{ij}$ as the ijth matrix element of operator **A**, we write Eq. (8.4.8) as

$$\frac{d\omega_{\gamma\kappa}}{ds} = \left(\frac{\partial \boldsymbol{\Omega}}{\partial s} \right)_{\gamma\kappa} + \frac{i}{\hbar}([\boldsymbol{K}, \boldsymbol{\Omega}])_{\gamma\kappa} \tag{8.4.9}$$

where we have used the commutator bracket $[\mathbf{K},\boldsymbol{\Omega}]$.

Equation (8.4.9) is the equation of motion of the matrix element $\omega_{\gamma\kappa}$ with respect to changes in s. A comparison of Eqs. (8.4.9) and (2.2.14) shows that the quantum and classical equations of motion are related by substituting the relativistic commutator bracket divided by $i\hbar$ for the relativistic Poisson bracket. A similar relationship exists in conventional nonrelativistic quantum theory.

CHAPTER 9
SPIN-0 PARTICLE

The mathematical foundation for PRQT was presented in Chapter 6 through Chapter 8, and several relationships between mathematical and physical concepts were introduced. We must extend the mathematics further into the physical world to test the validity of the mathematical foundation. This testing is essential to establishing or discrediting a new paradigm. We begin by considering the simplest case: the spin-0 particle [Fanchi, 1977; Fanchi and Collins, 1978; Collins and Fanchi, 1978].

§9.1 THE SPIN-0 STÜCKELBERG EQUATION

The Stückelberg equation for a spin-0 particle corresponds to Eqs. (8.1.2) and (8.1.3) for a single particle with 1 component:

$$i\frac{\partial \Psi}{\partial s} = \boldsymbol{K}_{(0)}\Psi = \left[\frac{\pi_\mu \pi^\mu}{2m} + V_{(0)}\right]\Psi \tag{9.1.1}$$

where

$$\pi_\mu = \frac{\hbar}{i}\partial_\mu - \frac{e}{c}A_\mu . \tag{9.1.2}$$

We continue to explicitly write the physical constants $\hbar$ and c for the sake of clarity. As we become more comfortable with the formalism, we will revert to natural units. The operator $\mathbf{K}_{(0)}$ is the spin-0 Hamiltonian and $V_{(0)}$ is a Hermitian function that represents interactions of the particle with its environment. The solution Ψ satisfies the normalization condition

$$\int \Psi^* \Psi d^4x = 1. \tag{9.1.3}$$

Equation (9.1.1) is the lowest order representation of the Stückelberg equation. Its form is highly suggestive of the parametrized Klein-Gordon equations introduced in §3.3. The following sections take advantage of this similarity -- while retaining the constraints of the mathematical foundation presented in Chapter 6 through Chapter 8 -- to establish a physical interpretation for the mathematical quantities. For simplicity, we first present the physical description of a non interacting ("free") particle before introducing interactions.

§9.2 THE FREE PARTICLE

The non interacting particle is defined by setting the four-vector A^μ to zero, or by setting the charge e to zero. For a single particle we have N=1 and write $\varepsilon_a = \varepsilon_1 = \varepsilon$. The spin-0 Stückelberg equation becomes

$$\begin{aligned} i\frac{\partial \Psi_f}{\partial s} &= -\varepsilon \frac{\hbar}{2} \partial_\mu \partial^\mu \Psi_f \\ &= -\varepsilon \frac{\hbar}{2} \left[\frac{\partial^2}{\partial (x^0)^2} - \nabla^2 \right] \Psi_f . \end{aligned} \tag{9.2.1}$$

Equation (9.2.1) has the general solution

$$\begin{aligned} \Psi_f (x,s) &= \int \psi_{f\,\kappa}(x,s) dk \\ &= \int \eta_{f\,\kappa} \exp[i\kappa(k)s + ik_\mu x^\mu] dk \end{aligned} \tag{9.2.2}$$

where

$$\kappa = -\varepsilon \frac{\hbar}{2} k_\mu k^\mu . \tag{9.2.3}$$

Rather than using the physical correspondence presented in Eq. (8.4.1), we have written the parameters as in Eq. (9.2.1). A physical motivation for the correspondence postulated in Eq. (8.4.1) is obtained by considering the expectation values of x^μ and p^μ.

DEFINITION OF MASS

The expectation value of the four-velocity V_f^μ of the non interacting particle is

$$\langle V_f^\mu \rangle = \frac{d\langle x_f^\mu \rangle}{ds} = \varepsilon \langle p_f^\mu \rangle. \tag{9.2.4}$$

from Eqs. (7.2.23), (7.2.24) and (7.3.9). The subscript f denotes a free particle. Substituting Eq. (9.2.2) into $<p_f^\mu>$ shows that $<p_f^\mu>$ is independent of s. Integrating Eq. (9.2.4) from s to s+δs yields

$$\delta \langle x_f^\mu \rangle = \varepsilon \langle p_f^\mu \rangle \delta s. \tag{9.2.5}$$

Equation (9.2.5) expresses the most probable trajectory of the relativistic non-interacting particle.

The observable world-line of the particle is given by

$$\delta \langle x_f^\mu \rangle \delta \langle x_{f\mu} \rangle = \varepsilon^2 \langle p_f^\mu \rangle \langle p_{f\mu} \rangle (\delta s)^2. \tag{9.2.6}$$

Recalling the definition of dispersion is

$$\Delta^2 p_f \equiv \langle p_f^\mu p_{f\mu} \rangle - \langle p_f^\mu \rangle \langle p_{f\mu} \rangle \tag{9.2.7}$$

lets us rewrite Eq. (9.2.6) as

$$\delta \langle x_f^\mu \rangle \delta \langle x_{f\mu} \rangle = \varepsilon^2 \langle p_f^\mu p_{f\mu} \rangle (\delta s)^2 - \varepsilon^2 \Delta^2 p_f (\delta s)^2. \tag{9.2.8}$$

In the classical limit $\Delta^2 p_f$ vanishes and Eq. (9.2.8) becomes

$$[\delta \langle x_f^\mu \rangle \delta \langle x_{f\mu} \rangle]_{cl} = \varepsilon^2 \langle p_f^\mu p_{f\mu} \rangle (\delta s)^2. \tag{9.2.9}$$

Equation (9.2.9) expresses the classical behavior of a noninteracting spinless particle. By comparing Eq. (9.2.9) with experiment -- as described in §2.3, especially Eq. (2.3.9) -- we are able to deduce the physical significance of ε.

Rearranging Eq. (9.2.9) gives

$$\frac{1}{\varepsilon^2} \equiv m^2 = \frac{\langle p_f^{\mu} p_{f\mu} \rangle}{c^2} \left\{ \frac{c^2 (\delta s)^2}{[\delta \langle x_f^{\mu} \rangle \delta \langle x_{f\mu} \rangle]_{cl}} \right\} \tag{9.2.10}$$

where we have inserted the speed of light c so that the term {...} is dimensionless, and introduced a real variable m as the inverse of ε as postulated in Eq. (8.1.1). We want ε^2 to be positive because we stipulated in §7.2 that ε was a real c-number. This stipulation was made to assure that the operators $\mathbf{p}^{\mu}$ of Eq. (7.2.24) and $\mathbf{K}$ of Eq. (8.1.2) would be Hermitian. If we recall that the evolution parameter s monotonically increases, then we see from Eq. (9.2.10) that the parameter ε is real in the case of timelike (bradyon or slower-than-light) motion because $<p_f^{\mu} p_{f\mu}>$ and $<\delta x_f^{\mu}><\delta x_{f\mu}>$ are positive. If the motion is spacelike (tachyon or faster-than-light), ε^2 is still positive because both $<p_f^{\mu} p_{f\mu}>$ and $<\delta x_f^{\mu}><\delta x_{f\mu}>$ are negative. Similar comments apply to its inverse m.

The value of m^2 is a measure of the ratio of the invariant evolution parameter interval to the classical proper time interval δs_{cl} [Goldstein, 1980] where

$$c^2 (\delta s_{cl})^2 \equiv [\delta \langle x_f^{\mu} \rangle \delta \langle x_{f\mu} \rangle]_{cl} . \tag{9.2.11}$$

When the absolute value of the ratio $\delta s/\delta s_{cl}$ equals unity, as it does in the noninteracting case, m^2 becomes

$$m^2 c^2 = \frac{\langle p_f^{\mu} p_{f\mu} \rangle}{sgn[\delta \langle x_f^{\mu} \rangle \delta \langle x_{f\mu} \rangle]} \geq 0 \tag{9.2.12}$$

where sgn(x) denotes the sign of x. Equation (9.2.12) applies to both timelike (bradyonic) and spacelike (tachyonic) motion. In this approach, Eq. (9.2.12) serves as the definition of mass for a spacelike particle and corresponds to the "rest" mass of a timelike particle. We do not refer to m as "rest" mass, however, because Eq. (9.2.12) applies to tachyons as well. The inclusion of tachyons within the PRQT framework is an important, experimentally testable prediction. An introduction to tachyon kinematics is presented in Part IV.

By identifying m as the mass of the particle, Eq. (9.2.9) becomes the light cone constraint of special relativity. Observe that the constraint is expressed in terms of average spacial and temporal displacements. The resulting constraint corresponds to Enatsu's "macroscopic causality" principle, i.e. the bradyon's most probable trajectory remains within the light cone. For a succinct review of Enatsu's work, see Fanchi [1993].

STÜCKELBERG-FEYNMAN INTERPRETATION

Equation (9.2.4) is the parametrized basis of the Stückelberg-Feynman interpretation. The space components of $<V_f^\mu>$ are the space components of the phase velocity, and $<V_f^0>$ is the temporal component. Recalling that s is monotonically increasing, Eq. (9.2.4) shows that the particle propagates into the temporal coordinate "future" when $d<x_f^0>$ is positive, and into the temporal coordinate "past" when $d<x_f^0>$ is negative. If we substitute Eq. (9.2.2) into Eq. (9.2.4) we find

$$\langle V_f^\mu \rangle = \frac{\hbar}{m} \langle k^\mu \rangle \tag{9.2.13}$$

where we now express $<V_f^\mu>$ in terms of wavenumbers $<k^\mu>$. Recognizing that k^0 is proportional to the frequency of the wave ($k^0 = \omega/c$), Eq. (9.2.13) says that a positive frequency particle propagates forward in time and a negative frequency particle propagates backward in time. This result is the Stückelberg-Feynman interpretation. Furthermore, in the nonrelativistic limit when

$$m^2 c^2 \gg \langle \vec{p} \cdot \vec{p} \rangle \tag{9.2.14}$$

we have the expected result

$$|\langle V_f^0 \rangle| \rightarrow c. \tag{9.2.15}$$

The Stückelberg-Feynman interpretation in the PRQT formulation provides an acceptable physical interpretation of positive and negative energy solutions within the framework of a single particle theory. Thus we see that solutions of field equations like the Klein-Gordon equation or the Stückelberg equation can be interpreted without resorting to the framework of many-body theories.

GAUSSIAN WAVE PACKET

Equation (9.2.12) shows that mass is an observable of the system. By examining Eq. (9.2.2), we see that Ψ_f is a superposition of states, each of which corresponds to a different energy-momentum four-vector $\{k^\mu\}$. Thus we obtain the reasonable result that measurements of the mass of the relativistic particle are not sharply defined, but have a distribution that depends on the probabilistic weight of each pure state, namely $|\eta_{f\kappa}|$. As an illustration, we consider the minimum wave packet representation of the noninteracting particle.

For clarity, suppose that a relativistic spin-0 particle is observed passing through a particle detector. We assume the particle does not interact after exiting the detector. If the uncertainties in the measurements of x^μ and p^μ have been minimized, we should be able to describe the particle leaving the detector as a free particle with the initial ($s=0$) condition that its wave function had the form of the minimum wave packet.

The expansion coefficients for the minimum wave packet are calculated using §7.4. We assume the initial condition

$$\begin{aligned}\Psi_{f0}(x) &\equiv \Psi_f(x,s=0)\\ &= \int \eta_{f\kappa} \exp[ik_\mu x^\mu]\,dk \qquad (9.2.16)\\ &= \Phi_{\min}(x)\end{aligned}$$

where $\Phi_{\min}$ is given by Eqs. (8.3.12) and (8.3.17). Multiplying Eq. (9.2.16) by $\exp(-k'_\mu x^\mu)$ and integrating over dx yields

$$\eta_f\,(k^\mu) = \frac{8\pi}{(2\pi)^4} \prod_{\mu=0}^{3} \{(\Delta x^\mu)^{1/2} \cdot \exp[-g^2_{\mu\mu}\langle\Delta k^\mu\rangle^2\langle\Delta x^\mu\rangle^2 - i k_\mu \langle x^\mu\rangle]\} \tag{9.2.17}$$

where the Einstein summation convention has been suspended. The general solution is found by substituting Eq. (9.2.17) into Eq. (9.2.2) to get

$$\Psi_{f\,\min}(x,s) \equiv \prod_{\mu=0}^{3} N_\mu \exp\{i\langle k_\mu\rangle(x^\mu - \langle x^\mu\rangle)\} \cdot \exp\left\{-\frac{i\hbar}{2m}\langle k_\mu\rangle\langle k^\mu\rangle s - \frac{\left[x^\mu - \langle x^\mu\rangle - \frac{\hbar}{m}\langle k^\mu\rangle s\right]^2}{4\left[\langle\Delta x^\mu\rangle^2 + i\frac{\hbar}{m} g_{\mu\mu} s\right]}\right\} \tag{9.2.18}$$

where subscript "min" reminds us of the choice of initial condition and $g_{\mu\mu}$ is needed for $\Psi_{f\,\min}$ to be a correct solution of the noninteracting particle case. The normalization constants $\{N^\mu\}$ do not play a role in the ensuing discussion and are therefore not exhibited here.

Let us now consider four physically interesting points. First, the expectation value of x_f^μ is

$$\langle x_f^\mu\rangle = \langle x_0^\mu\rangle + \frac{\hbar}{m}\langle k_0^\mu\rangle s. \tag{9.2.19}$$

Equation (9.2.19) is the expected result.

Second, by taking the absolute square of Eq. (9.2.17) we obtain the probability distribution in the momentum-energy representation. The result is a Gaussian distribution which indicates that the wave packet is a superposition of states. Each state corresponds to a particular mass eigenvalue associated with each four-momentum vector k^μ.

Third, the absolute square of Eq. (9.2.18) gives the Gaussian probability distribution in space and time. From this joint distribution a marginal probability of observing the particle somewhere in space during a specified interval of time may be constructed by integrating $\rho(x|s)$ over all space to yield the marginal probability density $\rho(x^0|s)$. Whenever $\rho(x^0|s)$ is zero the particle cannot be observed anywhere in space, i.e. the particle effectively does not exist at that instant. This capability -- not present in conventional theories -- is necessary for describing unstable particles while simultaneously retaining probability conservation. It is discussed in greater detail in Chapter 13.

Finally, we note that the wave packet $\Psi_{f\,min}$ is not Lorentz-invariant. This is a result of the choice of the minimum eigenfunction as the initial condition. Although Eq. (9.2.18) is not a Lorentz invariant, it does illustrate the concept of superposition of mass states.

HAMILTON-JACOBI FORM

Expressing the field equation for a noninteracting particle in Hamilton-Jacobi form gives us additional insight into the relationship between quantum theory, classical theory and probability theory. We begin with the wave equation

$$i\hbar\frac{\partial}{\partial s}\psi(x,s) = -\frac{\hbar^2}{2m}\partial_\mu\partial^\mu\psi(x,s). \tag{9.2.20}$$

Substituting Eq. (7.2.9) into (9.2.20), multiplying by $(\rho^{1/2}e^{-i\xi})$, and simplifying gives

$$\begin{gathered} i\hbar\left[i\rho\frac{\partial\xi}{\partial s}+\frac{1}{2}\frac{\partial\rho}{\partial s}\right] = -\frac{\hbar^2}{2m} \\ \cdot\left\{-\rho\,\partial^\mu\xi\,\partial_\mu\xi + i\partial_\mu(\rho\,\partial^\mu\xi) - \frac{1}{4\rho}\partial_\mu\rho\,\partial^\mu\rho + \frac{1}{2}\partial_\mu\partial^\mu\rho\right\}. \end{gathered} \tag{9.2.21}$$

Equation (9.2.21) can be rewritten in the form

$$-\rho\hbar\frac{\partial\xi}{\partial s}+\frac{\hbar^2}{2m}\left\{-\rho\,\partial^\mu\xi\,\partial_\mu\xi-\frac{1}{4\rho}\,\partial^\mu\rho\,\partial_\mu\rho+\frac{1}{2}\partial_\mu\partial^\mu\rho\right\}$$
$$+\frac{i\hbar}{2}\left\{\frac{\partial\rho}{\partial s}+\hbar\partial_\mu(\rho\,\partial^\mu\xi)\right\}=0. \tag{9.2.22}$$

Equation (9.2.22) is satisfied if the following two equations are satisfied:

$$\rho\hbar\frac{\partial\xi}{\partial s}+\frac{\hbar^2}{2m}\left\{\rho\,\partial^\mu\xi\,\partial_\mu\xi+\frac{1}{4\rho}\,\partial^\mu\rho\,\partial_\mu\rho-\frac{1}{2}\partial_\mu\partial^\mu\rho\right\}=0 \tag{9.2.23}$$

and

$$\frac{\partial\rho}{\partial s}+\hbar\partial_\mu(\rho\,\partial^\mu\xi)=0. \tag{9.2.24}$$

Dividing Eq. (9.2.23) by $-\rho$ and replacing ξ by the action functional $S_K=\hbar\xi$ gives

$$\frac{\partial S_K}{\partial s}+\frac{1}{2m}(\partial^\mu S_K)(\partial_\mu S_K)+Q_K=0, \tag{9.2.25}$$

where

$$Q_K=\frac{1}{4\rho^2}\partial^\mu\rho\,\partial_\mu\rho-\frac{1}{2\rho}\partial_\mu\partial^\mu\rho, \tag{9.2.26}$$

and Eq. (9.2.24) becomes

$$\frac{\partial\rho}{\partial s}+\partial_\mu(\rho\,\partial^\mu S_K)=0. \tag{9.2.27}$$

Equation (9.2.25) corresponds to a Hamilton-Jacobi equation with a quantum potential Q_K {Eq. (9.2.26)}, and Eq. (9.2.27) is a continuity equation with velocity

$$V^\mu=\hbar\;\partial^\mu S_K \tag{9.2.28}$$

from Eq. (7.2.10).

§9.3 ELECTROMAGNETIC INTERACTION

A useful constraint imposed on the formalism of modern physical theories is gauge invariance. We require that probability conservation be preserved under a local U(1) gauge transformation that depends on the space-time four-vector. As a result of the gauge invariance requirement, we obtain some properties of A^μ that reinforce its identification as the four-vector potential of the electromagnetic interaction. Once this is done, we can discuss charge conjugation within the PRQT framework.

GAUGE INVARIANCE

The probability density ρ and velocity V^μ are related to ψ and A^u by Eqs. (7.2.8) and (7.2.10). The value of ρ is unchanged by the transformation

$$\psi'_\lambda(x,s) = \psi_\lambda(x,s)\exp\left[-i\frac{\varepsilon''}{\varepsilon'}\Lambda_g(x)\right] \tag{9.3.1}$$

where Λ_g is a real scalar function of $\{x^\mu\}$.

A more general transformation in which Λ_g is allowed to depend on both $\{x_\mu\}$ and s, i.e. $\Lambda_g(x)\rightarrow\Lambda_g(x,s)$, has been studied by several authors, e.g. Greenberger [1970, 1974] and Kyprianidis [1987]. The dependence on evolution parameter requires the introduction of a fifth gauge potential in addition to the four-vector $\{A^\mu\}$. Horwitz, Arshansky and Elitzur [1988] called these five potentials "pre-Maxwell fields". Building on parametrized field theories of Fanchi [1979], Hostler [1980, 1981], and Kubo [1985], Saad, Horwitz and Arshansky [1989] constructed a Lagrangian density for the matter field and pre-Maxwell fields. They were able to construct Maxwell's theory of electromagnetism as a subset of their formalism. Preliminary considerations for second quantizing the theory were presented. Green's functions for the field strength tensor were subsequently derived by Land and Horwitz [1991]. They showed how to reproduce the propagator approach of Schwinger [1951a] and DeWitt [1975], and studied the causal properties of the propagators. For further details, the interested reader should consult the references. Equation (9.3.1) is sufficient for our purposes.

Equation (9.3.1) is a gauge transformation of the first kind. It corresponds to changing the phase ξ in Eq. (7.2.10) by

$$\xi'(x,s) = \xi(x,s) - \frac{\varepsilon''}{\varepsilon'} \Lambda_g(x) . \tag{9.3.2}$$

Equation (9.3.2) must be accompanied by the gauge transformation of the second kind

$$A'^{\mu}(x,s) = A^{\mu}(x,s) + \frac{\partial \Lambda_g(x)}{\partial x_{a\mu}} \tag{9.3.3}$$

so that the probability conservation equation {Eq. (7.1.4)} remains invariant with respect to gauge transformations. Thus A^{μ} and ξ are only specified to within a gauge transformation.

The transformation properties described in Eqs. (9.3.1) through (9.3.3) are familiar from electrodynamics. They serve to reinforce our identification of ε'' as -e/c in Eq. (8.1.1). Given this interpretation, the appropriate equation for the eigenfunction of a single particle moving in an electromagnetic field is

$$i\hbar \frac{\partial \Psi_{em}}{\partial s} = \frac{1}{2m} \left[\frac{\hbar}{i} \frac{\partial}{\partial x_{\mu}} - \frac{e}{c} A^{\mu} \right] \left[\frac{\hbar}{i} \frac{\partial}{\partial x^{\mu}} - \frac{e}{c} A_{\mu} \right] \Psi_{em} . \tag{9.3.4}$$

CHARGE CONJUGATION

The Stückelberg-Feynman interpretation can be combined with the interpretation of ε'' and A^{μ} to obtain a familiar result. The complex conjugate of Eq. (9.3.4) is

$$-i\hbar\frac{\partial\Psi^*_{em}}{\partial s} = \frac{1}{2m}\left[-\frac{\hbar}{i}\frac{\partial}{\partial x_\mu} - \frac{e}{c}A^\mu\right]\left[-\frac{\hbar}{i}\frac{\partial}{\partial x^\mu} - \frac{e}{c}A_\mu\right]\Psi^*_{em} \tag{9.3.5}$$

or

$$i\hbar\frac{\partial\Psi^*_{em}}{\partial s} = \frac{1}{2m}\left[\frac{\hbar}{i}\frac{\partial}{\partial x_\mu} + \frac{e}{c}A^\mu\right]\left[\frac{\hbar}{i}\frac{\partial}{\partial x^\mu} + \frac{e}{c}A_\mu\right]\Psi^*_{em}. \tag{9.3.6}$$

The wave function $\Psi_{em}{}^*$ is the solution of Eq. (9.3.4) with the charge e replaced by -e. Thus the probability density remains unchanged with respect to charge conjugation. What does change?

If we write Ψ_{em} as $Re^{i\Theta}$, then ρV^μ becomes

$$j^\mu \equiv \rho V^\mu = \left(\frac{\hbar}{m}\frac{\partial\Theta}{\partial x_\mu} - \frac{eA^\mu}{mc}\right)R^2 \tag{9.3.7}$$

where j^μ is the probability flux. Replacing e by -e and interchanging Ψ_{em} with $\Psi_{em}{}^*$ changes the sign of V^μ and $<V^\mu>$. This says that a charged particle and its oppositely charged antiparticle propagate in opposite spacial and temporal directions for a given electromagnetic configuration. These remarks apply to noninteracting particles as well. Thus a nontrivial interpretation of j^0/c is provided by PRQT for both charged and neutral particles. A significant result of the interpretation is discussed below.

§9.4 SCATTERING FROM A STEP POTENTIAL

Conventional theories predict that a relativistic spinless particle incident on a step potential of sufficient strength will be reflected by the barrier with a reflection coefficient that exceeds unity. This result is known as the Klein paradox and is a direct consequence of identifying ej^0/c as the charge density [Klein, 1929]. Even though total charge is conserved, the prediction that more particles will be reflected than were incident is a surprising consequence which has not been experimentally tested. The Klein paradox becomes an anomaly associated with the mainstream paradigm when we observe that the Klein paradox does not arise in PRQT, as we now show.

We are interested in solving the problem of a spinless particle scattering from a step potential. It is sufficient for our purposes to work in one spacial dimension x. A four-vector potential for this problem may be written as

$$\begin{aligned} A^0(x) &= 0, \ x<0 \\ &= \alpha, \ x>0 \\ \vec{A}(t,x) &= 0. \end{aligned} \tag{9.4.1}$$

We define the region of negative x (x < 0) as region I and the region of positive x (x > 0) as region II. We assume throughout the calculation that a positively charged, spin-0 bradyon (slower-than-light) particle is incident from x = -∞ with energy ω_1 and momentum k_1, where e, ω_1 and k_1 are real and positive, and we are resuming our use of natural units ($\hbar$=c=1). Invoking the boundary conditions of §7.2, the wave function and its first derivative with respect to x must be continuous across the boundary at x = 0.

Wave functions for regions I and II are solutions of the respective equations

$$2mi\frac{\partial \Psi_I}{\partial s} = \left(\frac{\partial^2}{\partial x^2} - \frac{\partial^2}{\partial t^2}\right)\Psi_I \tag{9.4.2}$$

and

$$2mi\frac{\partial \Psi_{II}}{\partial s} = \left[\frac{\partial^2}{\partial x^2} + \left(\frac{1}{i}\frac{\partial}{\partial t} - e\alpha\right)^2\right]\Psi_{II}. \tag{9.4.3}$$

The eigenvalue relation in region I is

$$q_I \equiv m_I^2 = (\omega_1^2 - k_1^2) > 0 \tag{9.4.4}$$

for the solution

$$\Psi_I = [a_1 e^{(-ik_1x)} + b_1 e^{(ik_1x)}] e^{(i\omega_1 t)} e^{(-iq_I s/2m)}. \tag{9.4.5}$$

The inequality in Eq. (9.4.4) is valid for incident bradyons, and the real quantity

m_I corresponds to the mass state of the particle in region I. In the case of negligible dispersion, the eigenvalue m_I equals the expected value m. For simplicity, we assume negligible dispersion so that m may be used in subsequent calculations. The coefficients a_1 and b_1 in Eq. (9.4.5) are the coefficients of the incident and reflected plane wave solutions, respectively. These identifications are established as follows.

First recall that probability flux is given by Eq. (7.2.26). For our case Eq. (7.2.26) simplifies to

$$\rho V^{\mu} = -\frac{i}{2m}\left(\psi^* \frac{\partial \psi}{\partial x_{\mu}} - \psi \frac{\partial \psi^*}{\partial x_{\mu}}\right) - \frac{eA^{\mu}}{m}\psi^* \psi \tag{9.4.6}$$

with the conditions

$$\begin{aligned} &\textit{region I: } (A^0, A^1) = (0,0), \\ &\textit{region II: } (A^0, A^1) = (\alpha, 0). \end{aligned} \tag{9.4.7}$$

For the t and x components we obtain

$$\rho V^{0} = -\frac{i}{2m}\left(\psi^* \frac{\partial \psi}{\partial t} - \psi \frac{\partial \psi^*}{\partial t}\right) - \frac{eA^{0}}{m}\psi^* \psi \tag{9.4.8}$$

$$\rho V^{1} = -\frac{i}{2m}\left(\psi^* \frac{\partial \psi}{\partial x} - \psi \frac{\partial \psi^*}{\partial x}\right) \tag{9.4.9}$$

respectively. The propagation directions for a given plane wave in Region I are tabulated in Table 9.1.

A description of the situation in region II is a little more involved. The solution of Eq. (9.4.3) is

$$\Psi_{II} = [a_2 e^{-ik_2 x} + b_2 e^{ik_2 x}] e^{i\omega_2 t} e^{-iq_{II} s/2m} . \tag{9.4.10}$$

To guarantee that the wave function and its first derivative with respect to x are continuous at x=0 for all values of t and s, we must have

$$\omega_1 = \omega_2$$
$$q_I = q_{II}. \tag{9.4.11}$$

The eigenvalue relation for region II is

$$k_2^2 = (\omega_1 - e\alpha)^2 - m^2 \tag{9.4.12}$$

where Eq. (9.4.11) has been used, and k_2 may be real or imaginary. Results are presented in Table 9.2.

Four possible scattering cases with corresponding spacial solutions are shown in Table 9.3. The quantities ω_1, k_1, k_2, q_I and $e\alpha$ are all taken to be real and positive.

TABLE 9.1 PROPAGATION DIRECTIONS OF PLANE WAVE SOLUTIONS IN REGION I

Plane Wave	V^0	V^1	Propagation Direction
$e^{i\omega t}$	$\omega/m>0$	--	Forward in t
$e^{-i\omega t}$	$-\omega/m<0$	--	Backward in t
e^{ikx}	--	$k/m<0$	Backward in x
e^{-ikx}	--	$k/m>0$	Forward in x

TABLE 9.2 PROPAGATION DIRECTIONS OF PLANE WAVE SOLUTIONS IN REGION II

Plane Wave	V^0	V^1	Propagation Direction
$e^{i\omega t}$	$(\omega\text{-}e\alpha)/m>0$	--	Forward in t
	$(\omega\text{-}e\alpha)/m<0$		Backward in t
	$(\omega\text{-}e\alpha)/m=0$		Stationary in t
$e^{-i\omega t}$	$(-\omega\text{-}e\alpha)/m<0$	--	Backward in t
e^{ikx}	--	$k/m<0$	Backward in x
e^{-ikx}	--	$k/m>0$	Forward in x

TABLE 9.3 SCATTERING CASES

Case	Conditions	Spacial Solution
1	$\omega_1>e\alpha$ $(\omega_1\text{-}e\alpha)^2\geq q_I$	$a_2\ \exp(-ik_2x)$
2	$\omega_1>e\alpha$ $(\omega_1\text{-}e\alpha)^2<q_I$	$b_2\ \exp(-ik_2x)$
3	$\omega_1<e\alpha$ $(\omega_1\text{-}e\alpha)^2\geq q_I$	$a_2\ \exp(-ik_2x)$
4	$\omega_1<e\alpha$ $(\omega_1\text{-}e\alpha)^2<q_I$	$b_2\ \exp(-ik_2x)$

Reflection and transmission coefficients are found in the same manner as that of nonrelativistic quantum mechanics [Schiff, 1968]. For example, Case 1 satisfies the boundary conditions

$$\begin{aligned} a_1 + b_1 &= a_2 \\ (a_1 - b_1)k_1 &= a_2 k_2 . \end{aligned} \tag{9.4.13}$$

Solving Eq. (9.4.13) for b_1 and a_2 in terms of a_1 gives

$$\begin{aligned} b_1 &= \frac{k_1 - k_2}{k_1 + k_2} a_1 , \\ a_2 &= \frac{2k_1}{k_1 + k_2} a_1 . \end{aligned} \tag{9.4.14}$$

The reflection coefficient R is calculated as the ratio of reflected probability flux divided by incident probability flux, with an analogous definition for the transmission coefficient T. For this case we have

$$\begin{aligned} R &= \frac{k_1/m \ |b_1|^2}{k_1/m \ |a_1|^2} = \left[\frac{k_1 - k_2}{k_1 + k_2} \right]^2 , \\ T &= \frac{k_2/m \ |a_2|^2}{k_1/m \ |a_1|^2} = \frac{4k_1 k_2}{(k_1 + k_2)^2} . \end{aligned} \tag{9.4.15}$$

The coefficient b_2 is zero in cases 1 and 3 because no particles are incident on the barrier from $x = +\infty$. For cases 2 and 4 the quantity k_2 is positive imaginary, thus the coefficient of the term $\exp(|k_2|x)$ must be zero because the wave function must be finite for all values of x. Results for all cases are summarized in Table 9.4. Are these results physically realistic?

The results show that the reflection coefficient for each case does not exceed unity and that the sum of the reflection and transmission coefficients for each case is unity. These are desired results that avoid the Klein paradox. All of the results for cases 1 and 4 are easy to accept because these particular cases correspond to the familiar results of nonrelativistic quantum mechanics. The details of cases 2 and 3 deserve further attention.

TABLE 9.4 REFLECTION AND TRANSMISSION RESULTS

Case	Reflection Coefficient R	Transmission Coefficient T	Does R+T=1?
1	$[(k_1-k_2)]/(k_1+k_2)]^2$	$4k_1k_2/(k_1+k_2)^2$	Yes
2	1	0	Yes
3	$[(k_1-k_2)]/(k_1+k_2)]^2$	$4k_1k_2/(k_1+k_2)^2$	Yes
4	1	0	Yes

To see that case 2 is physically realistic, we first observe that $\omega_1 > m$ because of Eq. (9.4.4). The eigenvalue condition for case 2 can be written as

$$|e\alpha| > \omega_1 - m > 0. \tag{9.4.16}$$

The kinetic energy of the particle in region I is ω_1-m, so Eq. (9.4.16) shows that the barrier height exceeds the kinetic energy of the incident particle and we therefore expect reflection at the barrier. Thus the results of case 2 are realistic.

Now consider case 3. The particles transmitted into region II are propagating backward in time, and the fraction of reflected particles is

$$R_3 = \left[\frac{(k_1 - k_2)}{(k_1 + k_2)}\right]^2 . \tag{9.4.17}$$

For this same case, conventional theories predict that antiparticles propagating forward in time are transmitted into region II, and that the fraction of reflected particles is greater than 1. The fraction of reflected particles in the conventional theory is $1/R_3$.

An experiment capable of measuring the fraction of particles reflected from an appropriate step potential could ascertain which, if either, result is correct. The smallest value of $|e\alpha|$ for which the conditions of case 3 are satisfied is 2m. This is a very large value -- on the order of 280 MeV for a charged pion -- and is not yet experimentally attainable. If the same predictions hold for a spin-½ particle, the strength of the step potential will decrease significantly (to about 1 MeV for a positron or electron). We are now justified in proceeding to a study of spin-½ particles.

CHAPTER 10
SPIN-½ PARTICLE

From a pedagogical perspective, it was best to introduce many physical ideas in the context of the spin-0 PRQT formalism in Chapter 9 because of its relative simplicity. From the perspective of an everyday practitioner, we also need to understand spin-½ particles. The purpose of this chapter is to present a PRQT model of spin-½ particles.

§10.1 THE SPIN-½ STÜCKELBERG EQUATION

The Stückelberg equation for a spin-½ particle corresponds to Eqs. (8.1.2) and (8.1.3) for a single particle with 2 components:

$$i\frac{\partial}{\partial s}\underline{u} = \underline{\underline{K}}\underline{u} = \left[\frac{\pi_\mu \pi^\mu}{2m}\underline{\underline{I}} + \underline{\underline{V}}\right]\underline{u}. \tag{10.1.1}$$

The matrices $\underline{\underline{I}}$, $\underline{\underline{K}}$, $\underline{\underline{V}}$ are 2x2 matrices and $\underline{u}$ is a 2x1 column vector that satisfies the normalization condition

$$\int \underline{u}^+ \underline{u}\, d^4x = 1. \tag{10.1.2}$$

Equation (10.1.1) is the 2x2 representation of the spin-½ Stückelberg equation. Conventional treatments of the spin-½ particle based on the Dirac equation use a 4x4 representation. Our interest here is to construct a 4x4 representation in the context of PRQT. The resulting representation provides additional insight into the nature of PRQT while simultaneously providing an alternative expression for the field equations associated with a spin-½ particle.

4x4 Representation of the Spin-½ Stückelberg Equation

Equation (10.1.1) corresponds to two equations containing second order space-time derivatives. Following a procedure introduced by Davidon [1955], we find a matrix representation of Eq. (10.1.1) by first writing four equations containing first order space-time derivatives:

$$i\pi_\mu \underline{\underline{A}}^\mu \underline{u} = \underline{v} \tag{10.1.3}$$

and

$$-i\pi_\nu \underline{\underline{B}}^\nu \underline{v} = 2mi\frac{\partial}{\partial s}\underline{u}. \tag{10.1.4}$$

The 2x2 matrices $\{\underline{\underline{A}}^\mu, \underline{\underline{B}}^\mu\}$ are as yet unspecified, and $\underline{v}$ is defined by Eq. (10.1.3). Substituting Eq. (10.1.3) into (10.1.4), gives

$$\pi_\nu \pi_\mu \underline{\underline{B}}^\nu \underline{\underline{A}}^\mu \underline{u} = 2mi\frac{\partial}{\partial s}\underline{u}. \tag{10.1.5}$$

Comparing Eq. (10.1.5) with (10.1.1) yields

$$\pi_\nu \pi_\mu \underline{\underline{B}}^\nu \underline{\underline{A}}^\mu = 2m\underline{\underline{K}} = \pi_\mu \pi^\mu \underline{\underline{I}} + 2m\underline{\underline{V}}. \tag{10.1.6}$$

Equation (10.1.6) expresses $\{\underline{\underline{A}}^\mu, \underline{\underline{B}}^\mu\}$ in terms of the Hamiltonian $\underline{\underline{\mathbf{K}}}$.

Equations (10.1.3) and (10.1.4) can be combined to form the matrix equation

$$i\pi_\mu \underline{\underline{\Gamma}}^\mu \underline{\Psi} = \underline{\underline{M}}\underline{\Psi} \tag{10.1.7}$$

where the 4x4 matrices $\underline{\underline{\Gamma}}^\mu$, $\underline{\underline{\mathbf{M}}}$ are

$$\underline{\underline{\Gamma}}^\mu = \begin{bmatrix} 0 & \underline{\underline{A}}^\mu \\ -\underline{\underline{B}}^\mu & 0 \end{bmatrix}, \tag{10.1.8}$$

$$\underline{\underline{M}}^{\mu} = \begin{bmatrix} \underline{\underline{I}} & 0 \\ 0 & 2mi\frac{\partial}{\partial s}\underline{\underline{I}} \end{bmatrix}, \tag{10.1.9}$$

and the 4x1 column vector $\underline{\Psi}$ is

$$\underline{\Psi} = \begin{bmatrix} \underline{v} \\ \underline{u} \end{bmatrix}. \tag{10.1.10}$$

Equation (10.1.7) is the 4x4 representation of the spin-½ Stückelberg equation.

STATIONARY STATES

A stationary mass state equation may be derived from Eq. (10.1.7) by performing a non-unitary transformation. In this case $\underline{\Psi}$ is an eigenfunction of $\underline{\underline{\mathbf{K}}}$ such that

$$\underline{\Psi}(x,s) \sim \underline{\Psi}(x) e^{-ims/2} \tag{10.1.11}$$

so that

$$\underline{\underline{M}}\underline{\Psi} = \begin{bmatrix} \underline{\underline{I}} & 0 \\ 0 & m^2\underline{\underline{I}} \end{bmatrix} \underline{\Psi}. \tag{10.1.12}$$

Using Davidon's [1955] transformation matrix

$$\underline{\underline{T}} \equiv \begin{bmatrix} m^{1/2} & 0 \\ 0 & m^{-1/2} \end{bmatrix}, \tag{10.1.13}$$

we multiply Eq. (10.1.7) on the right by $\underline{\underline{\mathrm{T}}}^* = \underline{\underline{\mathrm{T}}}$ and expand as

$$(i\underline{\underline{T}}^* \pi_\mu \underline{\underline{\Gamma}}^\mu \underline{\underline{T}} - \underline{\underline{T}}^* \underline{\underline{M}}\, \underline{\underline{T}}) \underline{\underline{T}}^{-1} \underline{\Psi} = 0. \tag{10.1.14}$$

In terms of the transformed variables

$$\underline{\underline{\Gamma}}^{\prime\mu} = \underline{\underline{T}}^{*}\underline{\underline{\Gamma}}^{\mu}\underline{\underline{T}} = \underline{\underline{\Gamma}}^{\mu},$$
$$\underline{\underline{M}}^{\prime} = \underline{\underline{T}}^{*}\underline{\underline{M}}\,\underline{\underline{T}} = m\underline{\underline{I}}, \tag{10.1.15}$$

and

$$\underline{\Psi}^{\prime} = \underline{\underline{T}}^{-1}\underline{\Psi} = \begin{bmatrix} m^{-1/2}\underline{v} \\ m^{1/2}\underline{u} \end{bmatrix}, \tag{10.1.16}$$

Eq. (10.1.14) becomes

$$(i\pi_{\mu}\underline{\underline{\Gamma}}^{\mu} - m\underline{\underline{I}})\underline{\Psi}^{\prime} = 0. \tag{10.1.17}$$

Equation (10.1.17) is a 4x4 representation of the stationary mass state equation.

DIRAC EQUATION

It is tempting to identify $\{\underline{\underline{\Gamma}}^{\mu}\}$ as the Dirac gamma matrices by writing

$$\underline{\underline{A}}_{D}^{\mu} \equiv [-i\underline{\underline{I}}, \underline{\underline{\sigma}}_{x}, \underline{\underline{\sigma}}_{y}, \underline{\underline{\sigma}}_{z}],$$
$$\underline{\underline{B}}_{D}^{\mu} \equiv -[i\underline{\underline{I}}, \underline{\underline{\sigma}}_{x}, \underline{\underline{\sigma}}_{y}, \underline{\underline{\sigma}}_{z}] \tag{10.1.18}$$

where $\{\underline{\underline{\sigma}}_{j},\ j=x,y,z\}$ are the Pauli matrices

$$\underline{\underline{\sigma}}_{x} = \begin{bmatrix} 0 & 1 \\ 1 & 0 \end{bmatrix}, \quad \underline{\underline{\sigma}}_{y} = \begin{bmatrix} 0 & -i \\ i & 0 \end{bmatrix}, \quad \underline{\underline{\sigma}}_{z} = \begin{bmatrix} 1 & 0 \\ 0 & -1 \end{bmatrix}. \tag{10.1.19}$$

The resulting gamma matrices are the Dirac gamma matrices identified, for example, as $\underline{\underline{\gamma}}^{\mu}$ by Arfken [pg. 190, 1970], and the resulting equation is the Dirac equation.

Davidon [1955] pointed out that defining $\{\underline{\underline{\Gamma}}^{\mu}\}$ by Eq. (10.1.18) is inconsistent with the scalar product we have adopted, namely Eqs. (7.3.11) and (10.1.2). Preferring to retain the Dirac equation in the form that made it so successful, Davidon introduced an inner product with an indefinite metric. The indefinite metric allows both positive and negative probabilities, which is the same difficulty suffered by conventional theories based on the Dirac equation.

We are at a dilemma. If we insist on preserving the probabilistic concepts in Chapter 7, we cannot accept the Dirac equation as part of PRQT. Is this a defect in PRQT, or is it a clue that the Dirac equation, even with its history of success, is not quite right?

Mayants [1974, 1984], and Wu and Hwang [1991] have pointed out flaws in the Dirac equation. Mayants [Sec. 14.4, 1984] notes that the Dirac equation predicts the eigenvalue of the velocity of any free, relativistic, spin-½ particle is the speed of light. This is a well-known consequence of Dirac theory [e.g. Bjorken and Drell, Sec. 3.3, 1964; Wu and Hwang, Sec. 2.3, 1991]. Mayants argues that this result contradicts both the principles of relativity theory and experimental data. He concludes that the Dirac equation must be incorrect. Wu and Hwang [Sec. 0, 1991] note that it is very difficult to apply the single particle Dirac equation to many particles. They further observe that the Dirac theory -- even though it was originally intended to be a single particle theory -- could not be interpreted within the confines of a single particle theory. Klein's paradox and the Stückelberg-Feynman interpretation are related to this problem.

The difficulties of the Dirac equation suggest that a better approach may exist, but we will never know if we are afraid to challenge cherished ideas. We know that we can construct a 4x4 representation of a spin-½ field equation. Furthermore, we have consistently rejected formulations that require unconventional probabilistic concepts such as negative probabilities. Therefore we must be prepared to follow PRQT to its logical conclusion. In the process we may obtain a new perspective that will help us better understand the conventional paradigm. For these reasons we conclude that the usual form of the Dirac equation is inappropriate for PRQT.

Does this mean that PRQT cannot calculate experimental results to the same degree of accuracy as the Dirac equation or its second order counterpart? It does not. What it does mean is that a new model of a spin-½ particle interacting with an electromagnetic field is needed.

THE HPR (HORWITZ, PIRON AND REUSE) MODEL

A model of the spin-½ particle interacting with a four-vector potential has been proposed by Horwitz, Piron and Reuse. When we speak of the HPR model, we mean the model introduced by Horwitz, Piron and Reuse in a series of articles published in the latter half of the 1970's [Horwitz, Piron and Reuse, 1975; Piron and Reuse, 1977; and Reuse, 1978]. The HPR model represents a spin-½ particle with a two-component wave function that satisfies the equation

$$i\frac{\partial \Psi}{\partial s} = \underline{\underline{K}}_{HPR}\Psi. \tag{10.1.20}$$

The operator $\mathbf{K}_{HPR}$ has the form

$$\begin{aligned}\underline{\underline{K}}_{HPR} &= \frac{\pi_\mu \pi^\mu}{2m}\underline{\underline{I}} + \frac{g_1\mu_0}{m}\pi^\mu \tilde{F}_{\mu\nu}\underline{\underline{W}}^\nu_n \\ &- \frac{(g_2)^2\mu_0^2}{8m}F_{\mu\nu}n^\nu F^\mu_\rho n^\rho \underline{\underline{I}} + g_3\mu_0 n^\mu \tilde{F}_{\mu\nu}\underline{\underline{W}}^\nu_n\end{aligned} \tag{10.1.21}$$

in natural units ($\hbar$=c=1). The tensor $F_{\mu\nu}$ and its dual are related by [Jackson, 1975]

$$\tilde{F}^{\mu\nu} = \frac{1}{2}\epsilon^{\mu\nu\rho\lambda}F_{\rho\lambda}. \tag{10.1.22}$$

The Levi-Civita type tensor $\epsilon^{\mu\nu\rho\lambda} = \pm 1$ for $\mu\nu\rho\lambda$ an even or odd permutation of 0,1,2,3 respectively, and $\epsilon^{\mu\nu\rho\lambda} = 0$ otherwise. The constants m and μ_0 denote the observable mass and the Bohr magneton (e/2m), respectively. The quantities g_1, g_2, g_3 are phenomenological constants with magnitudes of order 1. Their physical correspondence is given within the context of examples presented in §10.3 and §12.3.

Equation (10.1.21) has a form which is consistent with Eq. (10.1.1) when we identify the interaction potential $\underline{\underline{V}}$ as

$$\underline{V}_{HPR} = \frac{g_1\mu_0}{m}\pi^\mu \tilde{F}_{\mu\nu}\underline{\underline{W}}^\nu_n - \frac{(g_2)^2\mu_0^2}{8m}F_{\mu\nu}n^\nu F^\mu_\rho n^\rho \underline{\underline{I}} + g_3\mu_0 n^\mu \tilde{F}_{\mu\nu}\underline{\underline{W}}^\nu_n . \tag{10.1.23}$$

The first term in Eq. (10.1.21) represents the Lorentz forces acting on the particle. The remaining three terms corresponding to $\underline{\underline{V}}_{HPR}$ are perturbations describing spin-orbit coupling. Notice that each of the last three terms has as a factor either $F_{\mu\nu}$ or its dual. Both $F_{\mu\nu}$ and its dual are antisymmetric tensors, i.e.

$$F_{\mu\nu} = -F_{\nu\mu}, \quad \tilde{F}_{\mu\nu} = -\tilde{F}_{\nu\mu}. \tag{10.1.24}$$

Since $F_{\mu\nu}$ and its dual are antisymmetric tensors, the diagonal elements of both tensors are zero. Thus there are only six independent elements to be calculated. These elements are the magnetic field

$$\vec{B} = (F_{32}, F_{13}, F_{21}) = (\tilde{F}^{10}, \tilde{F}^{20}, \tilde{F}^{30}) \tag{10.1.25}$$

and the electric field

$$\vec{E} = (F_{01}, F_{02}, F_{03}) = (\tilde{F}^{23}, \tilde{F}^{31}, \tilde{F}^{12}). \tag{10.1.26}$$

Spin enters the theory through the four 2x2 matrices $\underline{\underline{W}}_n^\mu$ associated with a timelike unit four-vector n^μ. Letting $L(n)_\nu^\mu$ denote a boost [Jackson, §11.7, 1975], i.e.

$$L(n)^\mu_\nu \, n_0^\nu = n^\mu \tag{10.1.27}$$

with $n_0^\nu \equiv (1,0,0,0)$, the definition of $\underline{\underline{W}}_n^\mu$ is

$$\underline{\underline{W}}^\mu_n \equiv L(n)^\mu_\nu \underline{\underline{W}}^\nu_{n_0} , \quad \underline{\underline{W}}^\nu_{n_0} = \left(0\underline{\underline{I}}, \frac{1}{2}\underline{\underline{\sigma}}\right) \tag{10.1.28}$$

where $\{\underline{\underline{\sigma}}_i\}$ are the Pauli matrices. The validity of the spin-orbit coupling terms is evaluated by comparison with experiment in subsequent chapters.

§10.2 SCATTERING FROM A STEP POTENTIAL

We saw in §9.4 how a spin-0 particle behaves in the presence of a step potential. A similar calculation may be performed for the spin-½ particle. The calculation requires a knowledge of the direction of particle propagation, which we now determine.

STÜCKELBERG-FEYNMAN INTERPRETATION

The direction of particle propagation of a spin-½ particle is determined by evaluating the proper velocity: thus

$$\frac{d\langle x^\mu\rangle}{ds} = -\frac{i}{m}\sum_{j=1}^{2}\int \psi_j^* \frac{\partial}{\partial x_\mu}\psi_j d^4x \tag{10.2.1}$$

where

$$\underline{\Psi} = \begin{bmatrix}\psi_1\\ \psi_2\end{bmatrix} \equiv \begin{bmatrix}u_1\\ u_2\end{bmatrix} f(x^\mu, s). \tag{10.2.2}$$

The elements u_1, u_2 are scalars. We assume they are independent of the space-time coordinates $\{x^\mu\}$, but may depend on s. Noting that u_1 and u_2 satisfy the relation

$$\sum_{j=1}^{2} u_j^* u_j = 1 \tag{10.2.3}$$

leads to the result

$$\frac{d\langle x^\mu\rangle}{ds} = -\frac{i}{m}\int f^* \frac{\partial}{\partial x_\mu} f d^4x. \tag{10.2.4}$$

Equation (10.2.4) agrees with the spin-0 particle case for plane waves presented in Chapter 9. Consequently we conclude that the Stückelberg-Feynman interpretation applies to the spin-½ particle. This result considerably simplifies the following discussion of a spin-½ particle scattering from a step potential.

RESOLUTION OF THE KLEIN PARADOX

Proceeding as in §9.4, we take the four-vector potential in two dimensions ($x^0=t$, $x^1=x$) as

$$A^0(x,t) = \begin{Bmatrix} 0, & x<0 \\ \alpha, & x>0 \end{Bmatrix}, \qquad (10.2.5)$$
$$A^1(x,t) = 0.$$

Again region I refers to $x < 0$ and region II refers to $x > 0$. We assume that a positively charged, spin-½ bradyon is incident from $x = -\infty$ with energy ω_1 and momentum k_1, where e, ω_1 and k_1 are real and positive. The boundary conditions at $x = 0$ are that the wave function and its first derivative must be continuous.

Wave functions for regions I and II are solutions of the respective equations

$$2mi\frac{\partial \underline{\Psi}_I}{\partial s} = \left(\frac{\partial^2}{\partial x^2} - \frac{\partial^2}{\partial t^2}\right)\underline{\Psi}_I \qquad (10.2.6)$$

and

$$2mi\frac{\partial \underline{\Psi}_{II}}{\partial s} = \left[\frac{\partial^2}{\partial x^2} + \left(\frac{1}{i}\frac{\partial}{\partial t} - e\alpha\right)^2\right]\underline{\Psi}_{II}. \qquad (10.2.7)$$

Equations (10.2.6) and (10.2.7) do not allow either the mixing of spin states or spin-orbit interactions. To see that this is the case, we must show that the last three terms of $\mathbf{K}_{HPR}$ vanish.

The four-vector potential in region I is

$$(A^\mu)_I = 0, \quad \mu = 0,1,2,3 \qquad (10.2.8)$$

while the four-vector potential in region II is

$$(A^0)_{II} = \alpha ,$$
$$(A^i)_{II} = 0, \; i = 1,2,3. \tag{10.2.9}$$

Since α is a constant, all of the elements of the four-vector potentials in both regions I and II are constants. Consequently the elements of the corresponding tensors $(F_{\mu\nu})_I$ and $(F_{\mu\nu})_{II}$ and their duals are all zero. Thus the last three terms of $\mathbf{K}_{HPR}$ vanish in both regions I and II. The resulting field equations are Eqs. (10.2.6) and (10.2.7).

The scattering problem is further simplified by using Eq. (10.2.2) to write

$$\underline{\Psi}_I = \begin{bmatrix} u_{I1} \\ u_{I2} \end{bmatrix} f_I(x,t,s); \; \underline{\Psi}_{II} = \begin{bmatrix} u_{II1} \\ u_{II2} \end{bmatrix} f_{II}(x,t,s) \tag{10.2.10}$$

with the u's assumed independent of (x,t,s) in this case. We have already shown that all of the terms in $\mathbf{K}_{HPR}$ which might have mixed spin states are zero for the given four-vector potential. This fact and the continuity of the eigenfunction at the boundary x = 0 imply that the spin of the incident particle does not change, thus

$$u_{Ij} = u_{IIj}, \; j = 1,2. \tag{10.2.11}$$

Using Eqs. (10.2.10) and (10.2.11) in Eqs. (10.2.6) and (10.2.7) shows that the scattering problem is reduced to solving the following equations:

$$2mi\frac{\partial f_I}{\partial s} = \left(\frac{\partial^2}{\partial x^2} - \frac{\partial^2}{\partial t^2}\right) f_I, \tag{10.2.12}$$

and

$$2mi\frac{\partial f_{II}}{\partial s} = \left[\frac{\partial^2}{\partial x^2} + \left(\frac{1}{i}\frac{\partial}{\partial t} - e\alpha\right)^2\right] f_{II}. \tag{10.2.13}$$

The solution of this simplified problem is identical to that for spin-0 particles presented in §9.4.

Thus we see that the results of the spin-0 scattering problem also apply to spin-½ particles. Of particular interest is the case identified as case 3 in §9.4. Let us recall that the particles transmitted into region II are propagating backward in time, and the fraction of reflected particles R_3 is given by Eq. (9.4.17). For this case, conventional theories predict that the fraction of reflected particles is $R_3^{-1} > 1$. The conventional rationale [Bjorken and Drell, 1964; Wu and Hwang, 1991] for resolving the Klein paradox is to argue that we must use a many-particle theory, such as Lagrangian quantum field theory, to solve the scattering problem. We show in Part V that Lagrangian quantum field theory is a special case of a parametrized quantum field theory. Thus the conventional rationale is to appeal to a theory which is a special case of parametrized relativistic quantum field theory. The flaw in the conventional argument is to assert that we must use a many-particle theory. The work here and in Chapter 9 shows that the problem of relativistic scattering from a step potential can be solved using single-particle PRQT without encountering the Klein paradox.

In my opinion, an experiment designed to measure the fraction of particles reflected from an appropriate step potential is needed to ascertain which result is correct. The smallest value of $|e\alpha|$ for which the conditions of case 3 are satisfied is 2m, which is about 1 MeV for a positron or electron. Such an experiment could help choose among several internally consistent but fundamentally different approaches to the Klein paradox [Fanchi, 1993].

§10.3 SPIN STATES

Thus far we have used the HPR model of spin-½ particles to establish the Stückelberg-Feynman interpretation for spin-½ particles and to solve the problem of scattering from a step potential. We have not yet established the HPR model as a suitable model for describing spin. We do so here by applying the model to the Stern-Gerlach experiment and to a stationary particle in a static magnetic field.

STERN-GERLACH EXPERIMENT

In the previous section we wrote the 2x1 column vector as a product of a 2x1 column vector of complex numbers times a scalar that was dependent on space-time coordinates and the evolution parameter. The HPR model generalizes this scheme by associating a Hilbert space H_n of the form $C^2 \otimes L^2(R^4)$ with each two-component eigenfunction. Notice that this representation is consistent with the eigenfunction form allowed by PRQT in §7.2. The subscript n of H_n refers to the 4-vector n^μ introduced in §10.1. Mathematically, n^μ serves as a superselection rule for indexing the family of Hilbert spaces. Its physical meaning is clarified by considering a spin-½ particle in the presence of a magnetic field.

Piron and Reuse [1978] wrote a general representation of the spin eigenvalue as

$$\underline{\Psi}_n(x^\mu,s) = \begin{bmatrix} e^{-i\phi/2} \cos(\theta/2) \\ e^{i\phi/2} \sin(\theta/2) \end{bmatrix} f(x^\mu,s) \tag{10.3.1}$$

where θ,ϕ denote orientation of the spin vector in spherical coordinates. The state $\underline{\Psi}$ is normalized according to the definition in §7.3 as long as Eq. (10.2.3) and

$$\int f^* f \, d^4x = 1 \tag{10.3.2}$$

are satisfied. It is an eigenstate of the observable

$$\frac{1}{2}\vec{s}\cdot\underline{\underline{\vec{\sigma}}} = s_\mu \underline{\underline{W}}^\mu_{n_0} = s'_\mu \underline{\underline{W}}^\mu_n \tag{10.3.3}$$

where $\{\underline{\underline{\sigma}}^j\}$ are Pauli matrices and the spin vector is

$$s^\mu = (0,\ \sin\theta\ \cos\phi,\ \sin\theta\ \sin\phi,\ \cos\theta). \tag{10.3.4}$$

The spin vectors s^μ, s'^μ are related by the boost $L(n)_\nu{}^\mu$, i.e.

$$s'^\mu = L(n)^\mu_\nu \ s^\nu \tag{10.3.5}$$

from the unprimed to the primed reference frame. The operator $s'_\mu \underline{\underline{W}}_n^\mu$

corresponds to a measurement of spin using a Stern-Gerlach apparatus with the time and space directions of the magnetic field given by n^μ and s^μ respectively. To verify this, we calculate the eigenvalues of the operator $s'_\mu \underline{\underline{W}}_n^\mu$ acting on the state $\underline{\Psi}$.

The eigenvalue equation for the operator $s'_\mu \underline{\underline{W}}_n^\mu$ acting on the state $\underline{\Psi}$ is

$$\frac{1}{2}\vec{s}\cdot\vec{\underline{\underline{\sigma}}}\,\underline{\Psi} = \lambda\,\underline{\Psi} \tag{10.3.6}$$

or

$$\frac{1}{2}[\sin\theta\,(\cos\phi\;\underline{\underline{\sigma}}_x + \sin\phi\;\underline{\underline{\sigma}}_y) + \cos\theta\;\underline{\underline{\sigma}}_z]\underline{\Psi} = \lambda\,\underline{\Psi}. \tag{10.3.7}$$

The eigenvalue problem is simplified by orienting the spin state such that $\phi=\theta=0$. This corresponds to orienting the magnetic field in the z-direction. The eigenvalue problem becomes

$$\frac{1}{2}\,\underline{\underline{\sigma}}_z\,\underline{\Psi} = \lambda\,\underline{\Psi}. \tag{10.3.8}$$

The eigenvalues of Eq. (10.3.8) are $\pm$½ and the spin state of the spin-½ particle is either parallel or antiparallel to the direction of the magnetic field as expected.

SPIN-½ PARTICLE IN A MAGNETIC FIELD

If the formalism of the HPR model is applicable to a spin-½ particle, we should be able to calculate the contribution of the spin-magnetic field coupling for a spin-½ particle at rest in a static magnetic field. We begin by choosing a reference frame that lets us simplify the problem. In particular, we consider the case in which $n^\mu = n_0^\mu$ where n_0^μ = (1,0,0,0) as in §10.1.

The spin-dependent term with the g_1 coupling constant becomes

$$\frac{g_1\mu_0}{m}(p^\mu - eA^\mu)\tilde{F}_{\mu\nu}W_n^\nu = \frac{g_1\mu_0}{m}[(\vec{p}-e\vec{A})\cdot(\vec{W}_{n_0}\times\vec{E}) + (p^0 - eA^0)\vec{W}_{n_0}\cdot\vec{B}] \tag{10.3.9}$$

when expressed in terms of electric and magnetic field vectors. Similarly, the terms with g_2 and g_3 coupling constants become

$$\frac{g_2^2\mu_0^2}{8m}F_{\mu\nu}n^\nu F_\rho^\mu n^\rho = \frac{g_2^2\mu_0^2}{8m}\vec{E}\cdot\vec{E} \tag{10.3.10}$$

and

$$g_3\mu_0 n^\mu \tilde{F}_{\mu\nu}W_n^\nu = g_3\mu_0\vec{W}_{n_0}\cdot\vec{B} \tag{10.3.11}$$

respectively. Using Eq. (10.1.28) in Eqs. (10.3.9) and (10.3.11) lets us explicitly include spin in the above interaction terms. The resulting expressions are

$$\frac{g_1\mu_0}{m}(p^\mu - eA^\mu)\tilde{F}_{\mu\nu}W_n^\nu = \frac{g_1\mu_0}{2m}[(\vec{p}-e\vec{A})\cdot(\vec{\sigma}\times\vec{E}) + (p^0 - eA^0)\vec{\sigma}\cdot\vec{B}] \tag{10.3.12}$$

and

$$g_3\mu_0 n^\mu \tilde{F}_{\mu\nu}W_n^\nu = \frac{g_3\mu_0}{2}\vec{\sigma}\cdot\vec{B}. \tag{10.3.13}$$

We find an interpretation of the coupling constants g_1 and g_3 by considering a spin-½ particle at rest in a static magnetic field. In this case, the g_2 term vanishes and the remaining g_1 and g_3 terms simplify to a spin-orbit coupling operator with the form

$$K_{so} \equiv \frac{(g_1+g_3)\mu_0}{2}\vec{\sigma}\cdot\vec{B}. \tag{10.3.14}$$

Aligning the magnetic field along the z-axis such that

$$\vec{B} = B_0 \hat{k} \tag{10.3.15}$$

and evaluating the expectation value of $\mathbf{K}_{so}$ gives

$$\langle \boldsymbol{K}_{so} \rangle = \frac{(g_1 + g_3)\mu_0 B_0}{2}. \tag{10.3.16}$$

Recalling that μ_0 is the Bohr magneton e/2m, we have

$$\langle \boldsymbol{K}_{so} \rangle = \frac{(g_1 + g_3) e B_0}{4m}. \tag{10.3.17}$$

Equation (10.3.16) is in qualitative agreement with experiment, and agrees quantitatively when we identify the sum $g_1 + g_3$ as the g-factor of the electron magnetic moment.

CHAPTER 11
EVOLUTION PARAMETER MEASUREMENTS

Thus far in the development of PRQT we have used the evolution parameter s with only minimal discussion about its physical validity. Many authors in the past have argued that s is not measurable, but is merely a mathematical convenience [Fanchi, 1993]. These authors failed to conceptualize a physically meaningful experiment for measuring s and therefore assumed that s could not be measured. Indeed, by going beyond the confines of a single particle formulation, we can design an apparatus for measuring s [Fanchi, 1986; Pavsic, 1991a]. One advantage of the formalism presented in Chapter 7 and Chapter 8 is its applicability to many-particle systems. We demonstrate the usefulness of the many-particle formulation below by showing how it can be used to help us design an evolution parameter clock.

§11.1 NONINTERACTING MANY-BODY SYSTEMS

The Stückelberg equation for an N-body system is

$$i\frac{\partial \underline{\Psi}}{\partial s} = \boldsymbol{K}\,\underline{\Psi} = \left[\sum_{a=1}^{N}\left(\frac{\pi^{\mu}_{a}\pi_{a\mu}}{2m_a}\right)\underline{\underline{I}} + \underline{\underline{V}}\right]\underline{\Psi}. \tag{11.1.1}$$

We illustrate how to use Eq. (11.1.1) by considering the relatively simple problem of a noninteracting many-body system. Systems in which interactions are allowed are discussed in greater detail later.

In the case of a noninteracting system we have $\{A_a^{\mu}\}=0$ and $\underline{\underline{V}}=0$ so that Eq. (11.1.1) becomes

$$i\frac{\partial\psi_f}{\partial s}=\sum_{a=1}^{N}\left(-\frac{1}{2m_a}\frac{\partial^2\psi_f}{\partial y_a^{\mu}\partial y_{a\mu}}\right). \tag{11.1.2}$$

The column vector eigenfunction $\underline{\Psi}$ is replaced by a scalar ψ_f since spin effects are unimportant in this case. Equation (11.1.2) has the solution

$$\psi_f\,(y,s)=\prod_{a=1}^{N}\eta_a^{1/2}\,\exp\left[-\frac{is}{2m_a}(k_a^{\mu}k_{a\mu})+ik_{a\mu}y_a^{\mu}\right] \tag{11.1.3}$$

where η_a is the normalization constant for particle a.

The expectation value of the four-velocity for particle a is given by Eq. (7.3.9) for a scalar eigenfunction such that

$$\langle V_a^{\mu}\rangle=\frac{1}{m_a}\int\psi_f^*\left[\frac{1}{i}\frac{\partial}{\partial y_{a\mu}}\right]\psi_f\,dy \tag{11.1.4}$$

where Eq. (8.1.1) has been used. Substituting Eq. (11.1.3) into (11.1.4) gives

$$\frac{d\langle y_a^{\mu}\rangle}{ds}=\langle V_a^{\mu}\rangle=\frac{\langle k_a^{\mu}\rangle}{m_a}. \tag{11.1.5}$$

Integrating Eq. (11.1.5) from s to s+δs gives

$$\delta\langle y_a^{\mu}\rangle=\frac{\langle k_a^{\mu}\rangle}{m_a}\delta s. \tag{11.1.6}$$

Equation (11.1.6) expresses the most probable trajectory of the relativistic free particle a when m_a is identified as the particle mass. This identification is justified as follows.

The observable world-line of particle a is

$$\delta\langle y_a^{\mu}\rangle\delta\langle y_{a\mu}\rangle=\frac{\langle k_a^{\mu}\rangle\langle k_{a\mu}\rangle}{m_a^2}\delta s^2. \tag{11.1.7}$$

Recalling the definition of dispersion from Eq. (9.2.7) lets us rewrite Eq. (11.1.7) in the form

$$\delta\langle y_a^\mu\rangle\delta\langle y_{a\mu}\rangle = \frac{\langle k_a^\mu k_{a\mu}\rangle}{m_a^2}\delta s^2 - \frac{\Delta^2 k_a}{m_a^2}\delta s^2 . \tag{11.1.8}$$

In the classical limit $\Delta^2 k_a$ vanishes and Eq. (11.1.8) becomes

$$\delta\langle y_a^\mu\rangle\delta\langle y_{a\mu}\rangle = \frac{\langle k_a^\mu k_{a\mu}\rangle}{m_a^2}\delta s^2 . \tag{11.1.9}$$

As in the single particle case, we define the observable mass of particle a by

$$\langle p_a^\mu p_{a\mu}\rangle = \langle k_a^\mu k_{a\mu}\rangle \equiv m_a^2 \tag{11.1.10}$$

so that Eq. (11.1.9) reduces to the light cone constraint of classical special relativity:

$$\delta\langle y_a^\mu\rangle\delta\langle y_{a\mu}\rangle = \delta s^2 . \tag{11.1.11}$$

Equation (11.1.11) assures macroscopic causality for each particle of the N-body system in the sense that the classical trajectory of the ath particle is its most probable trajectory. Similar causal behavior on the microscopic level cannot be surmised from the present formalism and led Enatsu [1963] to the notion of "microncausal" particle behavior.

Another physical statement is implicit in Eq. (11.1.11). In particular, Eq. (11.1.11) implies that the rest frame clocks of every particle run at the same rate, i.e. they have the same classical proper time intervals. This calibration of rest frame clocks is a result of assuming a probability distribution with a single evolution parameter in §7.1. Fanchi and Wilson [1983] and Fanchi [1993] discuss multiparameter formalisms. One advantage of the single parameter formalism for many-body systems is its suitability for developing a physically meaningful evolution parameter clock, as we now show.

§11.2 TEMPORAL MEASUREMENTS

Time is one of the most accurately measured quantities in all of physics, yet the literature--ranging from science to art--abounds with temporal paradoxes [Wheeler, 1986]. Ever since Aristotle, who made time the measure of change, people have tried to understand its role in nature. An understanding of time is possible when a proper context is established.

Time appears in many of the equations of physics ranging from Newtonian and Einsteinian mechanics to Maxwell's equations and quantum mechanics. Does time have the same meaning in all of these contexts? A basic flaw in understanding time is a confusion stemming from treating time as if it was the same physical quantity in every context in which it was used. An analogous situation occurred with the concept of mass. With the advent of Einsteinian relativity, it was legitimate to ask if Newton's mass was the same as Einstein's mass? Only by assuming the masses could be different was a deeper insight into the concept of mass attained by the scientific community.

When a similar treatment is accorded time, we find that the concept of time is really three concepts with physically distinguishable characteristics. For purposes of identifying the concepts, I refer to three times: Galilean time; Minkowski time; and historical time. Galilean time is the usual nonrelativistic time of Newton and his predecessors. Minkowski time is named after Hermann Minkowski, formulator of the four-dimensional geometry adopted by Einstein. Historical time is the temporal parameter s of PRQT and should not be confused with a physical dimension. The best way to understand the three times is to develop an operational understanding, that is, an ability to make measurements and calculations with each of the three times.

In the following I describe some simple experiments. They are the type of experiments that can be easily visualized or performed. The concepts may be so obvious, especially in the more traditional discussion, that the description will appear a waste of effort. It is not. By developing the concepts of Galilean and Minkowski times, the times most familiar to the physics mainstream, we demonstrate that the new historical time is a simple and natural extension to the relativistic domain.

Galilean Time

Consider the following simple experiment. I want to perform a sequence of measurements on a nonrelativistic, classical system. The classical system consists of a single sphere rolling down an incline. I have a ruler and a pendulum (a ball on a string hanging from the table). Initially the sphere is at rest at the top of the incline, and the ruler is placed on the incline parallel to the expected direction of motion. The sphere is released and begins to roll. I perform a measurement of its position and call it E1(1) where E1(1) refers to the first measurement of experiment 1. After the measurement I catch the sphere and put it back on top of the incline. Again the sphere is released and a second measurement E1(2) is made. The process is repeated until I have N measurements E1(N).

A week later I decide to repeat the experiment. I realize, however, that exact replication of experiment 1 requires more information than I obtained in E1. I made N measurements, but I do not know the interval between the release of the sphere and the subsequent position measurement. To remedy the situation, I modify the experiment.

I still make N measurements, but this time I include the swinging pendulum ball in the process. I have previously observed that the duration of the pendulum swing from one turning point to the other is constant if the initial displacement of the pendulum ball from a position of rest is small. Furthermore, there is no noticeable interaction between the pendulum ball and the sphere. The sphere is released at the same time that the pendulum ball begins its swing. I make position measurement n when the swinging pendulum ball has changed direction n times. Thus measurement 1 is made after 1 direction change, measurement 2 after 2 direction changes, and so on.

The number of direction changes of the swinging pendulum is a parameter for ordering my position measurements. The swinging pendulum is a simple physical clock for measuring Galilean time. My unit of time is a complete pendulum swing from one turning point to the next. With the parametrization provided by the pendulum, I can easily repeat the procedure used in experiment 2 at any later date.

An analogous procedure is applicable to nonrelativistic quantum systems. Placing the sphere at the top of the incline is analogous to preparing a pure quantum ensemble. A pure quantum ensemble is an ensemble of systems in pure -- rather than mixed -- quantum states. It is prepared prior to each of the N measurements because the process of measurement can alter the original state of the quantum mechanical system.

If I want I can continue using my classical pendulum clock. Alternatively, I can use any noninteracting, physically observable periodicity as my ordering parameter. Examples of ordering parameters are planetary orbits and the resonance frequency of cesium clocks. The resulting ordering parameter is Galilean time. The requirement that a clock must not interact with the system of interest is an operational requirement: an interaction may actually exist but must be too small to be experimentally detectable (such as a gravitational interaction).

Historical Time

The idea of a nonrelativistic ordering parameter can be extended to the relativistic domain. The nonrelativistic domain consists of three space dimensions and a "time" dimension which is the ordering parameter I call Galilean time. Einstein's relativistic domain consists of three space dimensions and a temporal dimension I call Minkowski time to distinguish it from other times. Minkowski time is theoretically equivalent to the three space dimensions and should not be viewed as an ordering parameter. The difference between a coordinate and a parameter is that coordinate values can either increase or decrease, but parameter values can change in one direction only. Parameter values must change monotonically, while coordinate values are mathematically reversible. In practice, of course, this is not the case.

Unlike space measurements, successive time measurements always increase because of the nature of the measuring apparatus (the clock). Thus the measurement of a system evolving from the present to the past cannot be directly performed with existing clocks. This contrasts with the theoretical notion of

Minkowski time. For example, a positron propagating forward in time can be viewed theoretically as an electron propagating backward in time. Do not be misled by theoretical notions, however. Actual measurements of Minkowski time are monotonically increasing.

Despite the lack of operational equivalence between space and time measurements, an observer O can make independent space and Minkowski time measurements using rulers and a clock at rest in his unprimed frame of reference. A similar procedure can be performed by an observer O' in a primed reference frame that is moving with a constant velocity relative to the unprimed reference frame of observer O. Given the physical ability to make space-Minkowski time measurements in a particular reference frame, observers O and O' are asked to perform a series of measurements on a particular system S1. As in the nonrelativistic measurement procedure, the measurements in each reference frame are to be performed at equivalent intervals. If the duration between each space-Minkowski time measurement in both reference frames is to be the same, it is necessary to parametrize the series of measurements by a parameter that has the same value in both reference frames, and is also independent of the system being studied (S1). Such a parameter is the invariant historical time s.

Measurements of historical time -- a historical time clock -- can be made by introducing a second system S2 that does not interact with S1 but is observable by both observers O and O'. System S2 can be a classical, nonrelativistic pendulum such as the one described earlier. The zeroes of the Minkowski time clocks and the historical time clock (S2) are fixed when the origins of the primed and unprimed coordinate systems are coincident.

At this point each observer is measuring the space-Minkowski time coordinates of two systems: S1 and S2. These measurements require a set of rods and a clock for each system in each reference frame. Thus observer O has two sets of rods and clocks. The space-Minkowski time measurements on system S2 are used to construct historical time. In practice, the system S2 should be a classical system so that the progression of historical time can be determined with a single series of measurements.

If S2 is a quantum system, each measurement is capable of disturbing the state of system S2. In the quantum case, we need to worry about the purity of the quantum ensemble. Since the choice of S2 is up to the observers, it is wise to select a system S2 that is easy to work with. These concepts are quantified in the following sections.

§11.3 MEASURING NONRELATIVISTIC TIME

We claim that the invariant evolution parameter s is the relativistic analog of the nonrelativistic, Galilean time. For this to be true, Galilean time must be an ordering parameter associated with any physically observable periodicity. The periodicity must not be interacting with the system being studied. In this section we quantify these concepts in the context of nonrelativistic quantum mechanics.

The nonrelativistic field equation for N particles without spin may be written in the standard form

$$\left\{ \sum_{j=1}^{N} \left[-\frac{1}{2m_j} \nabla_j^2 \right] + V(x_1, \ldots, z_N, t) \right\} \Phi(x_1, \ldots, z_N, t) = i \frac{\partial}{\partial t} \Phi(x_1, \ldots, z_N, t) \tag{11.3.1}$$

where $V(x_1, \ldots, x_N, t)$ is the potential of the N-particle system. Notice that only one Galilean time t is used in Eq. (11.3.1) to characterize the motion of all N particles. Furthermore, Eq. (11.3.1) can be constructed using the procedure presented in Chapter 7 through Chapter 9 by assuming a Euclidean metric g_{ij} rather than the Minkowski metric $g_{\mu\nu}$. Time dependence is separated out of Eq. (11.3.1) by expressing the system eigenfunction as the functional product

$$\Phi(x_1, \ldots, z_N, t) = \phi(x_1, \ldots, z_N) \exp(-iEt) \tag{11.3.2}$$

such that Eq. (11.3.1) becomes

$$\left\{ \sum_{j=1}^{N} \left[-\frac{1}{2m_j} \nabla_j^2 \right] + V \right\} \phi = E\phi . \tag{11.3.3}$$

As a specific illustration of how t can be measured in principle, we consider the hydrogen atom problem. In this case N=2, we neglect spin, and the interaction potential is

$$V_{12} = \frac{-e^2}{[(x_2 - x_1)^2 + (y_2 - y_1)^2 + (z_2 - z_1)^2]^{1/2}} . \tag{11.3.4}$$

Usually a change of variables to center of mass and relative coordinates is performed before the hydrogen problem is solved. The Hamiltonian is

$$H_{12} = \sum_{j=1}^{2} \left[-\frac{1}{2m_j} \nabla_j^2 \right] + V_{12} \tag{11.3.5}$$

and the energy eigenvalue equation has the form

$$H_{12} \phi_{12} = E_{12} \phi_{12} . \tag{11.3.6}$$

An alternative approach to the hydrogen atom problem will clarify the role t plays in nonrelativistic quantum mechanics. The same approach is applied in §11.4 to PRQT.

We introduce a third particle into the hydrogen atom Hamiltonian such that the energy eigenvalue equation becomes

$$(H_{12} + H_3) \phi_{123} = E_{123} \phi_{123} \tag{11.3.7}$$

where

$$H_3 = -\frac{1}{2m_3} \nabla_3^2 . \tag{11.3.8}$$

Particle 3 is a noninteracting particle. We show now that its observable motion can be used to replace t in the description of the hydrogen atom.

Write the system eigenfunction as

$$\phi_{123} = \phi_{12} \phi_3 = \phi_{12}(x_1, \ldots, z_2) \phi_3(x_3, y_3, z_3) . \tag{11.3.9}$$

Dividing the field equation by ϕ_{123} yields

$$\frac{1}{\phi_{12}} H_{12}\phi_{12} + \frac{1}{\phi_3} H_3\phi_3 = E_{123} \equiv E_{12} + E_3 . \tag{11.3.10}$$

A separation constant α_{NR} is introduced such that the following two equations are obtained:

$$H_{12}\phi_{12} = (E_{12} + \alpha_{NR})\phi_{12}, \tag{11.3.11}$$

and

$$H_3\phi_3 = (E_3 - \alpha_{NR})\phi_3 . \tag{11.3.12}$$

Equation (11.3.11) is the hydrogen atom equation with energy modified by the additive separation constant. Equation (11.3.12) is the equation of a nonrelativistic free particle with its energy modified by the separation constant. The solution to Eq. (11.3.12) is

$$\phi_3 \sim \exp(i\vec{k}_3 \cdot \vec{x}_3) \tag{11.3.13}$$

with the expectation value of spacial location given by

$$\langle \vec{x}_3 \rangle = \frac{\langle \vec{k}_3 \rangle}{m_3} t . \tag{11.3.14}$$

We use Eq. (11.3.14) to express t in terms of expectation values, thus

$$t = \left[m_3^2 \frac{\langle \vec{x}_3 \rangle \cdot \langle \vec{x}_3 \rangle}{\langle \vec{k}_3 \rangle \cdot \langle \vec{k}_3 \rangle} \right]^{1/2} . \tag{11.3.15}$$

The expression in Eq. (11.3.15) can be used to replace all occurrences of t in the hydrogen atom problem with the expectation values of the position and momentum of the third noninteracting particle. This is, in essence, what a clock does. A clock is an object that must have two qualities. It must have a measurable property that changes in a regular (predictable) manner, and it must not interact with the system of interest. The clock is an agreed upon standard with which to study changes in a system of interest.

§11.4 A HISTORICAL TIME CLOCK

FIGURE 11.1

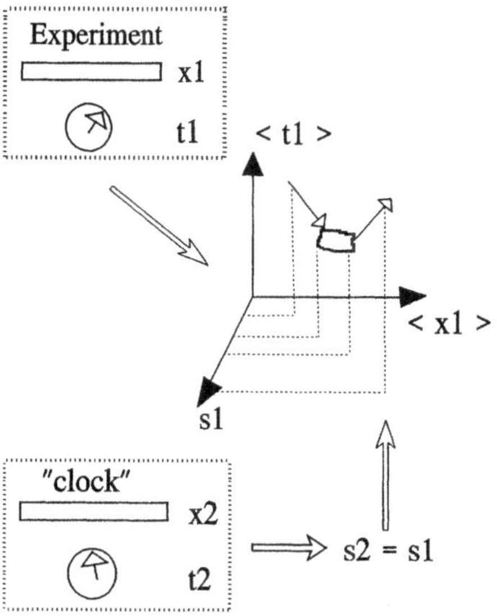

The procedure for designing an evolution parameter clock is straightforward given the information presented in §11.1 through §11.3. The basic concept is to introduce an autonomous system of "clock" particles whose trajectories can replace the evolution parameter of the particle system being studied. This concept is illustrated in Figure 11.1 and implemented using the following procedure.

Consider a system of three scalar particles. We are primarily interested in studying a system of two scalar particles. The third scalar particle is introduced because it serves to independently define the evolution parameter s. We make the following assumptions:

(a) The rest frame clocks of all three particles are calibrated.

(b) Particles 1 and 2 do not interact with particle 3, but may interact with each other through an interaction potential V_{12} or through the electromagnetic four-vector potentials A_1^μ and A_2^μ.

(c) Particle 3 may interact with an external potential defined by either V_3 or A_3^μ.

Assumption (a) is a fundamental assumption of the single evolution parameter formalism. It enhances the analogy we are drawing between the relativistic and nonrelativistic formalisms. The assumption appears schematically as "s2 = s1" in Figure 11.1. The remaining two assumptions serve to define the system of interest. Notice that particles 1 and 2 and their interactions comprise the physically interesting system (system S1 of §11.2). Particle 3 is an additional particle which serves as the evolution parameter clock (system S2 of §11.2).

Assumptions (b) and (c) simplify the problem. The interaction matrix $\underline{\underline{V}}$ of the Stückelberg equation is written as the sum of the two scalar potentials V_{12} and V_3. The appropriate PRQT equation for the three-particle system is

$$\{i\frac{\partial}{\partial s} - [\frac{\pi_1^\mu \pi_{1\mu}}{2m_1} + \frac{\pi_2^\mu \pi_{2\mu}}{2m_2} + V_{12} + \frac{\pi_3^\mu \pi_{3\mu}}{2m_3} + V_3]\}\,\psi(1,2,3,s) = 0 \tag{11.4.1}$$

where $\psi(1,2,3,s)$ denotes the s-dependent eigenfunction of the three-particle system. Equation (11.4.1) may be separated by writing

$$\psi(1,2,3,s) = \psi(1,2,s)\,\psi(3,s)\,. \tag{11.4.2}$$

Dividing Eq. (11.4.1) by (11.4.2), simplifying and using the separation constant α_R yields the two equations

$$\left\{i\frac{\partial}{\partial s} - \left[\frac{\pi_1^\mu \pi_{1\mu}}{2m_1} + \frac{\pi_2^\mu \pi_{2\mu}}{2m_2} + V_{12} + \alpha_R\right]\right\}\psi(1,2,s) = 0 \tag{11.4.3}$$

and

$$\left\{i\frac{\partial}{\partial s} - \left[\frac{\pi_3^\mu \pi_{3\mu}}{2m_3} + V_3 - \alpha_R\right]\right\}\psi(3,s) = 0\,. \tag{11.4.4}$$

As in the nonrelativistic case, the value of the separation constant α_R is determined in principle once the interactions and the boundary conditions of the physical system are specified.

Particle 3 was introduced as a means of removing the indefiniteness of the evolution parameter s. Carrying the nonrelativistic analogy further, we specify the physical environment in which particle 3 exists. The simplest possible environment, from the theoretical point of view, is an environment which contains no external or electromagnetic potentials that can influence particle 3, thus

$$V_3 = 0, \ \{A_3^\mu\} = 0. \tag{11.4.5}$$

Substituting Eq. (11.4.5) into (11.4.4) gives a slightly modified form of the field equation for a free particle:

$$i\frac{\partial \psi_f(3,s)}{\partial s} = \left[-\frac{1}{2m_3}\frac{\partial^2}{\partial y_3^\mu \partial y_{3\mu}} - \alpha_R \right] \psi_f(3,s). \tag{11.4.6}$$

Equation (11.4.6) has a particular solution

$$\psi_f(3,s) = \eta_3^{1/2} \exp\left[-i\left(\alpha_R + \frac{k_3^\mu k_{3\mu}}{2m_3} \right) s + i k_{3\mu} y_3^\mu \right] \tag{11.4.7}$$

where η_3 is a normalization constant. The free particle case is discussed at length in Chapter 9.

In the limit of negligible dispersion (the classical limit), the most probable trajectory of particle 3 is found from Eq. (9.2.9) to be

$$\delta\langle y_3^\mu \rangle \delta\langle y_{3\mu} \rangle = c^2 (\delta s)^2. \tag{11.4.8}$$

Equation (11.4.8) can be rearranged such that

$$(\delta s)^2 = (s - s_0)^2 = \frac{1}{c^2} \delta\langle y_3^\mu \rangle \delta\langle y_{3\mu} \rangle \tag{11.4.9}$$

where s_0 is the historical time at which the rest frame clocks of all three particles were calibrated. We can arbitrarily define the s axis so that $s_0 = 0$ and subsequently find from Eq. (11.4.9) an expression for historical time in terms of the measurable space-time trajectory of particle 3:

$$s^2 = \frac{1}{c^2} \delta\langle y_3^\mu \rangle \delta\langle y_{3\mu} \rangle. \tag{11.4.10}$$

By measuring the space-time trajectory of particle 3, we provide a determination of the historical time evolution parameter s. This parametrization can be used in the field equation describing the physically interesting two-particle system comprised of particles 1 and 2 {Eq. (11.4.3)}. This is the essence of an evolution parameter clock.

If particle 3 is a nonrelativistic particle, Eq. (11.4.10) simplifies to

$$s \simeq \delta \langle t_3 \rangle . \tag{11.4.11}$$

Equation (11.4.11) shows the expected result that the historical time is equivalent to the measured Galilean time t_3 in the nonrelativistic limit.

More complex clocks could be designed using the above procedure. For example, a particle moving in a static magnetic field [Fanchi and Collins, 1978] could be used as the basis of an historical time clock. Alternatively, a practical method using Atomic International Time -- Galilean time defined in a geocentric reference frame -- has been proposed [Horwitz, 1992]. The important point to remember is that historical time, which corresponds to the invariant evolution parameter, is measurable.

PART III:

APPLICATIONS OF PARAMETRIZED RELATIVISTIC QUANTUM MECHANICS

CHAPTER 12
BOUND STATE MODELS

Historically quantum theory achieved acceptance in large part because of its ability to correctly describe bound state problems, especially atomic spectra. One of the most important areas of application of quantum theory is to the solution of the bound state problem in which a charged particle is moving in a Coulomb potential. PRQT solutions to this problem are described in §12.1 through §12.3 for a particle moving in a field. Field equations for the relativistic two-body problem are then constructed in §12.4 and applied to the harmonic oscillator model of the strong interaction in §12.5.

§12.1 FIELD EQUATION WITH COULOMB POTENTIAL

The description of a single, charged particle moving in the presence of electromagnetic fields is described by the Stückelberg equation

$$i\frac{\partial \underline{\Psi}}{\partial s} = \left[\left(\frac{\pi^{\mu}\pi_{\mu}}{2m}\right)\underline{\underline{I}} + \underline{\underline{V}}\right]\underline{\Psi} \tag{12.1.1}$$

where the kinetic momentum operator is

$$\pi^{\mu} \equiv \frac{1}{i}\frac{\partial}{\partial x^{\mu}} - eA^{\mu}. \tag{12.1.2}$$

We are interested in calculating the behavior of a charged particle interacting with the Coulomb potential of an atomic nucleus. This system may be viewed as a single particle moving in the field generated by the nucleus. The four-vector potential for the system may be written in the form

$$A^1 = A^2 = A^3 = 0$$
$$eA^0 = -\frac{Ze^2}{r}, \quad r \equiv |\vec{x}| \tag{12.1.3}$$

where Z is the atomic number of the nucleus and r is the radial distance from the nucleus. The magnetic and electric fields corresponding to the Coulomb potential are

$$\begin{aligned}\vec{B} &\equiv (F_{32}, F_{13}, F_{21}) = (\tilde{F}^{10}, \tilde{F}^{20}, \tilde{F}^{30}) \\ &= (0,0,0)\end{aligned} \tag{12.1.4}$$

and

$$\begin{aligned}\vec{E} &= (F_{01}, F_{02}, F_{03}) = (\tilde{F}^{23}, \tilde{F}^{31}, \tilde{F}^{12}) \\ &= -Ze\frac{\vec{x}}{r^3}.\end{aligned} \tag{12.1.5}$$

The problem is most readily solved in spherical coordinates $\{x^1, x^2, x^3\} = \{r, \theta, \phi\}$. The form of $\underline{\underline{V}}$ depends on the model being used.

§12.2 SPIN-0 PARTICLE BOUND BY A COULOMB POTENTIAL

If we consider the problem of a spin-0 particle, we have the single-component field equation

$$i\frac{\partial \psi}{\partial s} = \left(\frac{\pi^\mu \pi_\mu}{2m}\right)\psi \tag{12.2.1}$$

where we have taken $\underline{\underline{V}} = 0$. Substituting Eq. (12.1.3) into (12.2.1) and expanding gives

$$i\frac{\partial\psi}{\partial s} = \frac{1}{2m}\left[\frac{1}{i}\frac{\partial}{\partial t} + \frac{Ze^2}{r}\right]\left[\frac{1}{i}\frac{\partial}{\partial t} + \frac{Ze^2}{r}\right]\psi$$
$$+ \sum_{j=1}^{3}\frac{1}{2m}\left[\frac{1}{i}\frac{\partial}{\partial x^j}\right]\left[\frac{1}{i}\frac{\partial}{\partial x_j}\right]\psi . \qquad (12.2.2)$$

Writing ψ as

$$\psi(x^0,x^1,x^2,x^3,s) = U_0(\vec{x})\,e^{i\omega t}e^{-iQs/2m} \qquad (12.2.3)$$

and substituting into Eq. (12.2.1) gives, after some rearranging, the result

$$-\nabla^2 U_0(\vec{x}) = \left\{\left[\omega + \frac{Ze^2}{r}\right]^2 - Q\right\}U_0(\vec{x}) . \qquad (12.2.4)$$

The Laplacian in spherical coordinates is

$$\nabla^2 = \frac{1}{r^2}\frac{\partial}{\partial r}\left(r^2\frac{\partial}{\partial r}\right) - \frac{\boldsymbol{L}\cdot\boldsymbol{L}}{r^2} . \qquad (12.2.5)$$

The operator **L·L** has the familiar form

$$\boldsymbol{L}\cdot\boldsymbol{L} \equiv -\left[\frac{1}{\sin\theta}\frac{\partial}{\partial\theta}\left(\sin\theta\frac{\partial}{\partial\theta}\right) + \frac{1}{\sin^2\theta}\frac{\partial^2}{\partial\phi^2}\right] \qquad (12.2.6)$$

and satisfies the eigenvalue equation

$$\boldsymbol{L}\cdot\boldsymbol{L}\; Y_{\ell m}(\theta,\phi) = \ell(\ell+1)Y_{\ell m}(\theta,\phi) \qquad (12.2.7)$$

where $Y_{\ell m}(\theta,\phi)$ are the orthonormal spherical harmonics. We write the operator **L** in terms of vector components as

$$\boldsymbol{L} = \vec{x}\times\vec{p} = \vec{x}\times\left(\frac{1}{i}\nabla\right) \qquad (12.2.8)$$

so that

$$\boldsymbol{L}\cdot\boldsymbol{L} = \boldsymbol{L}_x\cdot\boldsymbol{L}_x + \boldsymbol{L}_y\cdot\boldsymbol{L}_y + \boldsymbol{L}_z\cdot\boldsymbol{L}_z . \qquad (12.2.9)$$

The solutions of Eqs. (12.2.4) and (12.2.7) are well known [Schiff, 1968; Wu and Hwang, 1991]. They are

$$U_0(\vec{x}) = R_{n\ell}(r)\, Y_{\ell m}(\theta,\phi) \;\; for \left\{ \begin{matrix} \ell = 0, 1, 2, ... \\ m = 0, 1, ..., |\ell| \end{matrix} \right\}. \tag{12.2.10}$$

The radial dependent function $R_{n\ell}$ is a solution of the second order differential equation

$$\begin{gathered} \frac{1}{\rho^2}\frac{d}{d\rho}\left(\rho^2 \frac{d}{d\rho}\right) R_{n\ell}(\rho) + \\ \left[\frac{\lambda}{\rho} - \frac{1}{4} - \frac{[\ell(\ell+1) - \gamma^2]}{\rho^2}\right] R_{n\ell}(\rho) = 0 \end{gathered} \tag{12.2.11}$$

with

$$\begin{gathered} \rho \equiv \alpha r, \;\; \alpha^2 \equiv 4[Q - \omega^2], \\ \lambda \equiv \frac{2\omega\gamma}{\alpha}, \;\; \gamma \equiv Ze^2 \approx \frac{Z}{137}. \end{gathered} \tag{12.2.12}$$

A subsidiary condition arising from the solution of Eq. (12.2.11) is

$$n \equiv \lambda = n' + \frac{1}{2} + \left[\left(\ell + \frac{1}{2}\right)^2 - \gamma^2\right]^{1/2}, \;\; n' = 0, 1, 2, ... \tag{12.2.13}$$

together with the relation

$$Q = \omega^2 \left[1 + \frac{\gamma^2}{\lambda^2}\right]. \tag{12.2.14}$$

The value of λ is given by Eq. (12.2.13) where n is the total quantum number and n' is the radial quantum number.

We may simplify the solution of Eq. (12.2.11) by noting that γ is about Z/137, so γ^2 is less than 1% of $(\ell + ½)^2$ for Z as large as 6. Wu and Hwang [§1.3, 1991] have pointed out that n may acquire negative values for Z sufficiently large. They further observe that it is not possible to experimentally prepare a pionic atom with a single pion bound to a sufficiently large nucleus. Thus we may neglect γ^2 in Eq. (12.2.13). Then n is a nonzero, positive integer and the solution $R_{n\ell}(\rho)$ can be written as

$$R_{n\ell}(\alpha r) = N_{n\ell} e^{-\alpha r/2} (\alpha r)^{\ell} L_{n+\ell}^{2\ell+1}(\alpha r) \tag{12.2.15}$$

where $N_{n\ell}$ are normalization constants and $L_{n+\psi}^{2\ell+1}(\rho)$ are the associated Laguerre polynomials. Details of the derivation are given by Schiff [§16 and §51, 1968]. The normalization constants may be found from the orthonormality condition

$$\frac{p!}{(p+q)!}\int_0^\infty \rho^q e^{-\rho} L_p^q(\rho)\, d\rho = \begin{Bmatrix} 1, & p=s \\ 0, & p \neq s \end{Bmatrix} \tag{12.2.16}$$

where p, q, and s are integers. Another relation that is useful in evaluation integrals involving $R_{n\ell}(\rho)$ -- such as expectation values -- is the recurrence relation

$$\begin{aligned} \rho L_p^q(\rho) &= (2p+q+1) L_p^q(\rho) - (p+1) L_{p+1}^q(\rho) \\ &\quad - (p+q) L_{p-1}^q(\rho). \end{aligned} \tag{12.2.17}$$

Expectation Values

The expectation value defined in §7.3, denoted below as $<\cdots>_{PRQT}$ where the subscript refers to the parametrized theory, defines the scalar product of two vectors. According to this point of view, $<\mathbf{\Omega}>_{PRQT}$ is the transition probability from the state $\mathbf{\Omega}\psi$ into ψ, or conversely. We showed in §8.2 that the scalar product $<\cdots>_{PRQT}$ is a special case of the general scalar product defined by Wigner [1939]. It corresponds to replacing Wigner's measure $df(p,\zeta)$ by d^4k. Part of the physical significance of this replacement is that all volume elements of four-momentum space are weighted equally. The more familiar results of conventional quantum theory can be obtained by altering the weight distribution of the volume elements of four-momentum space, e.g. see §8.2 and §21.4. Our interest here is to obtain the nonrelativistic limit of $<\cdots>_{PRQT}$. A result of the calculation is the resolution of an anomaly introduced in Chapter 1.

One of the anomalies of the conventional theory discussed in Chapter 1 was the inability of the conventional expectation value Eq. (1.2.3) to reduce to the form Eq. (1.2.4) in the nonrelativistic limit. The pionic atom is an excellent example for quantifying the magnitude of the anomaly while showing how PRQT avoids the anomaly altogether [Fanchi, 1977].

We begin the analysis by observing that the joint probability density $\rho(x|s)$ can be written as the product of a marginal probability density $\rho(x^0|s)$ and a conditional probability density $\rho(x^1,x^2,x^3|x^0,s)$ as in Eq. (7.1.2), thus

$$\rho(x|s) = \rho(x^0|s)\,\rho(\vec{x}|x^0,s) \tag{12.2.18}$$

The expectation value $<g>_{PRQT}$ is

$$\langle g\rangle_{PRQT} = \int \rho(x^0|s)\left[\int g(\mathbf{x},t)\,\rho(\mathbf{x}|t,s)\,d^3x\right]dt \tag{12.2.19}$$

where **x** denotes a 3-vector with spacial components only, $x^0=t$, and $g(\mathbf{x},t)$ is, at present, unspecified. Equation (12.2.19) reduces to the 3-space definition of expectation value only for those cases in which the function g is independent of t or if the marginal probability density $\rho(t|s)$ is a delta function in t. This is a mathematical constraint. To completely define the region in which $<\cdots>_{PRQT}$ should agree with the nonrelativistic value $<\cdots>_{NRQM}$, we must also impose the physical constraint that $k^0=\omega\approx m_0$ where m_0 is the rest mass of the particle. For a meaningful comparison with $<\cdots>_{NRQM}$, the same physical requirement must be imposed on the expectation value of conventional theories $<\cdots>_{con}$ defined by Eq. (1.2.3).

A special case which satisfies all of the above conditions is that in which we take the expectation value of a function of the spacial coordinates $f(\mathbf{x})$ and the time dependence of ψ is $[\exp(i\omega t)]/(2T)^{1/2}$ where $-T<t<+T$ in the limit as $T\rightarrow\infty$. The resulting expectation values are

$$\langle f\rangle_{con} = \frac{\int_{-\infty}^{\infty} \psi^*\psi\,(\omega - eA^0) f(\mathbf{x})\, d^3x}{\int_{-\infty}^{\infty} \psi^*\psi\,(\omega - eA^0)\, d^3x} \tag{12.2.20}$$

and

$$\langle f\rangle_{PRQT} = \frac{\lim_{T\to\infty}\left\{\int_{-T}^{T}\int_{-\infty}^{\infty} \psi^*\psi f(\mathbf{x})\, d^3x\,dt\right\}}{\lim_{T\to\infty}\left\{\int_{-T}^{T}\int_{-\infty}^{\infty} \psi^*\psi\, d^3x\,dt\right\}} \tag{12.2.21}$$

where box normalization has been imposed in the time domain. Performing the integration over time in Eq. (12.2.21) lets us write

$$\langle f\rangle_{PRQT} = \frac{\int_{-\infty}^{\infty} \psi^*\psi f(\mathbf{x})\, d^3x}{\int_{-\infty}^{\infty} \psi^*\psi\, d^3x}\,. \tag{12.2.22}$$

The important difference between $<f>_{con}$ and $<f>_{PRQT}$ is due to the eA^0 term in $<f>_{con}$. The expectation values $<f>_{con}$ and $<f>_{PRQT}$ agree for free particles. They also agree when either A^0 or f is independent of the spacial coordinates, otherwise they are different. A test of the formulations is to see whether $<f>_{con}$ or $<f>_{PRQT}$ agree with $<f>_{NRQM}$ in the nonrelativistic limit.

It is well known that the Klein-Gordon equation reduces to the Schrödinger equation in the nonrelativistic limit, although the reduction violates Lorentz invariance and the gauge is altered. Nevertheless, under these circumstances the Klein-Gordon solutions are just the solutions of the Schrödinger equation. Thus $<f>_{PRQT}$ reduces to $<f>_{NRQM}$. Under the same circumstances, however, $<f>_{con}$

does not generally reduce to $<f>_{NRQM}$ because of the eA^0 term. The magnitude of the discrepancy is illustrated for the pionic atom case.

Using the solutions presented above, we determine the expectation value of the radius r to be

$$\langle r\rangle_{PRQT} = \frac{2n}{\alpha}\left[1+\frac{1}{2}\left(1-\frac{\ell(\ell+1)}{n^2}\right)\right],\ \alpha = \frac{2\omega Ze^2}{n} \tag{12.2.23}$$

and

$$\langle r\rangle_{con} = \frac{\frac{4n^2\omega}{\alpha}\left[1+\frac{1}{2}\left(1-\frac{\ell(\ell+1)}{n^2}\right)\right]+2nZe^2}{2n\omega+\alpha Ze^2}. \tag{12.2.24}$$

Dividing $<r>_{con}$ by $<r>_{PRQT}$ yields

$$\frac{\langle r\rangle_{con}}{\langle r\rangle_{PRQT}} = \frac{n^2\delta+Z^2}{[n^2+Z^2\beta^2]\delta} \approx 1+\frac{Z^2}{n^2\delta} \geq 1+\frac{Z^2}{n^2\delta} \tag{12.2.25}$$

where

$$\delta \equiv 1+\frac{1}{2}\left(1-\frac{\ell(\ell+1)}{n^2}\right),\ \beta \approx \frac{1}{137}. \tag{12.2.26}$$

The last step of Eq. (12.2.26) follows because $(Z\beta)^2$ is assumed negligible relative to a positive integer. The magnitude of the difference between $<r>_{con}$ and $<r>_{PRQT}$ is now easily obtained. For example, let us suppose a π^- particle is bound by the Coulomb potential of a Helium nucleus. If the π^- is in the 2P state, then $<r>_{con}$ is 80% larger than $<r>_{PRQT}$. For a π^--π^+ system, $<r>_{con}$ is 20% larger than $<r>_{PRQT}$. This substantial difference persists in the nonrelativistic limit.

In the nonrelativistic limit $\omega \approx m_0$ and the expectation values reduce to

$$\langle r\rangle_{PRQT} \to \frac{n^2\delta}{m_0 Ze^2} \tag{12.2.27}$$

and

$$\langle r \rangle_{con} \to \frac{n^2 \delta}{m_0 Z e^2}\left[1 + \frac{Z^2}{n^2 \delta}\right]. \tag{12.2.28}$$

Equation (12.2.27) agrees precisely with the nonrelativistic Schrödinger theory, while Eq. (12.2.28) differs substantially. The above comparison clearly shows that PRQT properly reduces to the nonrelativistic limit for stationary mass states while conventional theories do not.

§12.3 SPIN-½ PARTICLE BOUND BY A COULOMB POTENTIAL

We calculated the spectra associated with pionic atoms in §12.2. A more common problem is the calculation of the energy levels of hydrogenic atoms.

The field equation describing the behavior of an electron in the Coulomb field of an atomic nucleus is taken from the HPR model of §10.1 with the Coulomb potential given by Eq. (12.1.3). The resulting equation is

$$\begin{aligned} i\frac{\partial \underline{\Psi}}{\partial s} &= \frac{1}{2m}\left[\frac{1}{i}\frac{\partial}{\partial t} + \frac{Ze^2}{r}\right]\left[\frac{1}{i}\frac{\partial}{\partial t} + \frac{Ze^2}{r}\right]\underline{\Psi} \\ &+ \sum_{j=1}^{3}\frac{1}{2m}\left[\frac{1}{i}\frac{\partial}{\partial x^j}\right]\left[\frac{1}{i}\frac{\partial}{\partial x_j}\right]\underline{\Psi} \\ &+ \left[\frac{g_1 \mu_0}{2m}\vec{p}\cdot\underline{\underline{\sigma}}\times\vec{E} - \frac{g_2^2 \mu_0^2}{8m}\vec{E}\cdot\vec{E}\,\underline{\underline{I}}\right]\underline{\Psi} \end{aligned} \tag{12.3.1}$$

where we have chosen $n^\mu = n_0^\mu$ as in §10.3. We may rearrange the g_1 term by using the vector relation

$$\vec{A}\cdot\vec{B}\times\vec{C} = -\vec{B}\cdot\vec{A}\times\vec{C} \tag{12.3.2}$$

for the triple scalar product to get

$$
\begin{aligned}
i\frac{\partial \underline{\Psi}}{\partial s} = {} & \frac{1}{2m}\left[\frac{1}{i}\frac{\partial}{\partial t} + \frac{Ze^2}{r}\right]\left[\frac{1}{i}\frac{\partial}{\partial t} + \frac{Ze^2}{r}\right]\underline{\Psi} \\
& + \sum_{j=1}^{3}\frac{1}{2m}\left[\frac{1}{i}\frac{\partial}{\partial x^j}\right]\left[\frac{1}{i}\frac{\partial}{\partial x_j}\right]\underline{\Psi} \\
& - \left[\frac{g_1\mu_0}{2m}(\vec{p}\times\vec{E})\cdot\underline{\underline{\sigma}} + \frac{g_2^2\mu_0^2}{8m}\vec{E}\cdot\vec{E}\,\underline{I}\right]\underline{\Psi}.
\end{aligned}
\tag{12.3.3}
$$

We solve Eq. (12.3.3) by writing the 2x1 eigenfunction as

$$
\underline{\Psi} = U_{1/2}(\vec{x})\,e^{i\omega t}\,e^{-iQs/2m}\begin{bmatrix} u \\ v \end{bmatrix}
\tag{12.3.4}
$$

where constraints on the elements {u,v} have been discussed in Chapter 10.

Substituting Eq. (12.3.4) into (12.3.3) gives

$$
\begin{aligned}
\frac{Q}{2m}\underline{\Psi} = {} & \frac{1}{2m}\left[\omega + \frac{Ze^2}{r}\right]^2\underline{\Psi} + \frac{1}{2m}\nabla^2\underline{\Psi} \\
& - \left[\frac{g_1\mu_0}{2m}(\vec{p}\times\vec{E})\cdot\underline{\underline{\sigma}} + \frac{g_2^2\mu_0^2}{8m}\vec{E}\cdot\vec{E}\,\underline{I}\right]\underline{\Psi}.
\end{aligned}
\tag{12.3.5}
$$

Multiplying by 2m and rearranging lets us write

$$
\begin{aligned}
-\nabla^2\underline{\Psi} = {} & \left[\left(\omega + \frac{Ze^2}{r}\right)^2 - Q\right]\underline{\Psi} \\
& - \left[g_1\mu_0(\vec{p}\times\vec{E})\cdot\underline{\underline{\sigma}} + \frac{g_2^2\mu_0^2}{4}\vec{E}\cdot\vec{E}\,\underline{I}\right]\underline{\Psi}.
\end{aligned}
\tag{12.3.6}
$$

We may rewrite the cross product term as

$$\begin{aligned}\vec{p}\times\vec{E} &= \left(\frac{1}{i}\nabla\right)\times\left(-\frac{Ze\vec{x}}{r^3}\right) = -\frac{Ze}{i}\left(\nabla\times\frac{\vec{x}}{r^3}\right)\\ &= \frac{Ze}{i}\left(\frac{\vec{x}}{r^3}\times\nabla\right) = \frac{Ze}{r^3}(\vec{x}\times\vec{p})\\ &= \frac{Ze}{r^3}\vec{L}.\end{aligned} \tag{12.3.7}$$

Substituting Eq. (12.3.7) into (12.3.6) gives

$$\begin{aligned}-\nabla^2\underline{\Psi} &= \left[\left(\omega + \frac{Ze^2}{r}\right)^2 - Q\right]\underline{\Psi}\\ &\quad -\left[g_1\mu_0\frac{Ze}{r^3}\vec{L}\cdot\underline{\underline{\vec{\sigma}}} + \frac{g_2^2\mu_0^2}{4}\frac{Z^2e^2}{r^4}\underline{\underline{I}}\right]\underline{\Psi}.\end{aligned} \tag{12.3.8}$$

Making use of Eq. (12.2.5) yields the equation

$$\begin{aligned}-\left\{\frac{1}{r^2}\frac{d}{dr}\left(r^2\frac{d}{dr}\right) - \frac{\vec{L}\cdot\vec{L}}{r^2}\right\}\underline{\Psi} &= \left[\left(\omega + \frac{Ze^2}{r}\right)^2 - Q\right]\underline{\Psi}\\ &\quad -\left[g_1\mu_0\frac{Ze}{r^3}\vec{L}\cdot\underline{\underline{\vec{\sigma}}} + \frac{g_2^2\mu_0^2}{4}\frac{Z^2e^2}{r^4}\underline{\underline{I}}\right]\underline{\Psi}.\end{aligned} \tag{12.3.9}$$

Reuse [1978] found solutions of the form

$$\underline{\Psi} = R(r)\,\underline{Y}^m_{j,\ell}(\theta,\phi) \tag{12.3.10}$$

where the 2x1 column vector $\underline{Y}_{j,\ell}{}^{m}$ is

$$\underline{Y}^m_{j,\ell}(\theta,\phi) = \frac{1}{\sqrt{(2\ell+1)}}\begin{bmatrix} \left(\ell - m + \frac{1}{2}\right)^{1/2} Y_\ell^{m-\frac{1}{2}}(\theta,\phi) \\ \left(\ell + m + \frac{1}{2}\right)^{1/2} Y_\ell^{m+\frac{1}{2}}(\theta,\phi) \end{bmatrix} \quad (12.3.11)$$

for $\ell = j + ½$, and

$$\underline{Y}^m_{j,\ell}(\theta,\phi) = \frac{1}{\sqrt{(2\ell+1)}}\begin{bmatrix} \left(\ell + m + \frac{1}{2}\right)^{1/2} Y_\ell^{m-\frac{1}{2}}(\theta,\phi) \\ -\left(\ell - m + \frac{1}{2}\right)^{1/2} Y_\ell^{m+\frac{1}{2}}(\theta,\phi) \end{bmatrix} \quad (12.3.12)$$

for $\ell = j - ½$. We can simplify Eq. (12.3.9) by using the eigenvalue relation

$$\vec{L}\cdot\underline{\underline{\vec{\sigma}}}\,\underline{Y}^m_{j,\ell} = \begin{bmatrix} \boldsymbol{L}_z & \boldsymbol{L}_x - i\boldsymbol{L}_y \\ \boldsymbol{L}_x + i\boldsymbol{L}_y & \boldsymbol{L}_z \end{bmatrix}\underline{Y}^m_{j,\ell} = c(j,\ell)\,\underline{Y}^m_{j,\ell} \quad (12.3.13)$$

where

$$c(j,\ell) = \begin{Bmatrix} \ell - 1 \ \ for \ \ \ell = j + ½ \\ \ell \ \ for \ \ \ell = j - ½ \end{Bmatrix}. \quad (12.3.14)$$

Combining Eqs. (12.3.13) and (12.2.7) with (12.3.9) and factoring out the spherical harmonics yields the radial equation

$$-\left\{\frac{1}{r^2}\frac{d}{dr}\left(r^2\frac{d}{dr}\right) - \frac{\ell(\ell+1)}{r^2}\right\}R(r) = \left[\left(\omega + \frac{Ze^2}{r}\right)^2 - Q\right]R(r)$$
$$-\left[g_1\mu_0\frac{Ze}{r^3}c(j,\ell) + \frac{g_2^2\mu_0^2}{4}\frac{Z^2e^2}{r^4}\right]R(r). \quad (12.3.15)$$

Expanding the right hand side and rearranging gives

$$(\omega^2 - Q)R(r) = \left\{-\frac{1}{r^2}\frac{d}{dr}\left(r^2\frac{d}{dr}\right) + \frac{\ell(\ell+1)}{r^2} - 2\omega\frac{Ze^2}{r} - \frac{Z^2e^4}{r^2}\right\}R(r)$$
$$+\left\{g_1\mu_0\frac{Ze}{r^3}c(j,\ell) + \frac{g_2^2\mu_0^2}{4}\frac{Z^2e^2}{r^4}\right\}R(r). \tag{12.3.16}$$

Equation (12.3.16) is the HPR radial equation for the Coulomb potential. It is formally similar to the nonrelativistic Schrödinger equation with the potential

$$-2\omega\frac{Ze^2}{r} - \frac{Z^2e^4}{r^2} + g_1\mu_0\frac{Ze}{r^3}c(j,\ell) + \frac{g_2^2\mu_0^2}{4}\frac{Z^2e^2}{r^4}.$$

Terms on the order of r^{-2}, r^{-3} and r^{-4} are perturbations which serve as relativistic corrections to the nonrelativistic treatment.

Reuse pointed out that solutions of the radial equation look like solutions of the nonrelativistic Schrödinger model except in the region near the origin. In the region where $r \rightarrow 0$ the radial solution behaves like

$$R(r) \rightarrow e^{-r_0/r}O(r^{g_1c(j,\ell)/g_2}),\ r_0 = \frac{\alpha^2 g_2}{4}a_0 \tag{12.3.17}$$

where α is the fine structure constant, a_0 is the Bohr radius, and $r_0 \ll a_0$.

Comparison with Experiment

The energy spectrum is obtained by solving the HPR radial equation. It is

$$\omega_{n,\ell,j} = m\left[\ell - \frac{\alpha^2}{2n^2} - \frac{\alpha^4}{2n^4}\left(\frac{n}{j+1/2} - \frac{3}{4}\right) + \ldots\right]. \tag{12.3.18}$$

where Reuse assumed the coupling constants satisfy $g_1 = g_2 = 1$. Equation (12.3.18) is in agreement with the Dirac spectrum up to order α^4. Dropping the assumption of coupling constant equality yields the energy spectrum

$$\omega_{n,0,\frac{1}{2}} - \omega_{n,1,\frac{1}{2}} = \frac{m\alpha^4}{n^3}\left[\frac{g_1 - 1}{6} + \frac{g_2 - 1}{2}\right], \; n \geq 2. \tag{12.3.19}$$

Equation (12.3.19) represents the Lamb shift. The Lamb shift is usually obtained as a radiative correction to the Dirac spectrum, whereas here it follows from the HPR model when the coupling constants g_1 and g_2 are allowed to differ. Many more details are available in Piron and Reuse [1978], and Reuse [1978, 1979, 1980]. Especially noteworthy is Reuse's [1979] comparison of theoretical and experimental energy shifts for hydrogen-like atoms. An illustration of the level of accuracy of his results is shown in Tables 12.1 and 12.2. They indicate that the HPR model is capable of obtaining experimentally acceptable results. This should not be too surprising since the HPR model is an extension of the iterated (second-order) Dirac equation. An alternative treatment of hydrogen-like atoms using a parametrized formalism has been presented by Enatsu and Kawaguchi [1975].

TABLE 12.1 ENERGY SEPARATION FOR $^nS_{1/2}$ - $^nP_{1/2}$ (Z=1)

n	Theory (GHz)	Experiment (GHz) ^1H
2	1.058	1.058
3	0.314	0.315
4	0.132	0.133
5	0.068	0.065

TABLE 12.2 ENERGY SEPARATION FOR $^nP_{3/2}$ - $^nP_{1/2}$ (Z=1)

n	Theory (GHz)	Experiment (GHz)
2	10.973	10.969
3	3.251	3.248

§12.4 RELATIVISTIC TWO-BODY PROBLEM

One of the practical strengths of PRQT is its ability to deal with bound state problems. We have already seen examples of bound state models for describing mesic and hydrogenic atoms. These models have focused on the behavior of a single particle moving in an external field. Our next step in the study of bound state problems is to consider the behavior of two interacting particles.

The Stückelberg equation for an interacting N-body system is a combination of Eqs. (8.4.2) and (8.4.3) with Eq. (7.3.10) defining the expectation values of the observables. An interacting two-body system is described by the Stückelberg equation

$$i\frac{\partial \underline{\Psi}_{AB}}{\partial s} = \boldsymbol{K}_{AB}\underline{\Psi}_{AB} = \left[\sum_{a=1}^{2}\left(\frac{\pi^{\mu}_{a}\pi_{a\mu}}{2m_a}\right)\underline{I} + \underline{\underline{V}}_{AB}\right]\underline{\Psi}_{AB} \tag{12.4.1}$$

where the subscript AB signifies the two-body case for particles A and B. The kinetic momentum operator is

$$\pi^{\mu}_{a} \equiv p^{\mu}_{a} - eA^{\mu}_{a} = \frac{1}{i}\frac{\partial}{\partial y^{\mu}_{a}} - eA^{\mu}_{a}. \tag{12.4.2}$$

Applications depend on the form of $\{A_a^{\mu}\}$ and $\underline{\underline{V}}_{AB}$. Except for the Hermitian and Lorentz invariant requirements, no restrictions have been placed on $\underline{\underline{V}}_{AB}$. By defining a form for $\underline{\underline{V}}_{AB}$, the following examples let us compare PRQT results with the solutions of many standard bound state problems.

Suppose two scalar particles A and B are interacting via a real, scalar potential that depends only on the separation of A and B, i.e. a potential $V(y_A - y_B)$. The field equation becomes

$$i\frac{\partial \Psi_{AB}(y_A, y_B, s)}{\partial s} = \boldsymbol{K}_{AB}\Psi_{AB}(y_A, y_B, s) \tag{12.4.3}$$

with the Hamiltonian

$$K_{AB} = \frac{p^{\mu}_{A} p_{A\mu}}{2m_A} + \frac{p^{\mu}_{B} p_{B\mu}}{2m_B} + V(y_A - y_B) . \tag{12.4.4}$$

We have assumed $\{A_a^{\mu}\}=0$ for a=A,B. For completeness, we note that the normalization condition for the two-body problem is

$$\int_{D_{AB}} \Psi^*_{AB}(y_A, y_B, s) \, \Psi_{AB}(y_A, y_B, s) \, d^4 y_A \, d^4 y_B = 1 . \tag{12.4.5}$$

Center of Mass Coordinates

A particularly useful coordinate system for studying the two-body problem in nonrelativistic quantum theory is the center of mass (CM) coordinate system. The transformation to CM coordinates is a canonical transformation. By transforming to CM coordinates, we can directly study the behavior of the two particles relative to the motion of the composite system. A similar treatment may be accorded the relativistic two-body bound state model following the procedure introduced in §4.1 for the classical case. We begin by defining the CM variables.

The location of the composite system and the separation of its constituent particles is represented by the coordinates

$$X^{\mu} = \frac{m_A y^{\mu}_A + m_B y^{\mu}_B}{m_A + m_B} \tag{12.4.6}$$

and

$$x^{\mu} = y^{\mu}_A - y^{\mu}_B \tag{12.4.7}$$

respectively. The canonically conjugate four-momenta are

$$P^{\mu} = p^{\mu}_A + p^{\mu}_B \tag{12.4.8}$$

and

$$p^\mu = \frac{m_B p_A^\mu - m_A p_B^\mu}{m_A + m_B}. \tag{12.4.9}$$

The resulting Hamiltonian is

$$K_{AB} = \frac{P^\mu P_\mu}{2M} + \frac{p^\mu p_\mu}{2m} + V(x) \tag{12.4.10}$$

where total mass M and reduced mass m are defined by

$$M \equiv m_A + m_B,\ m \equiv \frac{m_A m_B}{m_A + m_B}. \tag{12.4.11}$$

The four-momentum operators are

$$P^\mu = \frac{1}{i}\frac{\partial}{\partial X_\mu},\ p^\mu = \frac{1}{i}\frac{\partial}{\partial x_\mu}. \tag{12.4.12}$$

We may simplify the field equation by recognizing that the interaction potential **V**(x) does not depend on the invariant evolution parameter s. Consequently we consider a state of the form

$$\Psi_{AB}(X,x,s) = \sum_\kappa \psi_\kappa(X,x) \exp\left[-\frac{i m_\kappa^2 s}{2M}\right] \tag{12.4.13}$$

where the sum is over the allowed eigenvalues κ. Although Eq. (12.4.13) implies that the eigenvalue spectrum is discrete, continuous eigenvalue spectra can be accommodated by replacing the sum with an integral. It is simply easier to proceed by assuming a particular type of spectrum, in this case a discrete spectrum.

Substituting Eqs. (12.4.10) and (12.4.13) into (12.4.3) gives the eigenvalue equation

$$\frac{m_\kappa^2}{2M}\psi_\kappa = \left[\frac{P^\mu P_\mu}{2M} + \frac{p^\mu p_\mu}{2m} + V(x)\right]\psi_\kappa. \tag{12.4.14}$$

We separate variables by using

$$\psi_\kappa(X,x) = f(X) g(x) \tag{12.4.15}$$

in Eq. (12.4.14) to obtain

$$\frac{1}{f(X)}\left[\frac{P^\mu P_\mu}{2M} - \frac{m_\kappa^2}{2M}\right] f(X) = -\frac{1}{g(x)}\left[\frac{p^\mu p_\mu}{2m} + V(x)\right] g(x) . \tag{12.4.16}$$

Defining a separation constant Λ yields the two partial differential equations

$$\left[\frac{P^\mu P_\mu}{2M} - \frac{m_\kappa^2}{2M}\right] f(X) = \Lambda f(X) \tag{12.4.17}$$

and

$$-\left[\frac{p^\mu p_\mu}{2m} + V(x)\right] g(x) = \Lambda g(x) . \tag{12.4.18}$$

The composite system is described by Eq. (12.4.17) while Eq. (12.4.18) describes the separation of the constituent particles.

Substituting

$$f(X) \sim e^{iK^\mu X_\mu} \tag{12.4.19}$$

in Eq. (12.4.17) yields

$$m_\kappa^2 = K^\mu K_\mu - 2M\Lambda . \tag{12.4.20}$$

The $K^\mu K_\mu$ term is the kinetic energy of the CM motion and the Λ term represents the contribution of the binding interaction to the mass eigenvalue. To calculate Λ, we must solve the binding interaction Eq. (12.4.18) for a specified interaction potential.

§12.5 QUARK MODEL OF HADRONS

Modern models of nuclei and hadrons use the strong interaction as the binding interaction. Numerous references describing these models exist. Some examples include Scadron [1979], Aitchison and Hey [1982], Hughes [1985], Cooper and West [1988], Kobzarev and Manin [1989], Williams [1991], and the Particle Data Group [1992]. We focus our attention here on the quark model of hadrons.

The typical model of a hadron treats a meson as a composite of a quark-antiquark pair, and a baryon as a composite of three quarks. Experiments have shown that hadron constituents behave like free particles inside the hadron, but cannot be freed from the binding potential. A simple interaction potential which represents these fundamental experimental results is the harmonic oscillator.

The idea of a relativistic harmonic oscillator has been around for decades, e.g. Dirac proposed a four-dimensional harmonic oscillator model as early as 1945. Feynman, Kislinger and Ravndal [1971] used a harmonic oscillator potential to describe the binding interaction of quark constituents in hadrons. Very similar ideas have been explored in detail by Kim and Noz [1986] based on a formulation of relativistic quantum mechanics proposed by Dirac [1949]. We show here that their solutions may be used as a PRQT model of the strong interaction.

COVARIANT HARMONIC OSCILLATOR

Substituting Eq. (12.4.12) into (12.4.18) gives

$$\left[-\frac{1}{2m}\frac{\partial^2}{\partial x^\mu \partial x_\mu} + V(x) + \Lambda\right] g(x) = 0 . \tag{12.5.1}$$

The interaction potential **V**(x) for a harmonic oscillator may be written in the form

$$V(x) = \frac{m\,\omega_0^2 x^2}{2} = \frac{m\,\omega_0^2 x^\mu x_\mu}{2} \qquad (12.5.2)$$

so that Eq. (12.5.1) becomes

$$\left[-\frac{1}{2m}\frac{\partial^2}{\partial x^\mu \partial x_\mu} + \frac{1}{2} m\,\omega_0^2 x^\mu x_\mu + \Lambda\right] g(x) = 0. \qquad (12.5.3)$$

Multiplying Eq. (12.5.3) by 2m and defining a dimensionless set of space-time coordinates $\{\xi^\mu\}$ by

$$\xi^\mu = \frac{x^\mu}{\chi} \qquad (12.5.4)$$

where χ is a constant lets us rewrite Eq. (12.5.3) as

$$\left[\frac{\partial^2}{\partial \xi^\mu \partial \xi_\mu} - \chi^4 m^2 \omega_0^2 \xi^\mu \xi_\mu - 2m\Lambda\chi^2\right] g(\xi) = 0. \qquad (12.5.5)$$

Equation (12.5.5) may be put in a familiar form by making the identifications

$$\chi = (m\,\omega_0)^{½}, \ \lambda = -\frac{2\Lambda}{\omega_0} \qquad (12.5.6)$$

with the result

$$\left[\frac{\partial^2}{\partial \xi^\mu \partial \xi_\mu} - \xi^\mu \xi_\mu + \lambda\right] g(\xi^\mu) = 0. \qquad (12.5.7)$$

Separation of variables is made possible by letting

$$\lambda = \eta^0 - (\eta^1 + \eta^2 + \eta^3) \qquad (12.5.8)$$

and

$$g(\xi^\mu) = h_0(\xi^0)h_1(\xi^1)h_2(\xi^2)h_3(\xi^3). \tag{12.5.9}$$

Substituting Eqs. (12.5.8) and (12.5.9) into (12.5.7) and separating yields an ordinary differential equation of the form

$$\left[\frac{d^2}{d\xi^2} - \xi^2 + \eta\right]h(\xi) = 0 \tag{12.5.10}$$

and we have suppressed the component index. There is one equation like Eq. (12.5.10) for each component $\{\mu=0,1,2,3\}$.

Equation (12.5.10) is the well-known harmonic oscillator equation. It has the solutions [Blokhintsev, 1964; Schiff, 1968; Bohm, 1979]

$$h_n(\xi) = e^{-\frac{1}{2}\xi^2}H_n(\xi), \ n = 0,1,2,... \tag{12.5.11}$$

where $H_n(\xi)$ are the Hermite polynomials of order n:

$$H_n(\xi) = \frac{(-1)^n}{(2^n n!\sqrt{\pi})^{\frac{1}{2}}} e^{\xi^2} \frac{d^n e^{-\xi^2}}{d\xi^n}. \tag{12.5.12}$$

The solutions of the harmonic oscillator equation satisfy the normalization condition

$$\int_{-\infty}^{\infty} h_n^*(\xi)h_n(\xi)d\xi = \int_{-\infty}^{\infty} e^{-\xi^2} H_n^*(\xi)H_n(\xi)d\xi = 1. \tag{12.5.13}$$

The eigenvalue relation for Eq. (12.5.10) is

$$\eta = 2n + 1. \tag{12.5.14}$$

Substituting Eqs. (12.5.11) and (12.5.12) into (12.5.9) gives the solution of Eq. (12.5.3), namely

$$g(\xi^\mu) = \prod_{\mu=0}^{3} e^{-\frac{1}{2}(\xi^\mu)^2} H_{n^\mu}(\xi^\mu). \tag{12.5.15}$$

The eigenvalue relation for Eq. (12.5.3) is obtained by combining the eigenvalue relations in Eq. (12.5.14) with (12.5.8) and solving for Λ to get

$$\Lambda = \omega_0 [n^1 + n^2 + n^3 + 1 - n^0] . \tag{12.5.16}$$

The minus sign in Eq. (12.5.16) means that the eigenvalues of the relativistic harmonic oscillator problem are infinitely degenerate. Kim and Noz [1986] have studied this problem and shown that the degeneracy can be removed by imposing a covariant subsidiary constraint of the form

$$p_\mu \left(x^\mu - \frac{\partial}{\partial x_\mu} \right) g(x) = 0 . \tag{12.5.17}$$

Equation (12.5.17) suppresses time-like oscillations by requiring $n^0 = 0$ and yields the covariant harmonic oscillator solution

$$\begin{aligned} g(\xi^\mu) &= e^{-\frac{1}{2}[(\xi^0)^2 + (\xi^1)^2 + (\xi^2)^2 + (\xi^3)^2]} \\ &\cdot H_0(\xi^0) H_{n^1}(\xi^1) H_{n^2}(\xi^2) H_{n^3}(\xi^3) \end{aligned} \tag{12.5.18}$$

with the resulting eigenvalue relation

$$\Lambda = \omega_0 [n^1 + n^2 + n^3 + 1] . \tag{12.5.19}$$

Combining Eq. (12.5.19) with (12.4.11) and (12.4.20) gives

$$m_\kappa^2 = K^\mu K_\mu - 2(m_A + m_B)\, \omega_0 [n^1 + n^2 + n^3 + 1] . \tag{12.5.20}$$

Equation (12.5.20) expresses the mass state eigenvalues of the composite system in terms of the four-momentum of the center of mass of the system and the binding interaction term.

Kim and Noz transformed Eq. (12.5.18) to spherical coordinates and combined the resulting covariant harmonic oscillator solution with spin interaction perturbations to obtain a meson mass formula. They applied a similar procedure to the three-body problem to develop a baryon mass formula.

An alternative approach to the covariant harmonic oscillator formalism has been proposed by Arshansky and Horwitz [1989]. Their approach restricts the allowed space-time domain to what they call the "restricted Minkowski space".

By imposing a restriction on the space-time domain of the problem, Arshansky and Horwitz eliminated the need for the subsidiary condition of Kim and Noz. In both approaches the final result is an eigenvalue relation analogous to Eq. (12.5.20) that depends linearly on harmonic oscillator eigenvalues. The interested reader should consult the references for details. Another mass model is sketched in §22.4.

CHAPTER 13
PARTICLE STABILITY

We saw in Chapter 12 that PRQT could satisfactorily describe a particle bound by a Coulomb potential. In addition, a physically significant model of the strong interaction was constructed within the context of PRQT. The ability to successfully reproduce important experimental results is a prerequisite for any theory purporting to replace the conventional paradigm. Another test of a proposed paradigm is its ability to remove anomalies associated with the conventional paradigm. One of the anomalies of the conventional paradigm discussed in Chapter 1 was its inability to describe particle decay. In this chapter we go beyond the derivation of familiar results and develop a simple model of particle stability and decay within the PRQT formalism.

§13.1 EMPIRICAL PARTICLE DECAY

Experimentally the rate of decay of a collection of unstable particles is proportional to the number of particles. It may be empirically described using an exponential (Poisson) probability distribution. The argument for an exponential distribution may be found in several places -- e.g. Frauenfelder and Henley [1974], Scadron [1979], and Williams [1991]. I reproduce it here for ease of reference in subsequent sections.

Let N(t') be the number of particles observed at time t' and write the total number of particles at t' =0 as N(0). Then the number of particles decaying in the time duration t' to t'+dt' is

$$dN(t') = -\lambda N(t') dt' \tag{13.1.1}$$

where λ is the constant probability of particle decay per unit time. Integrating Eq. (13.1.1) from t'=0 to t'=t gives

$$N(t) = N(0)\, e^{-\lambda t} \tag{13.1.2}$$

as the exponential description of particle decay. The decay rate is

$$R = \frac{dN(t)}{dt} = -\lambda N(t) \tag{13.1.3}$$

as required. The probability of observing a particle at time t is

$$P_{obs} = \frac{N(t)}{N(0)} = e^{-\lambda t}. \tag{13.1.4}$$

Although Eq. (13.1.4) is commonly used for describing phenomenological results, it implies a lack of probability conservation within the context of conventional quantum theories using a normalization condition over spacial volume only. Additional problems with this approach are discussed by Mayants [Sec. 14.6, 1984]. PRQT avoids these problems by extending the probability concept to space-time with a conditioning parameter.

§13.2 THEORY OF PARTICLE DECAY

The conditional probability distribution ρ defined in Chapter 7 is capable of representing the disappearance or reappearance of a particle in space-time. Although this capability has not been thoroughly explored, it is possible to show that the empirical results described in §13.1 can be obtained within the context of PRQT.

MARGINAL PROBABILITY DENSITY IN TIME

Our goal here is to develop a PRQT description of a single unstable particle with mass m. For simplicity we neglect spin. The corresponding equations for probability conservation and normalization are

$$\frac{\partial \rho(x|s)}{\partial s} + \frac{\partial}{\partial x^{\mu}} \rho(x|s) V^{\mu} = 0 \tag{13.2.1}$$

and

$$\int \rho(x|s) d^4x = 1 \tag{13.2.2}$$

respectively. Equations (13.2.1) and (13.2.2) are simplified forms of the equations presented in Chapter 7. The Stückelberg equation for this problem is

$$i\frac{\partial \Psi(x,s)}{\partial s} = \left[\frac{\pi^{\mu}\pi_{\mu}}{2m} + V_I\right] \Psi(x,s) . \tag{13.2.3}$$

where the kinetic four-momenta $\{\pi^{\mu}\}$ include the four-vector potential $\{A^{\mu}\}$ and $\mathbf{V}_I$ represents other interactions.

We simplify the problem further by recognizing that we are dealing with a particle of fixed mass. Therefore we assume a stationary mass state such that $\rho(x|s) \rightarrow \rho(x)$, thus

$$\Psi(x,s) = \psi(x) e^{-iM^2 s/2m} . \tag{13.2.4}$$

We saw in Chapter 9 that the mass expectation value equals m for a free particle. The probability density and normalization condition become

$$\begin{aligned} \rho(x|s) &= \Psi^*(x,s)\, \Psi(x,s) \\ &= \psi^*(x)\, \psi(x) \\ &\equiv \rho_S(x) \end{aligned} \tag{13.2.5}$$

and

$$\int \rho(x|s) d^4x = \int \rho_S(x) d^4x = 1 \tag{13.2.6}$$

respectively. Recall from Chapter 7 that a multivariable probability density can be written as a product of conditional and marginal probability densities. Since we wish to study the marginal probability density in time, we choose to write ρ_S as

$$\rho_S(x) = \rho_C(\vec{x}|t)\,\rho_T(t) \tag{13.2.7}$$

where we have adopted the notation

$$\begin{aligned} x^0 &= x_0 = t, \\ k^0 &= k_0 = \omega . \end{aligned} \tag{13.2.8}$$

The conditional probability density ρ_C is the probability density associated with the conventional paradigm. We know from probability theory that conditional and marginal probability densities must satisfy the normalization conditions

$$\int \rho_C(\vec{x}|t)\,d^3x = 1 \tag{13.2.9}$$

and

$$\int \rho_T(t)\,dt = 1 \tag{13.2.10}$$

respectively. The spacial volume element is denoted by d^3x in Eq. (13.2.9) and represents the normalization condition of the conventional paradigm. Integrating ρ_S over the spacial volume and using Eq. (13.2.9) gives

$$\begin{aligned} \int \rho_S(x)\,d^3x &= \int \rho_C(\vec{x}|t)\,\rho_T(t)\,d^3x \\ &= \rho_T(t)\int \rho_C(\vec{x}|t)\,d^3x \\ &= \rho_T(t) . \end{aligned} \tag{13.2.11}$$

Unlike the conventional paradigm, Eq. (13.2.11) shows that a constant normalization is not required by PRQT.

It is instructive to see how the probability conservation equation is affected by our assumptions. The probability conservation equation for a stationary mass state is

$$\frac{\partial}{\partial x^{\mu}}\rho_S(x)V^{\mu} = \frac{\partial}{\partial t}\rho_S V^0 + \nabla\cdot\rho_S\vec{V} = 0 . \tag{13.2.12}$$

Substituting Eq. (13.2.7) into (13.2.12) and simplifying yields

$$\frac{\partial}{\partial t}\rho_C V^0 + \nabla\cdot\rho_C\vec{V} = -\rho_C V^0 \frac{\partial \ell n\, \rho_T}{\partial t}. \tag{13.2.13}$$

In the nonrelativistic limit we have $V^0 \to 1$ corresponding to $\partial t/\partial s \to 1$. The nonrelativistic limit is verified by the free particle application in §9.2. Equation (13.2.13) becomes

$$\frac{\partial \rho_C}{\partial t} + \nabla\cdot\rho_C\vec{V} \approx -\rho_C \frac{d\ell n\, \rho_T}{dt}. \tag{13.2.14}$$

Equation (13.2.14) is a three-space continuity equation with sources or sinks depending on the sign of the right hand side. Writing the marginal probability in the exponential form

$$\rho_T(t) = e^{-\eta(t)} \tag{13.2.15}$$

where $\eta(t)$ is a real function lets us express Eq. (13.2.14) as

$$\frac{\partial \rho_C}{\partial t} + \nabla\cdot\rho_C\vec{V} \approx \rho_C \frac{d\eta(t)}{dt}. \tag{13.2.16}$$

In this form the right hand side of Eq. (13.2.16) is clearly seen to represent sources (sinks) if the sign of $d\eta(t)/dt$ is positive (negative). As further justification of the nonrelativistic limit stated above, we note that Collins [1977a] combined Eq. (13.2.14) -- without the source (sink) term -- and the three-space normalization Eq. (13.2.9) to derive Schrödinger's equation using the procedure in Chapter 7.

Equation (13.2.16) supports our contention that PRQT lets us consider phenomena in which the conditional probability density ρ_C is not conserved. Unlike the conventional paradigm which asserts that particle decay must be described using many-particle techniques, we are still working within a single particle formalism. We are now in a position to examine the quantum theoretic description of particle decay.

PRQT Model of Particle Decay

The Stückelberg equation for the stationary mass state given by Eq. (13.2.4) is

$$M^2 \psi(x,s) = \left[\pi^\mu \pi_\mu + 2m V_I\right]\psi(x,s). \qquad (13.2.17)$$

Expanding the kinetic four-momentum operators $\{\pi^\mu\}$ and regrouping terms gives

$$M^2 \psi(x,s) = \left[p^\mu p_\mu + 2m U_I\right]\psi(x,s), \qquad (13.2.18)$$

where the interaction terms are collected in the operator

$$U_I = -eA^\mu p_\mu - ep^\mu A_\mu + e^2 A^\mu A_\mu + 2m V_I, \qquad (13.2.19)$$

and $\{\mathbf{p}^\mu\}$ are given by Eq. (7.2.24). For simplicity, we make the adiabatic-type of approximation that the interaction operator $\mathbf{U}_I$ satisfies an equation of the form

$$U_I \psi(x) = u_I \psi(x) \qquad (13.2.20)$$

where u_I changes slowly in the space-time region of interest. Solutions to more complex problems may be tractable, but this approximation is sufficient for our purposes. The resulting field equation is

$$\begin{aligned}[M^2 - u_I]\,\psi(x) &= p^\mu p_\mu \,\psi(x) \\ &= \left[-\frac{\partial^2}{\partial t^2} + \nabla^2\right]\psi(x). \end{aligned} \qquad (13.2.21)$$

Our interest is in calculating both ρ_C and ρ_T. To get the correct dependence on space and time variables, we try the state function

$$\psi(x) = \zeta(\vec{x},t)\,\xi(t) \qquad (13.2.22)$$

with the identifications

$$\rho_C(\vec{x}|t) = \zeta^*(\vec{x},t)\,\zeta(\vec{x},t),$$
$$\rho_T(t) = \xi^*(t)\,\xi(t). \tag{13.2.23}$$

Substituting Eq. (13.2.22) into (13.2.21), dividing by $\xi(t)$ and rearranging yields

$$[M^2 - u_I]\,\zeta = \left[-\frac{\partial^2}{\partial t^2} + \nabla^2\right]\zeta - \frac{\zeta}{\xi}\frac{d^2\xi}{dt^2} - \frac{2}{\xi}\frac{d\xi}{dt}\frac{\partial\zeta}{\partial t}. \tag{13.2.24}$$

Further simplification of Eq. (13.2.24) is motivated by noting from §13.1 that ρ_T should have the form

$$\rho_T(t) \sim e^{-\lambda t},\ 0 \le t \le \infty. \tag{13.2.25}$$

Introducing a real, constant phase factor ε, we combine Eq. (13.2.25) and (13.2.23) to get the trial solution

$$\xi(t) \sim e^{i\varepsilon t - \lambda t/2}. \tag{13.2.26}$$

Substituting Eq. (13.2.26) into (13.2.24) gives

$$[M^2 - u_I]\,\zeta = \left[-\frac{\partial^2}{\partial t^2} + \nabla^2\right]\zeta - \left(i\varepsilon - \frac{\lambda}{2}\right)\zeta - (2i\varepsilon - \lambda)\frac{\partial\zeta}{\partial t}. \tag{13.2.27}$$

Equation (13.2.27) is a field equation for ζ. Substituting the plane wave solution

$$\zeta(\vec{x},t) \sim e^{i\omega t - i\vec{k}\cdot\vec{x}} \tag{13.2.28}$$

into Eq. (13.2.27) yields the characteristic equation

$$-\omega^2 - i\lambda\omega - 2\varepsilon\omega + \frac{\lambda^2}{4} - \varepsilon^2 - i\lambda\varepsilon + M^2 + \vec{k}\cdot\vec{k} - u_I = 0. \tag{13.2.29}$$

The real and imaginary terms of Eq. (13.2.29) are separated to give

$$-\omega^2 - 2\varepsilon\omega + \frac{\lambda^2}{4} - \varepsilon^2 + M^2 + \vec{k}\cdot\vec{k} - u_I = 0, \tag{13.2.30}$$

$$-i\lambda\omega - i\lambda\varepsilon = 0. \tag{13.2.31}$$

Equation (13.2.31) requires that $\omega = -\varepsilon$. Using this result in Eq. (13.2.30) lets us express the physically significant transition rate λ in the form

$$\lambda^2 = 4\left[u_I - (M^2 + \vec{k}\cdot\vec{k})\right]. \tag{13.2.32}$$

How should Eq. (13.2.32) be interpreted?

Before providing an interpretation of Eq. (13.2.32), we restrict our discussion to bradyons. An introduction to tachyon physics -- including a discussion of their creation and annihilation within the context of PRQT -- is presented in Part IV. Given this restriction, we consider the simplest case in which there is no interaction, that is $u_I = 0$ and Eq. (13.2.32) becomes

$$\lambda_0^2 = -4(M^2 + \vec{k}\cdot\vec{k}). \tag{13.2.33}$$

The term in parentheses is greater than zero for bradyons, which implies that λ_0 would have to be imaginary. If λ is allowed to be imaginary, the resulting marginal probability distribution would be uniform instead of exponentially declining. We therefore conclude that a bradyon cannot decay in the absence of an interaction. Furthermore, for a bradyon to decay, the interaction must satisfy

$$u_I > (M^2 + \vec{k}\cdot\vec{k}) \tag{13.2.34}$$

so that $\lambda^2 > 0$.

Finally, let us consider the state $\Psi(x,s)$ of the system. Combining Eqs. (13.2.4), (13.2.22), (13.2.26) and (13.2.28) gives

$$\Psi(x,s) \sim \exp\left[-i\frac{M^2 s}{2m} - \frac{\lambda}{2}t + i\vec{k}\cdot\vec{x}\right]. \tag{13.2.35}$$

The terms involving ω and ε cancel. Consequently the state $\Psi(x,s)$ decays in time but does not propagate in time. Nevertheless, the marginal probability density ρ_T has the exponential decay form we need to reproduce empirical results.

Energy Representation

The above description of particle decay is designed for relatively long-lived particles, where by long-lived or "stable" particles we mean particles which exist long enough to leave a measurable track in a detector [Williams, 1991]. In the case of excited states or resonance states, particle lifetimes must be inferred from measurements of energy distributions [Hughes, 1985]. In this case an expression of the marginal probability density $\rho_T(t)$ in the energy representation is needed.

Let us write $\rho_T(t)$ in terms of the state

$$\xi(t) = \xi_0 \exp\left(i\varepsilon t - \frac{\lambda}{2} t\right) \tag{13.2.36}$$

where ξ_0 is a normalization constant. Forming $\rho_T(t)$ gives

$$\rho_T(t) = \xi_0^2 e^{-\lambda t} \tag{13.2.37}$$

for $0 \leq t \leq \infty$. The Fourier transform of $\xi(t)$, namely

$$g(\varepsilon') = \frac{1}{2\pi} \int_0^{\infty} \xi(t) e^{-i\varepsilon' t} dt, \tag{13.2.38}$$

gives the energy representation $g(\varepsilon')$. Using the integral

$$\int_0^{\infty} e^{(i\alpha - \beta)t} dt = \frac{i}{\alpha + i\beta} \tag{13.2.39}$$

lets us find

$$g(\varepsilon') = \frac{\xi_0}{2\pi} \frac{i}{(\epsilon - \varepsilon') + i(\lambda/2)}. \tag{13.2.40}$$

Forming $g^* g$ yields the marginal probability density in the energy representation as

$$\rho_T(\varepsilon') = \frac{\xi_0^2}{(2\pi)^2} \frac{1}{(\varepsilon - \varepsilon')^2 + \lambda^2/4} \tag{13.2.41}$$

which has the familiar Lorentzian shape.

The line width Γ is the full width at half height of $\rho_T(\varepsilon')$. It is related to λ by

$$\lambda = \Gamma/\hbar \tag{13.2.42}$$

where we have explicitly written Planck's constant. The uncertainty in energy $\Delta\varepsilon'$ of the state is $\Gamma/2$. We may also estimate the uncertainty in energy of the state from the uncertainty principle for a particle with lifetime T, thus

$$T\Delta\varepsilon' \approx \hbar/2. \tag{13.2.43}$$

Substituting the estimate of uncertainty from the Lorentzian distribution into the uncertainty principle gives the desired relationship

$$\Delta\varepsilon' T = \frac{\Gamma}{2} T \approx \frac{\hbar}{2} \Rightarrow \Gamma T \approx \hbar. \tag{13.2.44}$$

Equation (13.2.44) is applied in §13.4 below.

§13.3 PARTICLE OSCILLATIONS

The ideas presented in §13.2 apply to particle decay. Particle instability may also manifest itself as a particle oscillation in which the particle periodically appears and disappears. Experimental studies of neutrino oscillations [Particle Data Group, 1992] comprise one such example. We see this type of behavior in the relatively simple case of a spin-0 particle in a space-time box [Collins and Fanchi, 1978].

Particle in a Space-Time Box

We define a space-time box as an infinite, four-dimensional well with boundaries

$$\begin{aligned} &0<x^0<T \\ &0<x^j<L \ for\ j=1,2,3 \end{aligned} \tag{13.3.1}$$

and we have reverted to natural units ($\hbar=c=1$). The normal component of probability flux ρV^μ was shown in §7.2 to be continuous at any surface in space-time. Recalling the relation between V^μ and four-vector potential A^μ in Eq. (7.2.10), we surmise that at any infinite discontinuity in A^μ, the probability distribution ρ must vanish. Thus the wave function must be zero outside the space-time box and A^μ is zero inside. Such a configuration represents a particle which exists only during $0<x^0<T$ within the spacial domain $\{0<x^j<L\}$. For this particular case the spin-0 Stückelberg equation (9.1.1) has square integrable solutions of the form

$$\Psi(x,s)=\frac{4}{(L^3T)^{1/2}}e^{-\frac{iQs}{2m}}\prod_{\mu=0}^{3}\sin(k^\mu x^\mu) \tag{13.3.2}$$

where the Einstein summation convention is not being used. The eigenvalue Q satisfies

$$Q=\pi^2\left\{\frac{(n^0)^2}{T^2}-\frac{1}{L^2}[(n^1)^2+(n^2)^2+(n^3)^2]\right\} \tag{13.3.3}$$

with

$$k^0=\frac{\pi n^0}{T},\ k^j=\frac{\pi n^j}{L} \tag{13.3.4}$$

for $j=1,2,3$. The quantum numbers $\{n^\mu\}$ are nonzero, positive integers.

The eigenvalue Q is infinitely degenerate because of the minus sign in Eq. (13.3.3). The degeneracy may be removed by imposing additional constraints on the system, as we discussed in §12.5 for a covariant harmonic oscillator model. Spacelike eigenvalues $Q<0$ correspond to faster-than-light tachyons and are considered in Part IV. By limiting our discussion to timelike eigenvalues $Q<0$, we can develop a constraint on the product of a particle mass and its lifetime. The resulting mass-lifetime relationship is validated for all known particles in §13.4. First we consider the behavior of the marginal probability distribution in time.

OSCILLATING PROBABILITY DISTRIBUTION

We wish to determine the probability that the particle will be observable during some interval 0 to t, where $0 \leq t \leq T$ and $t=x^0$. This probability is evaluated by first calculating the marginal probability density in time, i.e. $\rho(t)$. Integrating $\Psi^*\Psi$ over the spacial volume L^3 as prescribed in §7.1 yields

$$\begin{aligned} \rho(t) &= \int_{L^3} \Psi^*(x,s)\, \Psi(x,s)\, d^3x \\ &= \frac{2}{T} \sin^2 \frac{\pi n^0}{T} t \\ &= \frac{2}{T} \sin^2 \omega_Z t \end{aligned} \tag{13.3.5}$$

where ω_Z is the Zitterbewegung frequency

$$\omega_Z \equiv 2mc^2/\hbar \tag{13.3.6}$$

including the explicit expression of $\hbar$ and c. Returning to natural units ($\hbar = c = 1$) and using a trigonometric identity we rewrite Eq. (13.3.5) as

$$\rho(t) = \frac{1}{T}(1 - \cos \omega_z t). \tag{13.3.7}$$

Equation (13.3.7) predicts that at certain instants the probability of finding the particle somewhere in the box is zero. Is Eq. (13.3.7) a reasonable result?

The average values of quantum mechanical results are in agreement with classical values. For example, the expectation values of four-momentum and its scalar product are

$$\left\langle i\hbar \frac{\partial}{\partial x_\mu} \right\rangle = 0 \ \forall \ \mu = 0,1,2,3 \tag{13.3.8}$$

and

$$\left\langle -\hbar^2 \frac{\partial^2}{\partial x_\mu \, \partial x^\mu} \right\rangle = m^2 \tag{13.3.9}$$

respectively. The former result represents the fact that the average value of each component of the four-momentum is zero, while the latter expresses the result that the magnitude of the four-momentum is nonzero. Statistically the probability distribution is isotropic in space and time, therefore the components of $<\mathbf{p}^\mu>$ are zero even though the four-momentum has a nonzero magnitude. Classically this corresponds to a particle being reflected at the walls of a box; the particle four-momentum periodically reverses direction. As a result of this direction reversal in both space and time, the average value of the four-momentum components should be zero even though the four-momentum magnitude is non-zero, which agrees with our calculated results.

Based on the above discussion, we conclude that the PRQT calculation yields expected classical results. Upon closer inspection of the PRQT calculation, we find that it also contains significant new information about the temporal behavior of the system.

During the lifetime of the particle -- the interval $0 \le t \le T$ -- the particle does not interact. Consequently there is no classical reason to believe that the particle will be more observable at some time $0 \le t_1 \le T$ then at some other time $0 \le t_2 \le T$. In other words, the classical probability distribution would be uniform in $0 \le t \le T$. If we examine the quantum mechanical result in Eq. (13.3.7), we see that the probability distribution oscillates about the classical value with the Zitterbewegung frequency. These oscillations are so rapid for most massive particles ($\omega_Z \approx 2.2 \times 10^{23}$ s^{-1} for a charged pion) that they are not directly measurable. By contrast, very light particles -- such as neutrinos -- may have a sufficiently small Zitterbewegung frequency to make their oscillations detectable. Further work in this area is needed. For another approach to the problem of mass state oscillations, see §18.1.

Despite its simplicity, the particle in a space-time box model shows that particle disappearance and reappearance -- as in the case of neutrino oscillations -- may be consistently described within the context of PRQT. We use the model in §13.4 to find a tighter constraint on the mass-lifetime product than that predicted by the uncertainty principle. The constraint is verified by comparison with experimental data.

§13.4 MASS-LIFETIME RELATIONSHIP

Let us consider the case in which the term in L is negligible relative to the term in T. This situation occurs when the particle is essentially unbounded, i.e. $L \to \infty$; and when $(n^0)^2 \gg (n^1)^2 + (n^2)^2 + (n^3)^2$, which is the nonrelativistic case. Equation (13.3.3) becomes

$$m_n^2 \equiv Q \approx \left(\frac{\pi n^0}{T}\right)^2 \tag{13.4.1}$$

where m_n is the mass state eigenvalue. The minimum value of a spacially unconfined particle is obtained by setting $n^0 = 1$ in Eq. (13.3.3). Let us interpret T as the lifetime of the particle (see also Cooke [1968]). If we explicitly write the speed of light c and Planck's constant divided by 2π ($\hbar$), we obtain the following mass-lifetime constraint:

$$(mc^2)T \geq \hbar\pi \equiv L_b = 2.068 \times 10^{-21} MeV \cdot s. \tag{13.4.2}$$

The quantity L_b is a lower bound on the product mc^2T. It is 2π times larger than the lower bound L_u imposed by the uncertainty principle in §8.3, namely

$$(mc^2)T \geq \frac{\hbar}{2} \equiv L_u = 3.291 \times 10^{-22} MeV \cdot s. \tag{13.4.3}$$

Our model of an essentially unconfined particle in a temporal box has led to a stronger constraint L_b on the mass-lifetime product than obtained from considerations of uncertainty relations. Tables 13.1 through 13.3 list the mass-lifetime products of leptons, mesons and baryons respectively [Particle Data Group, 1992]. The particles shown represent all of the particles for which lifetimes are reported. Tables 13.1 through 13.3 show that Eq. (13.4.2) is in accord with experimental data for "stable" particles.

If we view resonance states as particles suitable for description by this simple model, we see that L_b is again satisfied when we calculate particle lifetime from the width of the resonance Γ using the relation $\Gamma = \hbar/T$ from §13.2. The corresponding mass-lifetime relationship becomes

$$\frac{mc^2}{\Gamma} \geq \pi \tag{13.4.4}$$

where both particle mass and resonance width are reported in energy units [Particle Data Group, 1992]. A comparison of Eq. (13.4.4) with experimental results shows that the mass-resonance width relationship is satisfied for all particles with reasonably reliable resonance width measurements. Equation (13.4.4) serves as a prediction of observations yet to be confirmed for particles such as the $a_1(1260)$ meson and several of the N series baryons. For example, a particle with an observed mass of 2600 MeV should have a resonance width of less than 828 MeV. By contrast, the uncertainty principle requires the resonance width to be less than 5200 MeV.

We should bear in mind that resonance states are restricted to volumes whose radii are on the order of a nuclear radius (about 10^{-13} to 10^{-12} cm.). This observation raises a question about the validity of our assumption that $L \gg cT$ for all particles. However, our comparison with experimental data has shown the mass-resonance width relation to be satisfactory. Therefore we are justified in viewing Eqs. (13.4.2) and (13.4.4) as "rules-of-thumb" that more severely constrain mass-lifetime and mass-resonance width relations than does the uncertainty principle.

TABLE 13.1 MASS-LIFETIME RELATION FOR LEPTONS

Particle	Mass (MeV)	Lifetime (sec.)	$(mc^2)T$ (MeV·s)
e	0.5	stable	--
μ	105.6	2.2×10^{-6}	2.3×10^{-4}
τ	1784.1	3.0×10^{-13}	5.4×10^{-6}

TABLE 13.2 MASS-LIFETIME RELATION FOR MESONS

Particle	Mass (MeV)	Lifetime (sec.)	$(mc^2)T$ (MeV·s)
$\pi^{\pm}$	139.6	2.6×10^{-8}	3.6×10^{-6}
π^0	135.0	8.4×10^{-17}	1.1×10^{-14}
$K^{\pm}$	493.6	1.2×10^{-8}	5.9×10^{-6}
K_S^0	497.7	8.9×10^{-11}	4.4×10^{-8}
K_L^0	497.7	5.2×10^{-8}	2.6×10^{-5}
$D^{\pm}$	1869.3	1.1×10^{-12}	2.0×10^{-9}
D^0	1864.5	4.2×10^{-13}	7.8×10^{-10}
$D_S^{\pm}$	1968.8	4.5×10^{-13}	8.9×10^{-10}
$B^{\pm}$, B^0	5278.6	1.3×10^{-12}	6.8×10^{-9}

TABLE 13.3 MASS-LIFETIME RELATION FOR BARYONS

Particle	Mass (MeV)	Lifetime (sec.)	$(mc^2)T$ (MeV·s)
p	938.3	stable	--
n	939.6	889	$8.4 \times 10^{+5}$
Λ	1115.6	2.6×10^{-10}	2.9×10^{-7}
Σ^+	1189.4	8.0×10^{-11}	9.5×10^{-8}
Σ^0	1192.6	7.4×10^{-20}	8.8×10^{-17}
Σ^-	1197.4	1.5×10^{-10}	1.8×10^{-7}
Ξ^0	1314.9	2.9×10^{-10}	3.8×10^{-7}
Ξ^-	1321.3	1.6×10^{-10}	2.1×10^{-7}
Ω^-	1672.4	8.2×10^{-11}	1.4×10^{-7}
Λ_c^+	2284.9	1.9×10^{-13}	4.3×10^{-10}
Ξ_c^+	2466.4	3.0×10^{-13}	7.4×10^{-10}
Ξ_c^0	2472.7	8.2×10^{-14}	2.2×10^{-10}

CHAPTER 14
PARAMETER INDEPENDENT PERTURBATION THEORY

There are very few realistic problems in modern physics that can be solved using analytic solutions. For that reason it is worthwhile to investigate approximation techniques that can extend the range of applicability of the formalism. We present approximation techniques for exactly solvable systems which are experiencing a perturbation that does not depend on the evolution parameter.

§14.1 FORMALISM

We assume the Stückelberg equation (8.1.1) may be written in the form

$$i\frac{\partial \underline{\Psi}}{\partial s} = \boldsymbol{K}\underline{\Psi} \tag{14.1.1}$$

where the operator **K** does not depend on the evolution parameter s. For the perturbation calculation, we further assume that **K** may be decomposed into a dominant part $\mathbf{K}_0$ and a perturbation $\delta\mathbf{K}$ such that

$$\boldsymbol{K} = \boldsymbol{K}_0 + \delta \boldsymbol{K}. \tag{14.1.2}$$

The unperturbed operator $\mathbf{K}_0$ satisfies the eigenvalue equation

$$\boldsymbol{K}_0 \underline{\Phi}_k = q_{0k} \underline{\Phi}_k \tag{14.1.3}$$

where $\{q_{0k}\}$ is the set of eigenvalues of the unperturbed system. The set of eigenvectors $\{\underline{\Phi}_k\}$ is a complete set and satisfy the orthonormality condition

$$\langle \underline{\Phi}_k | \underline{\Phi}_h \rangle = \delta_{kh}. \tag{14.1.4}$$

One way to solve the eigenvalue problem in Eq. (14.1.2) is to use the solution of the unperturbed problem to calculate successively more accurate values of the eigenvalues. This may be done by introducing a parameter λ that will track the order of the approximation. The total operator **K** becomes

$$\boldsymbol{K} = \boldsymbol{K}_0 + \lambda\, \delta \boldsymbol{K}. \tag{14.1.5}$$

Separating variables lets us write the eigenvalue problem in the form

$$\boldsymbol{K} \underline{\psi} = q \underline{\psi} \tag{14.1.6}$$

where

$$\underline{\Psi}(x,s) = \underline{\psi}(x) \exp(-iqs). \tag{14.1.7}$$

We assume the eigenvectors and eigenvalues of the eigenvalue problem can be obtained from the expansions

$$q = q_{0k} + \sum_{j=1}^{\infty} \lambda^j \delta q_j, \tag{14.1.8}$$

and

$$\underline{\psi} = \underline{\Phi}_k + \sum_{j=1}^{\infty} \lambda^j \delta \underline{\Phi}_j, \tag{14.1.9}$$

where $\lambda^j \delta \underline{\Phi}_j$ is the jth order correction to $\underline{\psi}$. Since $\{\underline{\Phi}_k\}$ is a complete set of basis vectors that spans the Hilbert space, we can write the jth order correction $\delta \underline{\Phi}_j$ as a linear superposition of the basis vectors, thus

$$\delta \underline{\Phi}_j = \sum_{i=1}^{\infty} \alpha_{ij} \underline{\Phi}_i \tag{14.1.10}$$

and the expansion coefficients α_{ij} do not depend on x, s or λ. Equation (14.1.9) becomes

$$\underline{\psi} = \underline{\Phi}_k + \sum_{j=1}^{\infty} \lambda^j \Big(\sum_{i=1}^{\infty} \alpha_{ij} \underline{\Phi}_i \Big). \tag{14.1.11}$$

Substituting Eqs. (14.1.5), (14.1.8) and (14.1.9) into (14.1.6) gives

$$\begin{aligned} &\boldsymbol{K}_0 \underline{\Phi}_k + \lambda\, \delta \boldsymbol{K}\ \underline{\Phi}_k + \boldsymbol{K}_0 \lambda \sum_{i=1}^{\infty} \alpha_{i1} \underline{\Phi}_i + \cdots \\ &= q_{0k} \underline{\Phi}_k + \lambda\, \delta q_1 \underline{\Phi}_k + q_{0k} \lambda \sum_{i=1}^{\infty} \alpha_{i1} \underline{\Phi}_i + \cdots . \end{aligned} \tag{14.1.12}$$

First Order Correct Calculation

The expansion in Eq. (14.1.12) is correct, in principle, to any order. In practice, we hope to obtain meaningful results by calculating corrections to the lowest order possible. Here we obtain the first order correction to the unperturbed eigenvalues.

Equation (14.1.12) may be simplified by using Eq. (14.1.3) and keeping terms only to first order in λ:

$$\begin{aligned} &\lambda \Big[\delta \boldsymbol{K}\ \underline{\Phi}_k + \boldsymbol{K}_0 \sum_{i=1}^{\infty} \alpha_{i1} \underline{\Phi}_i \Big] \\ &= \lambda \Big[\delta q_1 \underline{\Phi}_k + q_{0k} \sum_{i=1}^{\infty} \alpha_{i1} \underline{\Phi}_i \Big]. \end{aligned} \tag{14.1.13}$$

Setting the arbitrary parameter $\lambda = 1$ gives

$$(\delta \boldsymbol{K} - \delta q_1 \underline{\underline{I}})\ \underline{\Phi}_k + (\boldsymbol{K}_0 - q_{0k} \underline{\underline{I}}) \sum_{i=1}^{\infty} \alpha_{i1} \underline{\Phi}_i = 0. \tag{14.1.14}$$

We now form the scalar product of Eq. (14.1.14) with $\underline{\Phi}_k$ and simplify:

$$\langle \underline{\Phi}_k | (\delta \boldsymbol{K} - \delta q_1 \underline{\underline{I}}) |\ \underline{\Phi}_k \rangle + \sum_{i=1}^{\infty} \alpha_{i1} \langle \underline{\Phi}_k | (\boldsymbol{K}_0 - q_{0k} \underline{\underline{I}}) | \underline{\Phi}_i \rangle = 0. \tag{14.1.15}$$

The matrix element of $\mathbf{K}_0 - q_{0k}\underline{\underline{I}}$ can be shown to vanish by using Eqs. (14.1.3) and the orthonormality condition Eq. (14.1.4). The remaining matrix element becomes

$$\langle \underline{\Phi}_k | (\delta \boldsymbol{K} - \delta q_1 \underline{\underline{I}}) | \underline{\Phi}_k \rangle = 0 \tag{14.1.16}$$

or

$$\delta q_1 = \langle \underline{\Phi}_k | \delta \boldsymbol{K} | \underline{\Phi}_k \rangle . \tag{14.1.17}$$

Using Eq. (14.1.8) to first order in λ gives the first order correction to the eigenvalues:

$$q = q_{0k} + \langle \underline{\Phi}_k | \delta \boldsymbol{K} | \underline{\Phi}_k \rangle . \tag{14.1.18}$$

§14.2 CHARGED SPIN-0 PARTICLE IN A STATIC MAGNETIC FIELD

In preceding chapters a theory has been developed which defines a mass operator and asserts that the state of a particle is a superposition of mass states. Consequently the particle does not necessarily have a sharply defined mass. We may justifiably ask how this feature can be reconciled with the existence of, for example, beams of particles having sharply defined masses. The answer to this question, as well as the related question of why mass is quantized, is suggested by using PRQT to describe a particle beam interacting with a magnetic field [Fanchi and Collins, 1978]. The calculation techniques needed to answer the question require using parameter independent perturbation theory.

We begin by considering how a particle beam may be obtained from an ensemble of charged particles moving in a static magnetic field with magnitude B. Without loss of generality we choose a frame of reference that has an origin about which the particles execute circular motion, and orient the magnetic field parallel to the x^3 axis. For simplicity, we consider spin-0 bradyons only. This lets us avoid worrying about perturbations caused by particle spin-magnetic field interactions. We assume that bradyons with different rest masses have the same total energy, execute concentric trajectories, and do not interact with one another. Then the classical rest mass m_0 of a particle with charge e and total classical energy ω_c is given by

$$m_0^2 = \omega_c^2 - e^2 B^2 a^2 \tag{14.2.1}$$

where a is the radius of gyration and natural units ($\hbar = c = 1$) have been adopted. The usual interpretation of Eq. (14.2.1) is that the total classical energy is given in terms of the mass and momentum of the particle. In practice, however, the actual measured quantities are ω_c, e, B and a. Mass m_0 is calculated from these quantities. Thus, energy and momentum are the independent variables [Cooke, 1968].

The particle having mass m_0 can be observed by placing a detector at a or, alternatively, a beam of particles having mass m_0 can be obtained by putting collectors at all radii except a. Thus a beam of particles having a particular mass m_0 can be obtained by eliminating all particles having a mass other than m_0. Quantum mechanically, this is equivalent to preparing a pure mass state by eliminating contributions from all other mass states. Once the pure mass state is obtained, the wave function describing the particle beam-magnetic field system can be renormalized. Let us quantify these results.

The Stückelberg equation for a spin-0 particle is to be solved with the four-vector potential chosen such that

$$A_1 = -yB,\ A_0 = A_2 = A_3 = 0. \tag{14.2.2}$$

Writing

$$\psi(x,s) = \phi(x) \exp(-iqs/2m) \tag{14.2.3}$$

gives the eigenvalue equation

$$q\phi(x) = \pi^\mu \pi_\mu \phi(x) \tag{14.2.4}$$

where the four-vector potential is given by Eq. (14.2.2), and the kinetic momentum operator is

$$\pi_\mu = -i\partial_\mu - eA_\mu. \tag{14.2.5}$$

The solution to Eq. (14.2.4) with (14.2.2) has been derived by Lam (1971). Using the notation

$$\{x\} = \{x^0, x^1, x^2, x^3\} = \{t, x, y, z\},$$
$$\{k\} = \{k^0, k^1, k^2, k^3\} = \{\omega, k_x, k_y, k_z\}, \tag{14.2.6}$$

the solution may be written

$$\phi(x,y,z,t) = P(y)\exp\{-i[\omega t - (k_x x + k_z z)]\} \tag{14.2.7}$$

where

$$P(\xi) = g_j(i\alpha - i\xi^2)\exp\left(\frac{i\xi^2}{2}\right);\ j = 1,2 \tag{14.2.8}$$

$$\xi^2 = i\,|eB|\left(y + \frac{k_x}{|eB|}\right)^2, \tag{14.2.9}$$

and

$$\alpha = \frac{(m^2 - \omega^2)}{i|eB|}. \tag{14.2.10}$$

The g_j are expressed in terms of confluent geometric functions [Dennery and Krzywicki, 1967] as

$$g_1(b,\tau') = F\left(\frac{1}{4}(1+b), \frac{1}{2}, \tau'\right)$$
$$g_2(b,\tau') = \tau^{\frac{1}{2}} F\left(\frac{1}{4}(1+b) + \frac{1}{2}, \frac{3}{2}, \tau'\right). \tag{14.2.11}$$

The eigenvalue relation for our solution is

$$q = \omega^2 - (2n+1)|e|B;\ n = 0,1,2,... \tag{14.2.12}$$

and we have taken $k_z = 0$ as in the classical case. For a given energy ω, the mass spectrum is discrete. The general solution is the superposition

$$\psi(x,s) = \sum_{\lambda} \int A(\kappa,\lambda)\, \phi_{\kappa,\lambda}(x)\, e^{-iqs/2m}\, d\omega \tag{14.2.13}$$

where κ and λ represent the energy and momentum eigenvalues, respectively.

As a check on the quantum mechanical result, we demonstrate the classical correspondence. Finding $<y>$ using the definition of expectation value, Eq. (7.3.10), gives

$$2\,|e|\,B\langle y_n^2\rangle = 2n + 1 \tag{14.2.14}$$

which, when substituted into Eq. (14.2.12), yields

$$q = \omega^2 - 2e^2 B^2 \langle y_n^2\rangle \tag{14.2.15}$$

and the subscript n corresponds to the eigenvalue index. If we now recall that the classical, time-averaged value of y^2, denoted by $<y^2>_t$, is related to the gyration radius a by

$$\langle y^2\rangle_t = \frac{a^2}{2}, \tag{14.2.16}$$

we obtain the expected result that the eigenvalue q equals the classical mass m_0^2 of Eq. (14.2.1). Thus the classical and quantum results agree in the classical limit.

If a particle beam has been prepared experimentally as described at the beginning of this section, then a pure mass state exists and is treated analytically by defining

$$A(\kappa,\lambda) = B(\kappa,\lambda)\, \delta(\kappa - \omega)\, \delta_{\lambda n} \tag{14.2.17}$$

where $B(\kappa,\lambda)$ is a factor fixed by renormalizing the particle beam-magnetic field system. Thus a beam of particles having a sharply defined mass m_0 can exist because the system has been prepared so that all mass states other than m_0 are eliminated.

PERTURBATION OF THE MAGNETIC FIELD

The particle beam-magnetic field system illustrates an important aspect of PRQT. In particular, the square of the mass [q in Eq. (14.2.4)] is an observable that depends on the four-vector potential. If the electromagnetic potential is perturbed, then the observable rest mass can be altered. Furthermore, since the operator $\pi^\mu \pi_\mu$ is Hermitian on the Hilbert space $L^2(x)$ for an arbitrary four-vector potential, we can infer that the eigenvalues are real and that the set of eigenfunctions $\{\phi_{\kappa,\lambda}(x)\}$ of $\pi^\mu \pi_\mu$ span the space $L^2(x)$. Therefore $\{\phi_{\kappa,\lambda}(x)\}$ can be used as a basis set in terms of which a parameter independent perturbation calculation of the rest mass change can be made.

As an example, suppose that in addition to the experiment described previously an independent experiment is performed. The independent experiment is identical to the original experiment except for the magnitude of the magnetic field. It is important to emphasize that the location of the experimentalist's tools, e.g. the particle detectors or collectors, has also not changed. Letting the perturbing four-vector potential $\{a_\mu\}$ be

$$a_1 = y\delta B, \ a_0 = a_2 = a_3 = 0, \ |\delta B| < |B|, \tag{14.2.18}$$

gives the total four-vector potential $\{A^\mu + a^\mu\}$. We want to know what mass will be observed at the gyration radius a.

The first-order difference is found from Eq. (14.1.18) to be

$$q_{\delta B} = q + \int \phi^*_{\kappa,\lambda}(x) Q \phi_{\kappa,\lambda} dx \tag{14.2.19}$$

where

$$Q \equiv [-ie(a^\mu \partial_\mu - \partial_\mu a^\mu) + 2e^2 A^\mu a_\mu]. \tag{14.2.20}$$

The quantities q and $\phi_{\kappa,\lambda}(x)$ are given by Eqs. (14.2.12) and (14.2.7), respectively, and $q_{\delta B}$ is the mass squared eigenvalue of the perturbed system correct to first order. Performing the necessary integrations yields

$$q_{\delta B} = q + (2n+1)\,|e|\,\delta B, \quad n = 0, 1, 2, \ldots \tag{14.2.21}$$

Solving Eq. (14.2.21) for q and substituting into Eq. (14.2.12) gives

$$q_{\delta B} = \omega^2 - (2n+1)\,|e|\,(B - \delta B). \tag{14.2.22}$$

This result is the same as that obtained by solving Eq. (14.2.4) directly for the total four-vector potential $\{A^\mu + a^\mu\}$. Parameter independent perturbation theory has been used here primarily because it illustrates the usefulness of the completeness condition on the wave function, and also because perturbation theory is applicable for an arbitrary four-vector potential.

§14.3 INTRODUCTION TO RELATIVISTIC CHEMICAL PHYSICS

Grelland [1981] applied the HPR model to the study of atomic and molecular systems. His approach is based on an approximation technique called the self-consistent energy method. The validity of this approach and its applicability to realistic systems is demonstrated here as another example of a parameter independent perturbation theory.

We begin by considering an N-body problem satisfying a Stückelberg equation of the form

$$i\frac{\partial}{\partial s}\Psi(x_1,\ldots,x_N,s) = \boldsymbol{K}\,\Psi(x_1,\ldots,x_N,s). \tag{14.3.1}$$

An eigenvalue equation for the evolution operator **K** is found by writing the state Ψ as

$$\Psi(x_1,\ldots,x_N,s) = \psi(x_1,\ldots,x_N)\,e^{-iQs/2} \tag{14.3.2}$$

and substituting into Eq. (14.3.1) to get

$$\frac{Q}{2}\psi = \boldsymbol{K}\psi. \tag{14.3.3}$$

The binding energy ε of the system is defined as the difference between the total energy Ω and total mass Q:

$$\varepsilon = \Omega - Q \tag{14.3.4}$$

where

$$\Omega = \sum_{\eta=1}^{N} k_\eta^0 = N\omega \tag{14.3.5}$$

and

$$Q = \sum_{\eta=1}^{N} m_\eta = Nm. \tag{14.3.6}$$

We have made the approximation in the evaluation of the sums that $k_\eta^0 = \omega$ and $m_\eta = m$ for each particle. Comparing Eqs. (14.3.3) and (14.3.6) shows that Grelland identified the eigenvalue Q of **K** as Nm. The validity of these identifications and approximations is established by comparing the ensuing calculations with experiment.

We restrict the interactions of interest to those which satisfy the commutator

$$[\boldsymbol{K}, \boldsymbol{p}_\eta^0] = 0 \tag{14.3.7}$$

where $\mathbf{p}_\eta^0$ is the energy operator of particle η. Among the interactions which satisfy Eq. (14.3.7) are time-independent interactions. By assuming Eq. (14.3.7), we can separate the state ψ such that

$$\psi(x_1, \ldots, x_N) = \phi(\vec{x}_1, \ldots, \vec{x}_N) \exp\left(-i \sum_{\eta=1}^{N} k_\eta^0 x_\eta^0\right). \tag{14.3.8}$$

Grelland postulated that the bound states of the system correspond to the solutions of the eigenvalue equations

$$\boldsymbol{P}^0 \psi = \Omega \psi \tag{14.3.9}$$

and

$$K\psi = \frac{Q}{2}\psi \tag{14.3.10}$$

with the total energy operator defined as

$$P^0 = \sum_{\eta=1}^{N} p_\eta^0 = -\sum_{\eta=1}^{N} \partial_\eta^0 . \tag{14.3.11}$$

We seek an approximate solution of Eqs. (14.3.9) and (14.3.10) by replacing $\mathbf{p}_\eta^0$ with $\omega=\Omega/N$ in $\mathbf{K}$ to obtain

$$K(\Omega)\psi = \frac{Q}{2}\psi . \tag{14.3.12}$$

Self-Consistent Energy Method

A general evolution operator for an N-body system may be written as

$$K(\Omega) = \sum_{\eta=1}^{N} \frac{p_\eta^\mu p_{\eta\mu}}{2m} + V_I(\Omega) = K_0 + V_I(\Omega) \tag{14.3.13}$$

where $\mathbf{V}_I$ accounts for interactions. As a rule, analytical solutions of such complex operators are rarely found. Instead, approximation techniques are needed. To obtain Grelland's self-consistent energy method, we expand the first term of Eq. (14.3.13) and replace $\mathbf{p}_\eta^0$ with Ω/N to get

$$K_0 = \frac{N}{2m}\left(\frac{\Omega}{N}\right)^2 - \sum_{\eta=1}^{N} \frac{\vec{p}_\eta \cdot \vec{p}_\eta}{2m} . \tag{14.3.14}$$

Solving Eq. (14.3.4) for Ω and substituting into Eq. (14.3.14) lets us write

$$K_0 = \frac{(\varepsilon + Q)}{2mN} - \sum_{\eta=1}^{N} \frac{\vec{p}_\eta \cdot \vec{p}_\eta}{2m} . \tag{14.3.15}$$

Recalling that Q=Nm from Eq. (14.3.6) yields the result

$$\boldsymbol{K}_0 = \frac{\varepsilon^2}{2Q} + \varepsilon + \frac{Q}{2} - \sum_{\eta=1}^{N} \frac{\vec{p}_\eta \cdot \vec{p}_\eta}{2m} . \tag{14.3.16}$$

Substituting Eq. (14.3.16) into Eq. (14.3.12) and rearranging leads to Grelland's eigenvalue equation

$$\boldsymbol{H}_0 \psi \equiv \left[-\frac{\varepsilon^2}{2Q} - \boldsymbol{V}_I(\varepsilon) + \sum_{\eta=1}^{N} \frac{\vec{p}_\eta \cdot \vec{p}_\eta}{2m} \right] \psi = \varepsilon \psi \tag{14.3.17}$$

for the operator $\mathbf{H}(\varepsilon)$.

Equation (14.3.17) can be solved iteratively. First specify an initial value $\varepsilon^{(0)}$ and determine the next iterate $\varepsilon^{(1)}$ from

$$\boldsymbol{H}(\varepsilon^{(0)}) \psi = \varepsilon^{(1)} \psi . \tag{14.3.18}$$

The process can be repeated using $\varepsilon^{(1)}$ as the initial value to calculate $\varepsilon^{(2)}$. The iterative relation becomes

$$\boldsymbol{H}(\varepsilon^{(n-1)}) \psi = \varepsilon^{(n)} \psi \tag{14.3.19}$$

and is repeated until $|\varepsilon^{(n)}-\varepsilon^{(n-1)}|$ is less than a specified convergence criterion. This procedure is Grelland's self-consistent energy (SCE) method. It can be used to calculate the electronic states of an atom or molecule once an evolution operator is specified.

GRELLAND-REUSE EVOLUTION OPERATOR

Grelland selected an evolution operator based on Reuse's [1978] interaction model. It has the form

$$\boldsymbol{K} = \sum_{\eta=1}^{N} \left[\frac{(p_\eta^\mu + eA_\eta^\mu)(p_{\eta\mu} + eA_{\eta\mu})}{2m} + \boldsymbol{R}_\eta \right] \tag{14.3.20}$$

where -e is the electron charge. The four-vector potential consists of an external potential a_η^μ applied to the system and a sum of internal interaction potentials $a_{\xi\eta}^\mu$

representing the potential experienced by particle η as generated by particle ξ, thus

$$A_\eta^\mu = a_\eta^\mu + \sum_{\eta=1}^{N} \lambda_{\xi\eta} a_{\xi\eta}^\mu . \tag{14.3.21}$$

The coefficient $\lambda_{\xi\eta}$ is a "mass balance" weighting factor with the form

$$\lambda_{\xi\eta} = \frac{m_\xi}{m_\xi + m_\eta} = ½ . \tag{14.3.22}$$

The factor of ½ is a result of the previous mass assumption in which all particle masses were set equal to m. The resulting four-vector potential is

$$A_\eta^\mu = a_\eta^\mu + \frac{1}{2} \sum_{\eta=1}^{N} a_{\xi\eta}^\mu . \tag{14.3.23}$$

Spin terms are contained in $\mathbf{R}_\eta$. Grelland selected the Piron and Reuse [1978] model presented in §10.1. Consequently the state of the system must be an appropriate column vector.

Further simplification of the evolution operator is achieved by assuming the four-vector potential can be expressed in terms of temporal components only, i.e.

$$\begin{aligned} \{a_\eta^\mu\} &= (\phi_\eta, \vec{0}), \\ \{a_{\xi\eta}^\mu\} &= (\phi_{\xi\eta}, \vec{0}) \end{aligned} \tag{14.3.24}$$

so that

$$\{A_\eta^\mu\} = \left(\phi_\eta + \frac{1}{2} \sum_{\xi=1}^{N} \phi_{\xi\eta}, \vec{0} \right). \tag{14.3.25}$$

The evolution operator becomes

$$\boldsymbol{K} = \sum_{\eta=1}^{N} \left[-\frac{\vec{p}_\eta \cdot \vec{p}_\eta}{2m} + \frac{(p_\eta^0 + eA_\eta^0)^2}{2m} + \boldsymbol{R}_\eta \right]. \tag{14.3.26}$$

Applying the SCE method, we replace $\mathbf{p}_\eta^0$ with $(\varepsilon + Q)/N$ and use $Q = Nm$ to get

$$\boldsymbol{K} = \sum_{\eta=1}^{N}\left[-\frac{\vec{\boldsymbol{p}}_\eta \cdot \vec{\boldsymbol{p}}_\eta}{2m} + eA_\eta^0\right] + \frac{1}{2m}\sum_{\eta=1}^{N}\left[\frac{\varepsilon}{N} + eA_\eta^0\right]^2 + \varepsilon + \frac{Q}{2} + \sum_{\eta=1}^{N} \boldsymbol{R}_\eta . \tag{14.3.27}$$

Substituting Eq. (14.3.27) into (14.3.12) yields the operator

$$\boldsymbol{H}(\varepsilon) = \sum_{\eta=1}^{N}\left[\frac{\vec{\boldsymbol{p}}_\eta \cdot \vec{\boldsymbol{p}}_\eta}{2m} - eA_\eta^0 - \frac{1}{2m}\left(\frac{\varepsilon}{N} + eA_\eta^0\right)^2 - \boldsymbol{R}_\eta\right] \tag{14.3.28}$$

for use in the SCE method.

COMPARISON WITH THE DIRAC MODEL

Grelland used Eq. (14.3.28) and the SCE method to study an electron moving in a Coulomb potential. He validated the model by comparing his calculated binding energy with the binding energy calculated using the Dirac model. Letting Z equal atomic number, α equal the fine structure constant, and adopting Hartree atomic units ($m=1$, $e=1$, $\hbar=1$ and $c=1/\alpha$), we write Grelland's binding energy as

$$\varepsilon^G = -\frac{1}{2}Z^2 + \alpha^2 Z^4\left[-\frac{1}{8} + \frac{1}{2}(g_2 - 1)\right]. \tag{14.3.29}$$

The Dirac result with $g_2=1$ is

$$\varepsilon^D = \alpha^{-2}\left(1 + \frac{\alpha^2 Z^2}{1-\alpha^2 Z^2}\right)^{-1/2} - \alpha^{-2} \approx -\frac{1}{2}Z^2 - \frac{1}{8}\alpha^2 Z^4 - \frac{1}{16}\alpha^4 Z^6 + \cdots . \tag{14.3.30}$$

Comparing ε^G with ε^D shows that the Grelland model agrees with the Dirac model to order $\alpha^2 Z^4$. The agreement confirms that the SCE method can be used to calculate atomic and molecular binding energies with relativistic corrections. A more accurate model of the H-atom than that presented either here or in §12.3 -- but still within the context of PRQT -- has been given by Reuse [1980].

CHAPTER 15
INTRODUCTION TO SCATTERING THEORY

We wish to quantify a scattering experiment in which radiation from a source is directed toward a scattering region (Figure 15.1). The scattering region deflects some of the incident wave. Both scattered and incident waves are measured at large distances from the scattering region. The purpose of scattering theory is to predict the relative amounts of scattered and non-scattered waves.

One advantage of the parametrized formulation is its similarity to nonrelativistic Schrödinger wave mechanics. We make use of these similarities to develop a parameter dependent scattering formalism. An application of the formalism to electromagnetic scattering of charged, spin-0 particles is presented in Chapter 16.

FIGURE 15.1

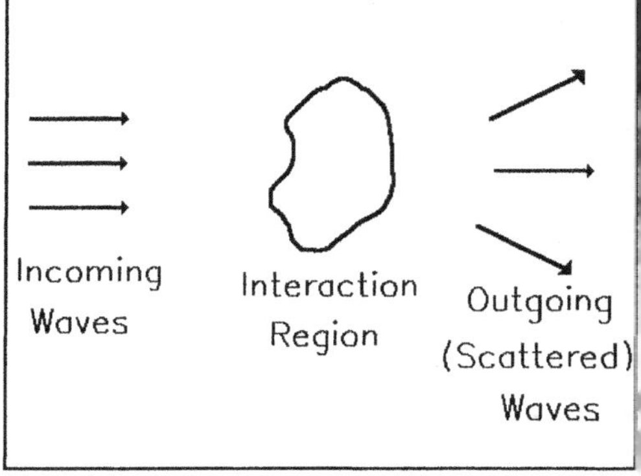

§15.1 SCHRÖDINGER PICTURE

The nonrelativistic Schrödinger picture [Schiff, 1968; Scadron, 1979] has the most direct correspondence to the Schrödinger equation of the three pictures: Heisenberg, interaction, and Schrödinger. It describes the evolution of nonrelativistic states which depend on Galilean time using the Galilean time dependent Schrödinger equation. Similarities between the Schrödinger equation and the Stückelberg equation suggest an obvious parallel.

We saw in Chapter 11 that it is important to distinguish between three types of time. Furthermore, we saw that the invariant evolution parameter s is the relativistic equivalent of Galilean time. We therefore expect to be able to apply Schrödinger picture concepts to PRQT.

EVOLUTION PARAMETER TRANSLATION OPERATOR

Parametrized scattering theory is concerned with the evolution of a state with respect to the invariant evolution parameter s. We introduced the concept of evolution parameter translation operator in Chapter 6. Those ideas are extended here by defining the operator for evolution parameter translations from s to s" as

$$\underline{\Psi}_S(s'') \equiv \boldsymbol{U}_S(s'',s)\,\underline{\Psi}_S(s). \tag{15.1.1}$$

Applying Eq. (15.1.1) to the translation from s" to s' gives

$$\underline{\Psi}_S(s') = \boldsymbol{U}_S(s',s'')\,\underline{\Psi}_S(s''). \tag{15.1.2}$$

Combining Eqs. (15.1.1) and (15.1.2) lets us write

$$\underline{\Psi}_S(s') = \boldsymbol{U}_S(s',s'')\,\boldsymbol{U}_S(s'',s)\,\underline{\Psi}_S(s). \tag{15.1.3}$$

By contrast, the definition in Eq. (15.1.1) gives the operator for evolution parameter translation from s to s' as

$$\underline{\Psi}_S(s') = \boldsymbol{U}_S(s',s)\,\underline{\Psi}_S(s). \tag{15.1.4}$$

Comparing Eqs. (15.1.3) and (15.1.4) gives us the constraint

$$\boldsymbol{U}_S(s',s) = \boldsymbol{U}_S(s',s'')\,\boldsymbol{U}_S(s'',s). \tag{15.1.5}$$

Equation (15.1.5) says that two successive evolution parameter translations from s to s" and then from s" to s' are equivalent to a single evolution parameter translation from s to s'. Furthermore, we have from Eq. (15.1.2) the identity element

$$U_S(s',s'') = 1 \ for \ s' = s''. \tag{15.1.6}$$

Equations (15.1.1), (15.1.5) and (15.1.6) are properties associated with an evolution parameter translation operator. To construct the operator explicitly, we need to consider a specific physical description of the evolution of a state.

NEUMANN SERIES SOLUTION OF INTEGRAL EQUATION

We asserted in Chapter 8 that the state $\underline{\Psi}_S(s)$ satisfies the Stückelberg equation

$$i\frac{\partial}{\partial s}\underline{\Psi}_S(s) = \boldsymbol{K}_S\underline{\Psi}_S(s). \tag{15.1.7}$$

Equation (15.1.7) may be converted to an integral equation by multiplying by -i and integrating from s_0 to s':

$$\underline{\Psi}_S(s') = \underline{\Psi}_S(s_0) - i\int_{s_0}^{s'} \boldsymbol{K}_S(s)\underline{\Psi}_S(s)\,ds. \tag{15.1.8}$$

The Neumann series method may be used to solve Eq. (15.1.8). Replacing $\underline{\Psi}_S(s)$ in the integrand by Eq. (15.1.8) gives

$$\begin{aligned}\underline{\Psi}_S(s') = \underline{\Psi}_S(s_0) - \left[i\int_{s_0}^{s'} \boldsymbol{K}_S(s)\,ds\right]\underline{\Psi}_S(s_0)\\ - \int_{s_0}^{s'} \boldsymbol{K}_S(s)\left[\int_{s_0}^{s} \boldsymbol{K}_S(s'')\underline{\Psi}_S(s'')\,ds''\right]ds.\end{aligned} \tag{15.1.9}$$

Equation (15.1.9) is formally an equality, but we still do not know how to solve the differential Stückelberg equation. A practical simplification is obtained if we assume the operator $\mathbf{K}_S(s)$ is an infinitesimal generator. Canonical transformation properties of the noninteracting particle presented in Chapter 6 suggest that this is a reasonable assumption. In this case, the state $\underline{\Psi}_S(s)$ may be approximated to first order in $\mathbf{K}_S(s)$ by

$$\underline{\Psi}_S(s') = \left[1 - i\int_{s_0}^{s'} \boldsymbol{K}_S(s)\,ds\right]\underline{\Psi}_S(s_0). \tag{15.1.10}$$

Notice that $\mathbf{K}_S(s)$ does not necessarily commute with $\underline{\Psi}_S(s)$. Referring to Figure 15.1, we observe that the term in brackets corresponds to the interaction region which transforms the incoming wave $\underline{\Psi}_S(s_0)$ into the outgoing wave $\underline{\Psi}_S(s')$. We denote "incoming" and "outgoing" relative to the monotonically increasing invariant evolution parameter axis.

Comparing Eq. (15.1.10) with Eq. (15.1.1) shows the evolution parameter translation operator may be written in the form

$$\boldsymbol{U}_S(s',s_0) = 1 - i\int_{s_0}^{s'} \boldsymbol{K}_S(s)\,ds. \tag{15.1.11}$$

A finite transformation is constructed as in §6.3 and yields

$$\boldsymbol{U}_S(s',s_0) = \exp\left(-i\int_{s_0}^{s'} \boldsymbol{K}_S(s)\,ds\right). \tag{15.1.12}$$

Comparing Eqs. (15.1.11) and (15.1.12) with Eqs. (15.1.1), (15.1.5) and (15.1.6) shows that the constraints on $\mathbf{U}_S(s',s)$ are satisfied.

§15.2 INTERACTION PICTURE

An alternative picture -- the interaction picture -- is constructed as follows. We write the Hamiltonian operator of Eq. (15.1.7) as

$$\boldsymbol{K}_S(s) = \boldsymbol{K}_{S0} + \boldsymbol{K}_{SI}(s) \tag{15.2.1}$$

where $\mathbf{K}_{S0}$ is the Hamiltonian in the absence of a perturbation and $\mathbf{K}_{SI}$ is the perturbation part of the total Hamiltonian $\mathbf{K}_S$. Notice that $\mathbf{K}_{S0}$ does not depend explicitly on s. In a typical scattering application $\mathbf{K}_{S0}$ is the Hamiltonian of noninteracting particles.

Let us consider the transformation

$$\underline{\Psi}_I(s) = e^{iK_{SO}s}\, \underline{\Psi}_S(s). \tag{15.2.2}$$

Differentiating Eq. (15.2.2) with respect to s and multiplying by i gives

$$i\frac{\partial \underline{\Psi}_I(s)}{\partial s} = i\left[i\boldsymbol{K}_{SO} e^{iK_{SO}s} \underline{\Psi}_S(s) + e^{iK_{SO}s} \frac{\partial \underline{\Psi}_S(s)}{\partial s} \right]. \tag{15.2.3}$$

Expanding Eq. (15.2.3) and using Eqs. (15.1.7) and (15.2.1) yields

$$i\frac{\partial \underline{\Psi}_I(s)}{\partial s} = -\boldsymbol{K}_{SO} e^{iK_{SO}s} \underline{\Psi}_S(s) + e^{iK_{SO}s} (\boldsymbol{K}_{SO} + \boldsymbol{K}_{SI}) \underline{\Psi}_S. \tag{15.2.4}$$

Making use of the relation

$$e^{-iK_{SO}s} e^{iK_{SO}s} = 1 \tag{15.2.5}$$

in Eq. (15.2.4) and simplifying gives

$$i\frac{\partial \underline{\Psi}_I(s)}{\partial s} = e^{iK_{SO}s} \boldsymbol{K}_{SI} e^{-iK_{SO}s} e^{iK_{SO}s} \underline{\Psi}_S. \tag{15.2.6}$$

If we define the Hamiltonian in the interaction picture as

$$\boldsymbol{K}_I \equiv e^{iK_{SO}s} \boldsymbol{K}_{SI} e^{-iK_{SO}s} \tag{15.2.7}$$

and use Eqs. (15.2.2) and (15.2.7) in Eq. (15.2.6) we obtain

$$i\frac{\partial \underline{\Psi}_I}{\partial s} = \boldsymbol{K}_I \underline{\Psi}_I. \tag{15.2.8}$$

Equation (15.2.8) gives the evolution parameter development of the interaction picture wave function $\underline{\Psi}_I$ in terms of an interaction picture Hamiltonian $\mathbf{K}_I$.

The Schrödinger picture and the interaction picture are connected by requiring that the matrix elements are equal in both pictures, thus

$$\langle \underline{\Psi}_I | \mathbf{\Omega}_I | \underline{\Psi}_I \rangle = \langle \underline{\Psi}_S | \mathbf{\Omega}_S | \underline{\Psi}_S \rangle . \tag{15.2.9}$$

Substituting Eq. (15.2.2) into the right hand side of Eq. (15.2.9) gives

$$\langle \underline{\Psi}_I | \mathbf{\Omega}_I | \underline{\Psi}_I \rangle = \langle \underline{\Psi}_S | e^{-i\boldsymbol{K}_{so}s} \mathbf{\Omega}_I e^{i\boldsymbol{K}_{so}s} | \underline{\Psi}_S \rangle . \tag{15.2.10}$$

Comparing Eq. (15.2.10) with the left hand side of Eq. (15.2.9) shows that

$$e^{-i\boldsymbol{K}_{so}s} \mathbf{\Omega}_I e^{i\boldsymbol{K}_{so}s} = \mathbf{\Omega}_S . \tag{15.2.11}$$

or

$$\mathbf{\Omega}_I = e^{i\boldsymbol{K}_{so}s} \mathbf{\Omega}_S e^{-i\boldsymbol{K}_{so}s} \tag{15.2.12}$$

where we have used Eq. (15.2.5). Equation (15.2.12) agrees with Eq. (15.2.7) and applies to both operators and observables.

§15.3 THE S-MATRIX

Our interest here is to find a formal solution for Eq. (15.2.8). We seek a Neumann series solution by converting the differential equation to an integral equation in the manner introduced in §15.1. First define the operator $\mathbf{U}(s,s_0)$ such that

$$\underline{\Psi}_I(s) = \boldsymbol{U}(s,s_0) \underline{\Psi}_I(s_0) \tag{15.3.1}$$

where $\underline{\Psi}_I(s_0)$ characterizes the system at a fixed evolution parameter value s_0 and $\mathbf{U}(s,s_0)$ satisfies the condition

$$\boldsymbol{U}(s_0,s_0) = 1 . \tag{15.3.2}$$

Using Eq. (15.3.1) in (15.2.8) gives

$$\left[i \frac{\partial}{\partial s} \boldsymbol{U}(s,s_0) - \boldsymbol{K}_I \boldsymbol{U}(s,s_0) \right] \underline{\Psi}_I(s_0) = 0 . \tag{15.3.3}$$

Equation (15.3.3) is satisfied if

$$i\frac{\partial}{\partial s}\boldsymbol{U}(s,s_0) = \boldsymbol{K}_I \boldsymbol{U}(s,s_0). \tag{15.3.4}$$

Integrating with respect to s yields

$$\boldsymbol{U}(s,s_0) = 1 - i\int_{s_0}^{s} \boldsymbol{K}_I(s')\,\boldsymbol{U}(s',s_0)\,ds' \tag{15.3.5}$$

where Eq. (15.3.2) has been used.

An iteration procedure is used to solve Eq. (15.3.5). We write the operator **U**(s',s_0) as

$$\boldsymbol{U}(s',s_0) = 1 - i\int_{s_0}^{s'} \boldsymbol{K}_I(s'')\,\boldsymbol{U}(s'',s_0)\,ds''. \tag{15.3.6}$$

Substituting Eq. (15.3.6) into (15.3.5) gives

$$\begin{aligned}\boldsymbol{U}(s,s_0) &= 1 - i\int_{s_0}^{s} \boldsymbol{K}_I(s')\,ds' \\ &+ (-i)^2\int_{s_0}^{s} \boldsymbol{K}_I(s')\left[\int_{s_0}^{s'} \boldsymbol{K}_I(s'')\,\boldsymbol{U}(s'',s_0)\,ds''\right]ds'.\end{aligned} \tag{15.3.7}$$

This procedure may be repeated indefinitely to any desired degree of accuracy, although calculation of higher order terms becomes progressively more difficult. We obtain the S-matrix by taking the limits

$$\begin{Bmatrix} s_0 \to -\infty \\ s \to +\infty \end{Bmatrix}.$$

The S-matrix becomes

$$
\begin{aligned}
S &\equiv U(\infty, -\infty) \\
&= 1 - i \int_{-\infty}^{\infty} K_I(s')\, ds' \\
&\quad + (-i)^2 \int_{-\infty}^{\infty} K_I(s') \left[\int_{-\infty}^{s'} K_I(s'')\, U(s'', s_0)\, ds'' \right] ds'.
\end{aligned}
\tag{15.3.8}
$$

The probability amplitude for scattering from an initial state $\underline{\Psi}_{Ii}$ to a final state $\underline{\Psi}_{If}$ due to the scattering matrix **S** is

$$
\langle \underline{\Psi}_{If} | S \underline{\Psi}_{Ii} \rangle = \int \underline{\Psi}_{If}^{+} S \underline{\Psi}_{Ii}\, d^4x. \tag{15.3.9}
$$

Substituting Eq. (15.3.8) into (15.3.9) gives

$$
\begin{aligned}
\langle \underline{\Psi}_{If} | S \underline{\Psi}_{Ii} \rangle &= \int \underline{\Psi}_{If}{}^{+} \underline{\Psi}_{Ii}\, d^4x \\
&\quad - i \int \left[\int_{-\infty}^{\infty} \underline{\Psi}_{If}^{+} K_I \underline{\Psi}_{Ii}\, ds \right] d^4x.
\end{aligned}
\tag{15.3.10}
$$

The first integral on the right hand side is just the orthonormality condition

$$
\int \underline{\Psi}_{If}^{+} \underline{\Psi}_{Ii}\, d^4x = \int \underline{\Psi}_{Sf}^{+} \underline{\Psi}_{Si}\, d^4x = \delta_{fi} \tag{15.3.11}
$$

where we have made use of Eqs. (15.2.2) and (15.2.5). The delta function is a Dirac delta function for continuum states and a Kronecker delta function for discrete states. Using Eq. (15.3.11) in (15.3.10) lets us write the scattering amplitude for the S-matrix element S_{fi} as

$$
\begin{aligned}
S_{fi} &\equiv \langle \underline{\Psi}_{If} | S \underline{\Psi}_{Ii} \rangle \\
&= \delta_{fi} - i \int_{-\infty}^{\infty} \left[\int \underline{\Psi}_{If}{}^{+} K_I \underline{\Psi}_{Ii}\, d^4x \right] ds
\end{aligned}
\tag{15.3.12}
$$

where we have interchanged the integrals.

We know from Eq. (15.2.9) that matrix elements in the interaction picture equal matrix elements in the Schrödinger picture. An alternative form of Eq. (15.3.12) is obtained by using Eqs. (15.2.2), (15.2.5) and (15.2.7) to get

$$S_{fi} = \delta_{fi} - i \int_{-\infty}^{\infty} \left[\int \underline{\Psi}_{Sf}^{+} K_{SI} \underline{\Psi}_{Si} d^4x \right] ds. \tag{15.3.13}$$

Equation (15.3.13) may be used directly with the Stückelberg equation represented by Eq. (15.1.7).

Transition Rates

Suppose the system of interest is known to be in state $\underline{\Psi}_{Ii}$ at s_0. The probability of finding the system in a final state $\underline{\Psi}_{If}$ at $s > s_0$ is

$$|\langle \underline{\Psi}_{If} | U(s,s_0) \underline{\Psi}_{Ii} \rangle|^2 = |U_{fi}(s,s_0)|^2 \tag{15.3.14}$$

where the matrix element $U_{fi}(s,s_0)$ is defined by

$$U_{fi}(s,s_0) \equiv \langle \underline{\Psi}_{If} | U(s,s_0) \underline{\Psi}_{Ii} \rangle. \tag{15.3.15}$$

Defining transition rate as transition probability per unit evolution parameter interval, we obtain an average transition rate for the transition from $\underline{\Psi}_{Ii}$ to $\underline{\Psi}_{If}$ as

$$w_{fi} = \frac{1}{s - s_0} |U_{fi}(s,s_0) - \delta_{fi}|^2. \tag{15.3.16}$$

The delta function δ_{fi} removes the case in which there is no change to a different state. In the interaction picture $\mathbf{U}(s,s_0)$ is composed only of the interacting part of the total Hamiltonian.

Transition rate may be expressed in terms of the S-matrix and the total evolution parameter interval Δs over which the scattering takes place. We replace $s-s_0$ in Eq. (15.3.16) with Δs and take the limits

$$\begin{Bmatrix} s_0 \rightarrow -\infty \\ s \rightarrow +\infty \end{Bmatrix}$$

to obtain the transition rate

$$w_{fi} = \frac{|S_{fi}|^2}{\Delta s}. \tag{15.3.17}$$

Another approach to scattering, transition probability and transition rate is given in Chapter 18.

SCATTERING CROSS-SECTION

Cooke [1968], Cook [1972], and Horwitz and Lavie [1982] defined a three-dimensional cross-section for a four-dimensional scattering region as the total number of observed scatterings N_{sc} into a four-momentum element d^4p divided by the number n_{inc} of wave packets per unit area per unit time, thus

$$\sigma = \frac{N_{sc}}{n_{inc}}. \tag{15.3.18}$$

An equivalent definition that more closely parallels the standard paradigm [e.g. Scadron, 1979; Aitchison and Hey, 1982; Peebles, 1992] would be to define the cross-section in terms of a rate divided by an incident flux. Horwitz and Lavie noted that such a definition is obtained by defining rate as $N_{sc}/\Delta s$ and flux as $n_{inc}/\Delta s$, where Δs is the evolution parameter duration of the perturbation. To put Eq. (15.3.18) into a more useful form, we express σ in terms of the S-matrix.

Suppose that a particle described by an incoming state ψ_{in}^a approaches an interaction region as in Figure 15.2. We are working in the laboratory frame where the accelerator, the interaction region, and the detector are at rest. The wave packets associated with the particles of the incident beam are assumed to satisfy the free particle equations of motion given in §2.3. The probability that a scattered particle will be found in the hypervolume

FIGURE 15.2

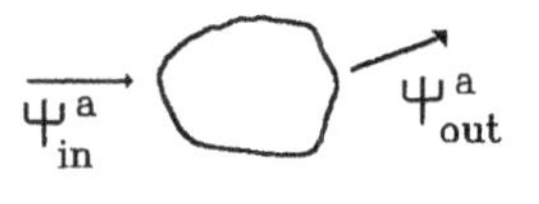

element d^4p around the four-momentum $\{p^\mu\}$ is

$$\omega_a(d^4p) = |\psi^a_{out}(p)|^2 d^4p \tag{15.3.19}$$

where the scattered particle is described by the outgoing state ψ_{out}^a. We write the outgoing state as a linear superposition of the complete set of incoming states, i.e.

$$\psi^a_{out}(p) = \int S(p',p)\,\psi^a_{in}(p')\,d^4p' \tag{15.3.20}$$

where S(p',p) is the probability amplitude for scattering from a state with four-momentum p' to a state with four-momentum p. An expression for S(p',p) may be found as follows.

Multiply Eq. (15.3.20) by $\psi_{in}^{a*}(p'')$ and integrate over space-time to find S(p',p):

$$\begin{aligned}\int \psi^{a*}_{in}(p'')\,\psi^a_{out}(p)\,d^4x &= \int\int \psi^{a*}_{in}(p'')\,S(p',p)\,\psi^a_{in}(p')\,d^4p'\,d^4x \\ &= \int S(p',p)\left[\int \psi^{a*}_{in}(p'')\,\psi^a_{in}(p')\,d^4x\right]d^4p' \\ &= \int S(p',p)\,\delta(p'-p'')\,d^4p' \\ &= S(p'',p).\end{aligned} \tag{15.3.21}$$

The outgoing state is related to the incoming state by Eq. (15.1.1) in the form

$$\psi_{out}(p,s') = \boldsymbol{U}(s',s_0)\,\psi_{in}(p,s_0) \tag{15.3.22}$$

and the unitary operator $\mathbf{U}(s',s_0)$ is given by Eq. (15.1.10). As a result of these relationships we obtain

$$\psi^a_{out}(p,s') = \left[1 - i\int_{s_0}^{s'} \boldsymbol{K}_S(s)\,ds\right]\psi^a_{in}(p,s_0). \tag{15.3.23}$$

Substituting Eq. (15.3.23) into (15.3.21) gives

$$S(p'',p) = \int \psi_{in}^{a*}(p'',s_0) U(s',s_0) \psi_{in}^{a}(p,s_0) d^4x \\ = \langle \psi_{in}^{a}(p'',s_0) | U(s',s_0) \psi_{in}^{a}(p,s_0) \rangle. \tag{15.3.24}$$

An expression of S(p',p) in terms of the S-matrix is obtained by taking the usual limits

$$\begin{Bmatrix} s_0 \to -\infty \\ s \ \to +\infty \end{Bmatrix}.$$

to get

$$S(p'',p) = \langle \psi_{in}^{a}(p'') | S \psi_{in}^{a}(p) \rangle. \tag{15.3.25}$$

Substituting Eq. (15.3.25) into (15.3.20) gives us our final result

$$\psi_{out}^{a}(p) = \int \langle \psi_{in}^{a}(p'') | S \psi_{in}^{a}(p) \rangle \psi_{in}^{a}(p'') d^4p''. \tag{15.3.26}$$

An analogous result is obtained in the conventional theory [e.g. Jauch and Rohrlich, pg. 132, 1976]. The physical interpretation of Eq. (15.3.26) is that the S-matrix transforms incoming states into outgoing states.

We now wish to calculate a cross-section that retains a functional dependence only on the energy p^0 and direction of propagation of the outgoing particle. Therefore we write the hypervolume as

$$d^4p = d^3p\,dp^0 \\ = d\Omega\, dp^0 p^2 dp \tag{15.3.27}$$

where $d\Omega$ is the solid angle of the scattered particle and p is the magnitude of the three-vector $\{p^j: j=1,2,3\}$. Substituting Eq. (15.3.27) into (15.3.19) and integrating over p gives

$$\omega_a(d\Omega\, dp^0) = d\Omega\, dp^0 \cdot \int_0^\infty |\psi_{out}^{a}(p)|^2 p^2 dp. \tag{15.3.28}$$

The total number of particles scattered into $d\Omega \cdot dp^0$ is calculated by summing the probabilities for each incident particle:

$$N_{sc}(d\Omega\, dp^0) = \sum_a \omega_a(d\Omega\, dp^0). \tag{15.3.29}$$

If we assume the number of incident particles per unit area per unit time is essentially uniform over the range needed to cover the target, the sum in Eq. (15.3.29) may be approximated by

$$N_{sc}(d\Omega\, dp^0) \approx n_{inc} \int d^2\rho \int dx^0\, \omega_a(d\Omega\, dp^0) \tag{15.3.30}$$

where $d^2\rho$ is an incremental spacial cross-section normal to the direction of propagation of the incident beam and dx^0 is an incremental duration of the beam pulse. Using the definition of cross-section in Eq. (15.3.18) gives

$$\begin{aligned} \sigma(d\Omega\, dp^0) &= \frac{N_{sc}(d\Omega\, dp^0)}{n_{inc}} \\ &= \int d^2\rho \int dx^0 \omega_a(d\Omega\, dp^0) \\ &= \int d^2\rho \int dx^0 d\Omega\, dp^0 \int_0^\infty |\psi^a_{out}(p)|^2 p^2 dp. \end{aligned} \tag{15.3.31}$$

The differential scattering cross-section becomes

$$\frac{d\sigma(d\Omega\, dp^0)}{d\Omega\, dp^0} = \int d^2\rho \int dx^0 \int_0^\infty |\psi^a_{out}(p)|^2 p^2 dp. \tag{15.3.32}$$

Combining Eqs. (15.3.31) and (15.3.32) with (15.3.26) formally expresses both the cross-section and the differential scattering cross-section in terms of incoming states and matrix elements of the S-matrix. An application of these definitions is provided by Horwitz and Lavie [1982] to Rutherford scattering. Results of their calculation are briefly discussed in §16.3. For a discussion of scattering based on phase shift analysis, see Arshansky and Horwitz [1989].

CHAPTER 16
ELECTROMAGNETIC SCATTERING OF A SPIN-0 PARTICLE

We illustrate the ideas of Chapter 15 for the case of a spin-0 particle scattering from an electromagnetic potential. In addition to demonstrating how the theory is applied, we obtain a new perspective on important ad hoc assumptions of the conventional, nonparametrized theory for both spin-0 and spin-½ particles.

§16.1 THE INTERACTION TERM AND THE S-MATRIX

The Stückelberg equation for a charged pion is

$$i\frac{\partial\Psi}{\partial s} = \boldsymbol{K}_{(0)}\Psi = \left[\frac{\pi_\mu\pi^\mu}{2m} + V_{(0)}\right]\Psi \tag{16.1.1}$$

where

$$\pi_\mu = \frac{1}{i}\partial_\mu - \frac{e}{c}A_\mu . \tag{16.1.2}$$

as in Eqs. (9.1.1) and (9.1.2). Equation (16.1.1) has the general solution

$$\Psi(x,s) = \int \eta_k e^{-i\kappa(k)s}\psi_k(x)\,dk \tag{16.1.3}$$

where $\{\eta_k\}$ are normalization factors and κ(k) satisfies the eigenvalue equation

$$\kappa\Psi = \boldsymbol{K}_{(0)}\Psi . \tag{16.1.4}$$

For the present application, it is sufficient to take $V_{(0)}=0$ so that Eq. (16.1.4) becomes

$$\kappa \Psi = \frac{\pi_\mu \pi^\mu}{2m} \Psi . \tag{16.1.5}$$

Equation (16.1.5) is the Klein-Gordon equation. Expanding Eq. (16.1.5) and rearranging gives

$$\begin{aligned}[\partial_\mu \partial^\mu + \kappa] \psi(x) &= \left[-ie\left(\partial_\mu A^\mu + A^\mu \partial_\mu\right) + e^2 A_\mu A^\mu\right] \psi(x) \\ &\equiv K_{SI}(x)\, \psi(x)\end{aligned} \tag{16.1.6}$$

where $\mathbf{K}_{SI}$ represents the interaction of the charged pion with the electromagnetic field in the Schrödinger picture.

The S-matrix element for Eq. (16.1.6) is obtained from Eq. (15.3.13). It is

$$S_{fi} = \delta_{fi} - i \int_{-\infty}^{\infty} \left[\int \psi_f^* K_{SI} \psi_i d^4x\right] ds \tag{16.1.7}$$

where the subscripts i and f denote initial and final respectively. If the interaction $\mathbf{K}_{SI}$ is independent of the evolution parameter s, which it is for the electromagnetic interaction considered here, the integration over s may be performed and yields

$$S_{fi} = \delta_{fi} - i\,\Delta s \left[\int \psi_f^* K_{SI} \psi_i d^4x\right]. \tag{16.1.8}$$

The evolution parameter interval Δs was introduced in §15.3 and includes the evolution parameter duration over which the scattering occurred.

Combining Eq. (16.1.6) and (16.1.8) gives

$$\begin{aligned}S_{fi} = \delta_{fi} &- e\,\Delta s \int \psi_f^* (\partial_\mu A^\mu + A^\mu \partial_\mu) \psi_i d^4x \\ &- ie^2 \Delta s \int \psi_f^* A_\mu A^\mu \psi_i d^4x .\end{aligned} \tag{16.1.9}$$

The term $\int \psi_f^* \partial_\mu (A^\mu \psi_i) d^4x$ may be integrated by parts with the result

$$\int \psi_f^* \partial_\mu (A^\mu \psi_i) d^4x = \psi_f^* A^\mu \psi_i \big|_R - \int (\partial_\mu \psi_f^*) A^\mu \psi_i d^4x \tag{16.1.10}$$

where R is a three dimensional hypersurface enclosing the four-dimensional hypervolume of integration. The hypersurface of the S-matrix approaches $\pm\infty$. To keep the results finite, the four-vector potential A^μ must vanish at R. Consequently the term $\psi_f^* A^\mu \psi_i$ vanishes on the hypersurface R and Eq. (16.1.10) becomes

$$\int \psi_f^* \partial_\mu (A^\mu \psi_i) d^4x = -\int (\partial_\mu \psi_f^*) A^\mu \psi_i d^4x. \tag{16.1.11}$$

Substituting Eq. (16.1.11) into (16.1.9) gives

$$\begin{aligned} S_{fi} = \delta_{fi} - e\Delta s \int \left[-(\partial_\mu \psi_f^*) A^\mu \psi_i + \psi_f^* A^\mu (\partial_\mu \psi_i) \right] d^4x \\ - ie^2 \Delta s \int \psi_f^* A_\mu A^\mu \psi_i d^4x. \end{aligned} \tag{16.1.12}$$

Factoring out A^μ and rearranging yields

$$\begin{aligned} S_{fi} = \delta_{fi} - e\Delta s \int \left[\psi_f^* (\partial_\mu \psi_i) - (\partial_\mu \psi_f^*) \psi_i \right] A^\mu d^4x \\ - ie^2 \Delta s \int \psi_f^* A_\mu A^\mu \psi_i d^4x. \end{aligned} \tag{16.1.13}$$

Equation (16.1.13) is valid as long as $\psi_f^* A^\mu \psi_i$ vanishes on the hypersurface R. If S-matrix elements correct only to first order in e are desired, as they are below, Eq. (16.1.13) becomes

$$S_{fi} = \delta_{fi} - e\Delta s \int \left[\psi_f^* (\partial_\mu \psi_i) - (\partial_\mu \psi_f^*) \psi_i \right] A^\mu d^4x. \tag{16.1.14}$$

§16.2 ON MASS SHELL SCATTERING

We can reproduce the results of conventional scattering theory by taking appropriate limits, such as requiring the use of on mass shell solutions. The on mass shell criterion is satisfied by states that are stationary with respect to s. In this case the states correspond to the distribution $\{\eta_k\}$ being sharply peaked around the observable pion rest mass m_π. The Dirac delta function is an example of such a distribution. The eigenvalue spectrum of the operator on the right hand side of

Eq. (16.1.5) has a line spectrum with a single value.

We want to calculate the scattering behavior of a pion with sharply defined rest mass in the presence of an electromagnetic field. Applying conventional techniques [e.g. Bjorken and Drell, 1964], the integral solution of Eq. (16.1.6) is

$$\psi(x) = \phi_f(x) + \phi_s(x), \qquad \phi_s(x) \equiv \int G(x-y)\, V(y)\, \psi(y)\, dy \tag{16.2.1}$$

where $\phi_f(x)$ is the incident wave and $\phi_s(x)$ is the scattered wave. The incident wave $\phi_f(x)$ is calculated as the free particle solution of Eq. (16.1.6), i.e. the solution when $A^\mu=0$, and G(x-y) is the Green's function solution of

$$[\partial_\mu \partial^\mu + \kappa] G(x'-x) = -\delta(x'-x). \tag{16.2.2}$$

To calculate the scattered wave $\phi_s(x)$, we must find the Green's function. We do this by letting

$$G(x'-x) = e^{ik_\mu (x'^\mu - x^\mu)} \tag{16.2.3}$$

and substituting into Eq. (16.2.2) to get

$$G(x'-x) = \frac{\delta(x'-x)}{k_\mu k^\mu - \kappa}. \tag{16.2.4}$$

Using the delta function representation

$$\delta(x'-x) = \frac{1}{(2\pi)^4} \int_{-\infty}^{\infty} e^{-ik_\mu (x'^\mu - x^\mu)} d^4k \tag{16.2.5}$$

in Eq. (16.2.4) yields the integral solution

$$G(x'-x) = \frac{1}{(2\pi)^4} \int_{-\infty}^{\infty} \frac{e^{-ik_\mu (x'^\mu - x^\mu)}}{k_\mu k^\mu - \kappa} d^4k. \tag{16.2.6}$$

The integrand of the Green's function has a singularity at the poles

$$k^0 = \pm(\vec{k}\cdot\vec{k} + \kappa)^{1/2}. \tag{16.2.7}$$

The conventional procedure for solving the Green's function integral is to use the calculus of residues. Several contours are possible [Appendix A, Urban, 1970]. The choice of contour in the conventional treatment represents an arbitrary imposition of a boundary condition [Bjorken and Drell, 1964]. This boundary condition is just the Stückelberg-Feynman interpretation, i.e. the positive frequency solution propagates forward in time and the negative frequency solution propagates backward in time. While the Stückelberg-Feynman interpretation is an ad hoc addition to the conventional theory, we showed in Chapter 9 that it is an implicit requirement in PRQT.

The Stückelberg-Feynman interpretation is incorporated into the mathematics by adding a small imaginary part $i\epsilon$ to the denominator of Eq. (16.2.6) and then letting ϵ go to zero after the integration over k^0. Thus Eq. (16.2.6) becomes the Feynman propagator

$$G_F(x'-x) = \lim_{\epsilon \to 0} \frac{1}{(2\pi)^4} \int_{-\infty}^{\infty} e^{i\vec{k}\cdot(\vec{x}'-\vec{x})} d^3k \cdot \int_{-\infty}^{\infty} \frac{e^{-i\omega(t'-t)}}{\omega^2 - \omega_k^2 + i\epsilon} d\omega \tag{16.2.8}$$

where $\omega = k^0$, $t = x^0$, and

$$\omega_k^2 = \vec{k}\cdot\vec{k} + \kappa . \tag{16.2.9}$$

Evaluating the energy integral and taking the limit gives

$$G_F(x'-x) = -\frac{i}{(2\pi)^3} \int_{-\infty}^{\infty} \frac{e^{i\vec{k}\cdot(\vec{x}'-\vec{x}) - i\omega_k t^+}}{2\omega_k} d^3k - \frac{i}{(2\pi)^3} \int_{-\infty}^{\infty} \frac{e^{i\vec{k}\cdot(\vec{x}'-\vec{x}) - i\omega_k t^-}}{2\omega_k} d^3k \tag{16.2.10}$$

where

$$t^+ = t' - t \geq 0, \qquad t^- = t' - t < 0. \tag{16.2.11}$$

We simplify notation by writing

$$\psi_k^{\pm}(x) = \frac{e^{\pm i(\vec{k}\cdot\vec{x} - \omega_k t)}}{\sqrt{(2\pi)^3 2\omega_k}} \tag{16.2.12}$$

and noting that the unit step function

$$\theta(t'-t) = \begin{Bmatrix} 1, & t' > t \\ 0, & t' < t \end{Bmatrix} \tag{16.2.13}$$

can be represented as

$$\theta(t) = \lim_{\epsilon \to 0} \left[-\frac{1}{2\pi i} \int_{-\infty}^{\infty} \frac{e^{-i\omega t}}{\omega + i\epsilon} d\omega \right]. \tag{16.2.14}$$

Combining these equations lets us write the Green's function as the desired result

$$\begin{aligned} G_F(x'-x) = & -i \int_{-\infty}^{\infty} \psi_k^+(x') \psi_k^{+*}(x) \theta(t'-t) d^3k \\ & -i \int_{-\infty}^{\infty} \psi_k^-(x') \psi_k^{-*}(x) \theta(t-t') d^3k. \end{aligned} \tag{16.2.15}$$

Substituting Eq. (16.2.15) into (16.2.1) we express the scattered wave $\phi_s(x)$ in the form

$$\begin{aligned} \phi_s(x) = & \int \theta(t'-t) \psi_k^+(x) T_+(k) d^3k \\ & + \int \theta(t-t') \psi_k^-(x) T_-(k) d^3k \end{aligned} \tag{16.2.16}$$

where the transition matrix elements are defined by

$$T_{\pm}(k) \equiv \int [-i \psi_k^{\pm *}(y) K_I(y) \psi(y)] d^4y. \tag{16.2.17}$$

The transition matrix elements T_+ and T_- apply to positive and negative frequency waves respectively. From here we can calculate the usual quantities associated with the conventional theory, e.g. scattering amplitudes and differential cross-sections. As an illustration, we outline the calculation of the scattering amplitude for a charged pion in a Coulomb potential.

§16.3 SCATTERING AMPLITUDE

Scattering amplitudes may be calculated using either the S-matrix element developed in §16.1 or §16.2. We begin by continuing the analysis of §16.2. For comparison, the scattering amplitude based on §16.1 is then calculated. We show the expected result that the amplitudes contain equivalent physical information.

CONVENTIONAL APPROACH

The following scattering amplitude calculation parallels that of Bjorken and Drell [§9.5, 1964]. For simplicity, we consider only the case of an incident positive frequency particle being scattered forward in time by the Coulomb potential.

Initial and final plane wave solutions are assumed to have the form of Eq. (16.2.12); thus

$$\psi_i(y) \approx \psi^+_{k_i}(y) = \frac{1}{(2\pi)^{3/2}} \frac{e^{-ik_i^\nu y_\nu}}{\sqrt{2\omega_i}} \tag{16.3.1}$$

and

$$\psi_f(y) \approx \psi^+_{k_f}(y) = \frac{1}{(2\pi)^{3/2}} \frac{e^{-ik_f^\nu y_\nu}}{\sqrt{2\omega_f}} . \tag{16.3.2}$$

Equations (16.3.1) and (16.3.2) are a natural assumption within the context of

PRQT. To obtain these plane wave solutions in the conventional, nonparametrized theory, the normalization condition must change from assuming a single particle represented by state ψ_S in a spacial volume V:

$$\int \psi_S^* \psi_S d^3x = 1, \tag{16.3.3}$$

to assuming $2\omega_i$ particles represented by state ψ_M in the same volume V:

$$\int \psi_M^* \psi_M d^3x = 2\,\omega_i. \tag{16.3.4}$$

A similar change must be made to the normalization condition for spin-½ spinors, as is clearly discussed by Aitchison and Hey [Sec. 2.5, 1982]. By replacing Eq. (16.3.3) with (16.3.4), we showed in §8.2 that the conventional theory has essentially adopted a four-dimensional normalization condition with the on mass shell assumption discussed in §16.2.

The scattering amplitude to lowest order in e is

$$\begin{aligned} S_{k_f k_i} &= T_+(k) \\ &= -\frac{i}{(2\pi)^3}\int \frac{1}{\sqrt{2\omega_i}\,\sqrt{2\omega_f}} e^{ik_f^\nu y_\nu} \\ &\cdot ie\left[\frac{\partial}{\partial y^\mu}A^\mu(y) + A^\mu(y)\frac{\partial}{\partial y^\mu}\right] e^{-ik_i^\nu y_\nu} d^4y. \end{aligned} \tag{16.3.5}$$

We have assumed that $k_i \neq k_f$ and the second order term $e^2 A_\mu A^\mu$ is negligible to first order in e. The four-vector potential is given by Eq. (12.1.3). Using Eq. (12.1.3) in (16.3.3) yields

$$\begin{aligned} S_{k_f k_i} &= -\frac{i}{(2\pi)^3}\int \frac{e^{iq^\nu y_\nu}}{\sqrt{4\omega_i\omega_f}} (k_{i\mu} + k_{f\mu}) A^\mu(y)\, d^4y \\ &= -\frac{ie}{(2\pi)^3}\frac{(k_{i\mu} + k_{f\mu}) A^\mu(q)}{\sqrt{4\,\omega_i\,\omega_f}} \end{aligned} \tag{16.3.6}$$

where the four-vector q is defined by

$$q \equiv k_f - k_i \neq 0, \tag{16.3.7}$$

and $A^\mu(q)$ is the Fourier transform

$$A^\mu(q) = \int e^{iq^\nu y_\nu} A^\mu(y)\, d^4y. \tag{16.3.8}$$

Substituting Eq. (12.1.3) into (16.3.8) gives

$$\begin{aligned} A^\mu(q) &= \int\int e^{iq^\nu y_\nu} A^\mu(y)\, dy^0 d^3y \\ &= g^{\mu 0} \int \frac{-Ze}{4\pi |\vec{y}|} e^{-\vec{q}\cdot\vec{y}} d^3y \int e^{i(\omega_f - \omega_i)y^0} dy^0 . \end{aligned} \tag{16.3.9}$$

Evaluating the integrals yields

$$A^\mu(q) = \frac{Ze}{|\vec{q}|^2} g^{\mu 0} 2\pi\, \delta(\omega_f - \omega_i), \tag{16.3.10}$$

and the scattering amplitude is the expected result

$$\begin{aligned} S_{k_f k_i} &= \frac{-ie}{(2\pi)^3} \frac{k_{i\mu} + k_{f\mu}}{\sqrt{4\omega_f \omega_i}} g^{\mu 0} \frac{Ze}{|\vec{q}|^2} 2\pi\, \delta(\omega_f - \omega_i) \\ &= \frac{-ie}{(2\pi)^3} \frac{\omega_i + \omega_f}{\sqrt{4\omega_i \omega_f}} \frac{Ze}{|\vec{q}|^2} 2\pi\, \delta(\omega_f - \omega_i) . \end{aligned} \tag{16.3.11}$$

PRQT APPROACH

We now wish to calculate the scattering amplitude from Eq. (16.1.14), which is the more direct PRQT approach to the problem. Rather than assuming the plane wave solutions in Eqs. (16.3.1) and (16.3.2), we recognize that the unperturbed Hamiltonian -- the left hand side of Eq. (16.1.6) -- is the Hamiltonian of a free particle with plane wave solutions

$$\psi_i(x) = \eta_i e^{-ik_{i\mu}x^\mu},$$
$$\psi_f(x) = \eta_f e^{-ik_{f\mu}x^\mu} \tag{16.3.12}$$

where η_i, η_f are normalization constants. Again we assume $k_f \neq k_i$ and consider terms to first order in A^μ only so that Eq. (16.1.14) becomes

$$S_{fi} = -e\,\Delta s \int \left[\psi_f^* (\partial_\mu \psi_i) - (\partial_\mu \psi_f^*) \psi_i \right] A^\mu d^4x. \tag{16.3.13}$$

Substituting Eq. (16.3.12) into (16.3.13) gives

$$\begin{aligned} S_{fi} = &-e\,\Delta s \int \left[\eta_f^* e^{ik_{f\mu}x^\mu} (-ik_{i\mu}) \eta_i e^{-ik_{i\mu}x^\mu} \right] A^\mu d^4x \\ &-e\Delta s \int \left[\eta_f^* (ik_{f\mu}) e^{ik_{f\mu}x^\mu} \eta_i e^{-ik_{i\mu}x^\mu} \right] A^\mu d^4x. \end{aligned} \tag{16.3.14}$$

Equation (16.3.14) may be simplified by factoring terms and using Eqs. (16.3.7) and (16.3.8) to get

$$S_{fi} = ie\,\Delta s (k_{i\mu} + k_{f\mu}) \eta_f^* \eta_i A^\mu(q). \tag{16.3.15}$$

Equation (16.3.15) is the same result as Eq. (16.3.6) except for normalization factors. The equations exhibit the same dependence on the four-vector potential and the sum of initial and final four-momenta. The remaining constants, such as the additional energy terms in the denominator of Eq. (16.3.4) and the normalization constants in Eq. (16.3.13), are factored out when experimentally meaningful quantities like differential scattering cross-sections are calculated.

A procedure for calculating differential scattering cross-sections within the context of PRQT was outlined in §15.3. To obtain the familiar Rutherford scattering cross-section, we need the relations

$$p_f dp_f = \omega_f d\omega_f \tag{16.3.16}$$

for this on mass shell example in which mass is fixed, and

$$|\vec{q}|^2 = 4p^2 \sin^2 \frac{\theta}{2} \tag{16.3.17}$$

where the scattering angle θ is defined in Figure 16.1. Equation (16.3.17) recognizes that the magnitude p of spacial momentum is

$$p = |\vec{p}_i| = |\vec{p}_f| . \tag{16.3.18}$$

The result of the calculation is

$$\frac{d\sigma}{d\Omega} = t\left(\frac{Ze^2}{2\pi}\right)^2 \frac{\omega_f^2}{\left(4p^2 \sin^2\frac{\theta}{2}\right)^2} \tag{16.3.19}$$

where the time factor t comes from an integration over dx^0. Except for the factor of t, Eq. (16.3.19) is the Rutherford cross-section for Coulomb scattering of a spin-0 particle off a heavy target such as a nucleus. A correction due to recoil is provided by Horwitz and Lavie [1982]. Additional scattering references may be found in the review by Fanchi [1993].

FIGURE 16.1

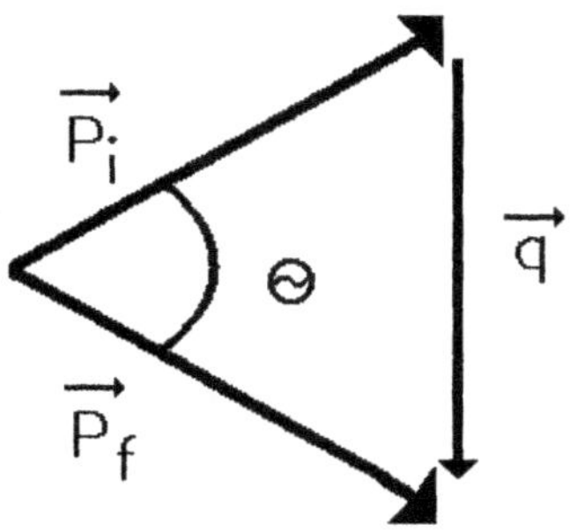

CHAPTER 17
STATISTICAL PRQT

We saw in Chapter 4 that the calculation of the free energy of a Gibbs ensemble in statistical PRCM led to a difference between the conventional theory and the parametrized theory. The difference was shown in Chapter 5 to have potentially measurable consequences in cosmology. The study of the statistical properties of an ideal quantum gas also leads to differences with experimental consequences, as is shown below. In addition, we find that the parametrized theory gives us a new perspective on an ancient enigma: the arrow of time.

We begin our study of statistical PRQT with a study of ensembles. The resulting background sets the stage for an analysis of an ideal quantum gas and the derivation of distributions for three important systems: Maxwell-Boltzmann, Fermi-Dirac, and Bose-Einstein. We end the chapter with a discussion of Boltzmann's H-theorem and its interpretation in PRQT.

§17.1 GENERALIZED ENSEMBLES

FIGURE 17.1

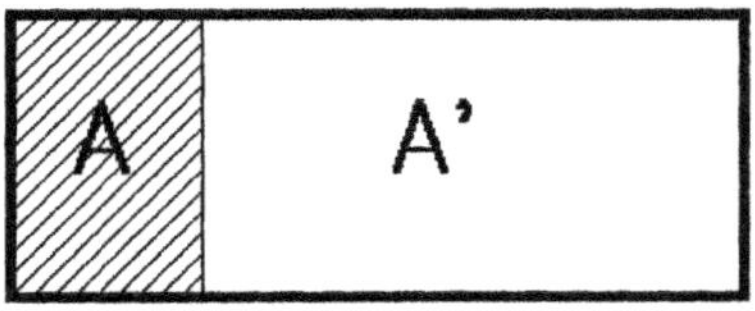

We wish to develop expressions for describing an ensemble of statistical objects. Using familiar arguments from, for example, Reif [§6.9, 1965] and Chandler [§3.5, 1987], we consider a small system A that is separated from a considerably larger system A' by a movable, perforated partition (Figure 17.1). The systems A and A' can exchange energy and an additional N_m extensive variables $\{X_i\}$. The extensive variables may include the number of particles, the volume of the system, or the mass of the system. The exchanged quantities are subject to the constraints

$$\begin{aligned} \omega + \omega' &= \omega_T \\ X_i + X'_i &= X_{iT} \ \forall\ i = 1,\ldots,N_m \end{aligned} \tag{17.1.1}$$

where ω_T and $\{X_{iT}\}$ are constants for the total system $A_T = A + A'$.

Let us denote the number of states accessible to reservoir A' by $\Omega'(\omega', \{X_i'\})$. If A is in a state r characterized by the quantities $(\omega_r, \{X_{ir}\})$, then the number of states accessible to the total system A_T is the number of states accessible to A', namely $\Omega'(\omega', \{X_i'\})$.

The probability P_r of finding A in A_r is proportional to $\Omega'(\omega', \{X_i'\})$:

$$P_r = \zeta\, \Omega'(\omega', \{X'_i\}) \tag{17.1.2}$$

where the proportionality constant ζ is found by normalizing the probability. Substituting Eq. (17.1.1) into (17.1.2) gives

$$P_r = \zeta\, \Omega'(\omega_T - \omega_r, \{X_{iT} - X_{ir}\}) . \tag{17.1.3}$$

We assumed at the outset that A' acts like a reservoir relative to A, i.e. $A \ll A'$. Therefore we have

$$\begin{aligned} \omega_r &\ll \omega_T \\ \{X_{ir}\} &\ll \{X_{iT}\} . \end{aligned} \tag{17.1.4}$$

Given Eq. (17.1.4), it is reasonable to approximate the number of states Ω' by expanding Ω' in a Taylor series about $(\omega_T, \{X_{iT}\})$ and keeping lower order terms. To minimize the number of terms used in the Taylor series expansion, we make the approximation more accurate by expanding $\ell n\ \Omega'$ rather than Ω' since $\ell n\ \Omega'$ changes more slowly than Ω'. The resulting Taylor series expansion is

$$\ell n\ \Omega'(\omega_T - \omega_r, \{X_{iT} - X_{ir}\}) \approx \ell n\ \Omega'(\omega_T, \{X_{iT}\}) - \beta\, \omega_r - \sum_{i=1}^{N_m} \gamma_i X_{ir} \tag{17.1.5}$$

where the expansion coefficients are

$$\beta \equiv \frac{\partial \ell n\ \Omega'}{\partial \omega_r}\Big|_{\omega_r=\omega_T} \tag{17.1.6}$$

and

$$\gamma_i \equiv \frac{\partial \ell n\ \Omega'}{\partial X_{ir}}\Big|_{X_{ir}=X_{iT}} \tag{17.1.7}$$

for all values of $i=1,\ldots,N_m$. The derivatives characterize state r. They may be expressed in terms of entropy by recalling that entropy S is related to the number of accessible states Ω by

$$S(\omega,\{X_i\}) \equiv k_B \Omega(\omega,\{X_i\}) \tag{17.1.8}$$

where k_B is Boltzmann's constant. Substituting the definition of entropy into Eqs. (17.1.6) and (17.1.7) gives

$$k_B \beta = \frac{\partial S}{\partial \omega_r}\Big|_{\omega_r=\omega_T} \tag{17.1.9}$$

and

$$k_B \gamma_i = \frac{\partial S}{\partial X_{ir}}\Big|_{X_{ir}=X_{iT}}. \tag{17.1.10}$$

We find the proportionality constant ζ of Eq. (17.1.2) by first taking the exponent of Eq. (17.1.5)

$$\Omega'(\omega',\{X'_i\}) = \Omega'(\omega_T,\{X_{iT}\})\, e^{-\beta\omega_r - \sum_{i=1}^{N_m} \gamma_i X_{ir}}. \tag{17.1.11}$$

Substituting Eq. (17.1.11) into (17.1.2) and normalizing probability by summing over all possible states R gives

$$\sum_R P_r = 1 = \sum_R \zeta\, \Omega'(\omega', \{X'_i\}) = \zeta\, \Omega'(\omega_T, \{X_{iT}\}) \sum_R e^{-\beta\omega_r - \sum_{i=1}^{N_m} \gamma_i X_{ir}} . \tag{17.1.12}$$

For simplicity, we have assumed the states are discrete. Continuous states may be treated by replacing sums with integrals and using analogous arguments. Performing the normalization gives

$$\zeta = \frac{1}{\Omega'(\omega_T, \{X_{iT}\})\, \Xi} \tag{17.1.13}$$

where the generalized partition function Ξ is defined as

$$\Xi \equiv \sum_R e^{-\beta\omega_r - \sum_{i=1}^{N_m} \gamma_i X_{ir}} . \tag{17.1.14}$$

The probability of finding A in state r can now be written in terms of the coefficients characterizing state r, thus

$$P_r = \zeta\, \Omega'(\omega', \{X'_i\}) = \frac{e^{-\beta\omega_r - \sum_{i=1}^{N_m} \gamma_i X_{ir}}}{\Xi} . \tag{17.1.15}$$

Mean values of physical quantities may be readily obtained from Eq. (17.1.15).

The mean value of energy is

$$\langle \omega \rangle = \sum_R \omega_r P_r = \frac{\sum_R \omega_r e^{-\beta\omega_r - \sum_{i=1}^{N_m} \gamma_i X_{ir}}}{\Xi} . \tag{17.1.16}$$

We rewrite the summand by introducing a partial derivative with respect to β to get

$$\langle\omega\rangle = \frac{1}{\Xi}\sum_R\left[-\frac{\partial}{\partial\beta}e^{-\beta\omega_r - \sum_{i=1}^{N_m}\gamma_i X_{ir}}\right]. \tag{17.1.17}$$

The β derivative may be factored out of the summation with the result

$$\langle\omega\rangle = \frac{1}{\Xi}\left[-\frac{\partial}{\partial\beta}\left(\sum_R e^{-\beta\omega_r - \sum_{i=1}^{N_m}\gamma_i X_{ir}}\right)\right]. \tag{17.1.18}$$

Making use of the definition of the partition function we find the formally simpler expression

$$\langle\omega\rangle = \frac{1}{\Xi}\left[-\frac{\partial}{\partial\beta}\Xi\right] = -\frac{\partial}{\partial\beta}\ell n\,\Xi. \tag{17.1.19}$$

A similar calculation for $\{<X_i>\}$ yields

$$\langle X_i\rangle = \sum_R X_{ir}P_r = -\frac{\partial}{\partial\gamma_i}\ell n\,\Xi \tag{17.1.20}$$

for all i. If we have a situation where only the means are known, Eqs. (17.1.19) and (17.1.20) would be used to solve for $\beta,\{\gamma_i\}$.

We are now in a position to establish Boltzmann's principle. Boltzmann's principle is needed for our subsequent discussion of Boltzmann's H-theorem and the arrow of time.

Boltzmann's Principle

When A is a very small macroscopic system relative to A', the fluctuations of $\omega,\{X_i\}$ around the means $<\omega>,\{<X_i>\}$ are relatively small. The entropy of such a system may be approximated as a function of the means only. We calculate the change in entropy with respect to changes in mean values as

$$dS = \frac{\partial S}{\partial\langle\omega\rangle} d\langle\omega\rangle + \sum_{i=1}^{N_m} \frac{\partial S}{\partial\langle X_i\rangle} d\langle X_i\rangle. \tag{17.1.21}$$

Making use of the relationship between the coefficients $\beta, \{\gamma_i\}$ and the partial derivative of entropy with respect to an extensive variable given in Eqs. (17.1.9) and (17.1.10) lets us rewrite Eq. (17.1.21) in the form

$$dS = k_B \beta\, d\langle\omega\rangle + k_B \sum_{i=1}^{N_m} \gamma_i d\langle X_i\rangle \tag{17.1.22}$$

where $\beta, \{\gamma_i\}$ are determined from mean values.

Following Chandler, [§3.5, 1987], we introduce the function

$$\sigma = -\sum_R P_r \ell n P_r. \tag{17.1.23}$$

Making use of Eq. (17.1.15) we obtain

$$\ell n\ P_r = -\ell n\ \Xi - \beta\omega_r - \sum_{i=1}^{N_m} \gamma_i X_{ir}. \tag{17.1.24}$$

Substituting Eq. (17.1.24) into (17.1.23) gives

$$\begin{aligned} \sigma &= -\sum_R P_r [-\ell n\ \Xi - \beta\omega_r - \sum_{i=1}^{N_m} \gamma_i X_{ir}] \\ &= \ell n\ \Xi [\sum_R P_r] + \beta [\sum_R P_r \omega_r] + \sum_{i=1}^{N_m} \gamma_i [\sum_R P_r X_{ir}]. \end{aligned} \tag{17.1.25}$$

Using probability normalization and the definition of means we get

$$\sigma = \ell n\ \Xi + \beta\langle\omega\rangle + \sum_{i=1}^{N_m} \gamma_i \langle X_{ir}\rangle. \tag{17.1.26}$$

The differential of σ is

$$d\sigma = d\ln\,\Xi + \beta\, d\langle\omega\rangle + \langle\omega\rangle d\beta + \sum_{i=1}^{N_m} [\gamma_i d\langle X_{ir}\rangle + \langle X_{ir}\rangle d\gamma_i]. \tag{17.1.27}$$

This expression may be simplified by recognizing that the partition function is a function of the expansion coefficients and has the differential

$$d\ln\Xi = \frac{\partial \ln\Xi}{\partial\beta} d\beta + \sum_{i=1}^{N_m} \frac{\partial \ln\,\Xi}{\partial\gamma_i} d\gamma_i. \tag{17.1.28}$$

Recalling the relationships between the means and the partition function from Eqs. (17.1.19) and (17.1.20) gives

$$d\ln\Xi = -\langle\omega\rangle d\beta - \sum_{i=1}^{N_m} \langle X_i\rangle d\gamma_i. \tag{17.1.29}$$

Substituting Eq. (17.1.29) into $d\sigma$ completes the Legendre transformation and yields

$$d\sigma = \beta\, d\langle\omega\rangle + \sum_{i=1}^{N_m} \gamma_i d\langle X_i\rangle. \tag{17.1.30}$$

Comparing Eq. (17.1.30) with (17.1.22) shows that σ is proportional to entropy, thus

$$S = k_B \sigma. \tag{17.1.31}$$

Substituting Eq. (17.1.23) into (17.1.31) gives Boltzmann's principle

$$S = -k_B \sum_R P_r \ln P_r \tag{17.1.32}$$

which is also known as Gibbs entropy.

ADDITIVITY OF ENTROPY

An important property of entropy is that it is additive. To show this, we use Eq. (17.1.26) to write

$$S = k_B \sigma = k_B \left[\ell n\, \Xi + \beta \langle \omega \rangle + \sum_{i=1}^{N_m} \gamma_i \langle X_i \rangle \right]. \tag{17.1.33}$$

We are interested in calculating the entropy for two weakly interacting subsystems $A_r, A_{r'}$ whose states are labelled by the indices r,r' and have extensive variables $\omega_r, \{X_{ir}\}$ and $\omega_{r'}, \{X_{ir'}\}$ respectively. We approximate the extensive variables of the combined weakly interacting systems by

$$\begin{aligned} &\omega_r + \omega_{r'} = \omega_T \\ &X_{ir} + X_{ir'} = X_{iT} \;\forall\; i = 1,...,N_m . \end{aligned} \tag{17.1.34}$$

The partition function of the combined system is

$$\begin{aligned} \Xi_T &= \sum_{R,R'} e^{-\beta(\omega_r + \omega_{r'}) - \sum_{i=1}^{N_m} \gamma_i (X_{ir} + X_{ir'})} \\ &= \left[\sum_R e^{-\beta \omega_r - \sum_{i=1}^{N_m} \gamma_i X_{ir}} \right] \left[\sum_{R'} e^{-\beta \omega_{r'} - \sum_{i=1}^{N_m} \gamma_i X_{ir'}} \right] \\ &= \Xi\, \Xi' \end{aligned} \tag{17.1.35}$$

where Ξ, Ξ' are the partition functions of subsystems $A_r, A_{r'}$ respectively. Recognizing that the means are additive, we have the relations

$$\begin{aligned} &\ell n\, \Xi_T = \ell n\, \Xi + \ell n\, \Xi', \\ &\langle \omega_T \rangle = \langle \omega_r \rangle + \langle \omega_{r'} \rangle, \\ &\langle X_{iT} \rangle = \langle X_{ir} \rangle + \langle X_{ir'} \rangle \;\forall\; i = 1,...,N_m . \end{aligned} \tag{17.1.36}$$

The entropy of the combined system is

$$S_T = k_B\left[\ell n\,\Xi_T + \beta\langle\omega_T\rangle + \sum_{i=1}^{N_m}\gamma_i\langle X_{i_T}\rangle\right]. \tag{17.1.37}$$

Substituting Eq. (17.1.36) into (17.1.37) gives

$$\begin{aligned} S_T &= k_B\left[\ell n\,\Xi_r + \beta\langle\omega_r\rangle + \sum_{i=1}^{N_m}\gamma_i\langle X_{ir}\rangle\right] \\ &\quad + k_B\left[\ell n\,\Xi_{r'} + \beta\langle\omega_{r'}\rangle + \sum_{i=1}^{N_m}\gamma_i\langle X_{ir'}\rangle\right] \\ &= S_r + S_{r'}. \end{aligned} \tag{17.1.38}$$

Equation (17.1.38) demonstrates the expected additivity of entropy.

§17.2 IDEAL QUANTUM GAS

One useful, albeit simple, application of quantum statistics is to the description of an ideal gas. An ideal quantum gas consists of N identical, non-interacting particles treated quantum mechanically. We consider such a gas in a space-time box of dimension

$$-\frac{\Upsilon}{2} \le x^0 \le \frac{\Upsilon}{2};\quad -\frac{L}{2} \le x^j \le \frac{L}{2}\ \ for\ j=1,2,3. \tag{17.2.1}$$

The field equation for N free particles was introduced in §11.1 and may be rewritten in the form

$$i\frac{\partial\Psi}{\partial s} = \sum_{a=1}^{N}\frac{p_{\mu a}p_a^{\mu}}{2M}\Psi \tag{17.2.2}$$

where M is the rest mass of each bradyon. Letting

$$\Psi(x,s) = e^{-iQs/2M}\psi(x) \tag{17.2.3}$$

yields the eigenvalue equation

$$Q_\lambda \psi_\lambda = \sum_{a=1}^{N} p_{\mu a} p_a^\mu \psi_\lambda . \tag{17.2.4}$$

The resulting eigenvalues are

$$Q_\lambda = (k_\lambda^0)^2 - \vec{k}_\lambda \cdot \vec{k}_\lambda \tag{17.2.5}$$

where λ denotes the state. Employing box normalization in the ranges defined by Eq. (17.2.1) gives

$$k_\lambda^0 = \frac{2\pi}{\Upsilon}\lambda^0;\ k_\lambda^j = \frac{2\pi}{L}\lambda^j \ for\ j=1,2,3 \tag{17.2.6}$$

and $\{\lambda^\mu\}=0, \pm 1, \pm 2, ...$ To derive familiar results, we work only with bradyons ($Q_\lambda > 0$) and neglect antiparticles corresponding to $k_\lambda^0 < 0$. Processes such as creation and annihilation are not considered because of our initial free particle assumption.

We are interested in determining the distribution of states of the ideal quantum gas. To proceed, we let n_λ be the occupation number of state λ, i.e. n_λ is the number of particles in state ψ_λ. The total mass and energy of the system are

$$Q_T = \sum_\Lambda n_\lambda Q_\lambda \tag{17.2.7}$$

and

$$k_T^0 = \sum_\Lambda n_\lambda k_\lambda^0 \tag{17.2.8}$$

where Λ denotes a sum over all states. Although there are many similarities between the argument presented here and the conventional argument given, for example, by Reif [1965] or Chandler [1987], it is important to remember that energy in this context is total relativistic energy. Furthermore, the extensive quantities that must be conserved are N, k_T^0, Q_T.

The particle in the λth state is viewed as a small system A which may interact weakly with all other particles $n' = N - n_\lambda$ comprising reservoir A'. From this perspective, we may invoke the results of §17.1. Thus, the equilibrium probability that the λth quantum state of system A contains n_λ particles is

$$P(n_\lambda) = \zeta e^{-\beta \omega_\lambda - \gamma_1 n_\lambda - \gamma_2 q_\lambda} \tag{17.2.9}$$

where ζ is determined by normalizing probability.

The extensive variables ω_λ and q_λ correspond to the mass and energy of state λ in system A. For this particular case, they satisfy the relations

$$\begin{aligned} \omega_\lambda &= n_\lambda k_\lambda^0 \\ q_\lambda &= n_\lambda Q_\lambda . \end{aligned} \tag{17.2.10}$$

The expansion coefficients γ_1 and γ_2 are associated with occupation number and mass respectively. They may be expressed in terms of entropy and the number of accessible states as

$$\begin{aligned} \gamma_1 &= \frac{1}{k_B}\frac{\partial S}{\partial n} = \frac{\partial \ln \Omega'}{\partial n}, \\ \gamma_2 &= \frac{1}{k_B}\frac{\partial S}{\partial Q} = \frac{\partial \ln \Omega'}{\partial Q}. \end{aligned} \tag{17.2.11}$$

Coefficient γ_2 was first introduced by Horwitz, Schieve and Piron [1981]. They dubbed γ_2 the "mass potential" in accordance with the interpretation of γ_1 as the "chemical potential".

Using Eq. (17.2.10) in (17.2.9) gives

$$P(n_\lambda) = \zeta e^{-\alpha_\lambda n_\lambda}, \ \alpha_\lambda \equiv \beta k_\lambda^0 + \gamma_1 + \gamma_2 Q_\lambda . \tag{17.2.12}$$

where ζ is determined by normalization to be

$$\frac{1}{\zeta} = \Xi = \sum_\Lambda e^{-\alpha_\lambda n_\lambda}. \tag{17.2.13}$$

Following the procedure presented in §17.1, we determine the mean value of the occupation number of state λ from

$$\langle n_\lambda \rangle = -\frac{\partial}{\partial \alpha_\lambda} \ln \Xi \tag{17.2.14}$$

where Ξ is the partition function given by Eq. (17.2.13).

MAXWELL-BOLTZMANN DISTRIBUTION

The Maxwell-Boltzmann distribution is calculated for a system of N distinguishable particles. The appropriate partition function is

$$\Xi_{MB} = \sum_{\lambda_1, \lambda_2, \ldots, \lambda_N} e^{-(\alpha_{\lambda_1} + \alpha_{\lambda_2} + \cdots + \alpha_{\lambda_N})} \tag{17.2.15}$$

where we are summing over all possible states for each of the N particles [Reif, §9.4, 1965] and the subscript MB denotes Maxwell-Boltzmann. Factoring yields

$$\begin{aligned} \Xi_{MB} &= \left[\sum_{\lambda_1} e^{-\alpha_{\lambda_1}}\right] \left[\sum_{\lambda_2} e^{-\alpha_{\lambda_2}}\right] \cdots \left[\sum_{\lambda_N} e^{-\alpha_{\lambda_N}}\right] \\ &= \left[\sum_{\lambda} e^{-\alpha_\lambda}\right]^N \end{aligned} \tag{17.2.16}$$

since all sums are equal for an ideal gas and λ is a dummy index. Given the partition function, we calculate from Eq. (17.2.14) the mean occupation number of state λ for the Maxwell-Boltzmann distribution:

$$\begin{aligned} \langle n_\lambda \rangle_{MB} &= -\frac{\partial}{\partial \alpha_\lambda} \ell n\, \Xi_{MB} \\ &= -\frac{\partial}{\partial \alpha_\lambda} N \ell n \left[\sum_{\ell} e^{-\alpha_\ell}\right] \end{aligned} \tag{17.2.17}$$

or

$$\langle n_\lambda \rangle_{MB} = \frac{N e^{-\alpha_\lambda}}{\sum_{\lambda} e^{-\alpha_\lambda}} \tag{17.2.18}$$

where α_λ is given by Eq. (17.2.12). Equation (17.2.18) is the PRQT analog of the Maxwell-Boltzmann distribution.

BOSE-EINSTEIN DISTRIBUTION

The Bose-Einstein distribution is calculated for the case in which every particle may occupy every state and all particles are indistinguishable. Thus the sum over occupation number must include all non-negative integers from 0 to ∞. To perform the calculation, we will need the geometric series representation [Reif, App. A, 1965]

$$\sum_{\eta=0}^{\infty} \chi^{\eta} = \frac{1}{1-\chi}. \tag{17.2.19}$$

The Bose-Einstein partition function becomes

$$\Xi_{BE} = \sum_{n_\lambda=0}^{\infty} e^{-\alpha_\lambda n_\lambda} = \sum_{n_\lambda=0}^{\infty} \chi^{n_\lambda} = \frac{1}{1-\chi}; \ \chi \equiv e^{-\alpha_\lambda} \tag{17.2.20}$$

or

$$\Xi_{BE} = \frac{1}{1-e^{-\alpha_\lambda}}. \tag{17.2.21}$$

The mean occupation number in state λ for the Bose-Einstein distribution is

$$\begin{aligned} \langle n_\lambda \rangle_{BE} &= \frac{1}{e^{-\alpha_\lambda} - 1} \\ &= \frac{1}{e^{-(\beta k_\lambda^0 + \gamma_1 + \gamma_2 Q_\lambda)} - 1}. \end{aligned} \tag{17.2.22}$$

FERMI-DIRAC DISTRIBUTION

Only one particle may occupy any given state λ in a Fermi-Dirac distribution. Consequently the occupation number may be either 0 or 1. The corresponding Fermi-Dirac partition function is

$$\Xi_{FD} = \sum_{n_\lambda=0}^{1} e^{-\alpha_\lambda} = 1 + e^{-\alpha_\lambda}. \tag{17.2.23}$$

Calculating the mean occupation number of the Fermi-Dirac distribution directly from the definition of the mean gives

$$\begin{aligned}\langle n_\lambda \rangle_{FD} &= \sum_{n_\lambda=0}^{1} P(n_\lambda) n_\lambda \\ &= \sum_{n_\lambda=0}^{1} \frac{e^{-\alpha_\lambda n_\lambda}}{1 + e^{-\alpha_\lambda}} n_\lambda \\ &= \frac{e^{-\alpha_\lambda}}{1 + e^{-\alpha_\lambda}}.\end{aligned} \tag{17.2.24}$$

Simplifying the fraction yields the more familiar form of the Fermi-Dirac distribution:

$$\begin{aligned}\langle n_\lambda \rangle_{FD} &= \frac{1}{e^{\alpha_\lambda} + 1} \\ &= \frac{1}{e^{(\beta k_\lambda^0 + \gamma_1 + \gamma_2 Q_\lambda)} + 1}.\end{aligned} \tag{17.2.25}$$

The distributions in Eqs. (17.2.18), (17.2.22) and (17.2.25) have the same form as their counterparts in the non-parametrized theory, but their details differ. Our distributions include total relativistic energy and a mass potential. These differences may have measurable consequences. For example, the Fermi energy k_F^0 is the energy at which a single Fermi-Dirac state is just as likely to be found occupied as unoccupied. The mean occupation number at the Fermi energy $<n_{F\lambda}>_{FD}$ must satisfy

$$[\langle n_{F\lambda} \rangle_{FD}] = ½. \tag{17.2.26}$$

Equation (17.2.26) is satisfied when $\alpha_\lambda = 0$, thus

$$\alpha_{F\lambda} = \beta k_F^0 + \gamma_1 + \gamma_2 Q_\lambda = 0 \tag{17.2.27}$$

with the result

$$k_F^0 = -\frac{(\gamma_1 + \gamma_2 Q_\lambda)}{\beta}. \tag{17.2.28}$$

The mean occupation number may be rewritten in terms of the Fermi energy as

$$\langle n_\lambda \rangle_{FD} = \frac{1}{e^{\beta(k_\lambda^0 - k_F^0)} + 1}. \tag{17.2.29}$$

Although Eq. (17.2.29) has the same form as the non-parametrized theory, the Fermi energy has now been shifted by the mass potential. If the number of accessible states does not depend on mass -- as in a uniform distribution in which mass is constant -- the mass potential will be negligible or zero by Eq. (17.2.11). The corresponding shift in Fermi energy will also be negligible. Efforts to design experimental tests of the parametrized theory using the above distributions are yet to be made.

§17.3 THE ARROW OF HISTORICAL TIME

To Newton, time marched inexorably from the past to the future; it was an arrow that many modern scientists have tried to link to the second law of thermodynamics: the law of ever-increasing and never-decreasing entropy. Entropy is a measure of the disorder of a system: the greater the entropy, the greater the disorder. It has been known since the 19th century that entropy, or the disorder of a system, can be constant or increase, but never decrease. This is the content of the second law of thermodynamics and follows readily from the statistical interpretation of entropy embodied in Eq. (17.1.8).

Einstein's views require a physical equivalence between coordinate time and coordinate space. In this view, which is widely accepted today, time should be a reversible coordinate in the same manner as space. Evidence for such reversibility has come indirectly from particle physics, where antiparticles are interpreted as particles moving backward in time. Newton's absolute time has slipped into disfavor. Time is a coordinate, say most physicists today, not an arrow. Yet is it? Can time possibly be two things at once? How can it be an irreversible arrow tied to entropy and a reversible coordinate in the Einsteinian sense?

Time, as a single concept, cannot. Herein lies the crux of a temporal paradox, the mystery of time, and a clue to its resolution. Time, as we showed in Chapter 11, is not a single concept. It is not just a coordinate or a parameter. It is three concepts with physically distinguishable characteristics. Minkowski time is the coordinate time of Einstein, while Galilean time and historical time are the irreversible parameters that are linked to entropy. These distinctions make it possible to have an arrow of (parametrized) time and an equivalence between (coordinate) time and space. In the following we show how parametrized time can be linked to the second law of thermodynamics by deriving Boltzmann's H-theorem within the context of PRQT.

BOLTZMANN'S H-THEOREM

A link between time and thermodynamics is usually made through Boltzmann's H-theorem [Kittel, 1961; Reif, 1965; Feynman, 1972] or, equivalently, the temporal rate of entropy change. An analogous theorem can be derived within the context of PRQT. The derivation requires two assumptions: the principle of detailed balance, and a master equation.

The principle of detailed balance asserts that the transition probabilities for the transition from state λ to λ', and the inverse transition from state λ' to λ, are equal; thus

$$W_{\lambda\lambda'} = W_{\lambda'\lambda} . \tag{17.3.1}$$

The PRQT analog of the master equation [Reif, 1965] is obtained as follows. Let $P_\lambda(s)$ be the probability that an isolated system is in state λ at evolution parameter s. The rate of change of P_λ with respect to s increases when systems in other states make transitions to state λ, and decreases when systems in state λ make transitions to other states; thus

$$\frac{dP_\lambda}{ds} = \sum_{\lambda'} P_{\lambda'} W_{\lambda'\lambda} - \sum_{\lambda'} P_\lambda W_{\lambda\lambda'}. \tag{17.3.2}$$

Equations (17.3.1) and (17.3.2) are viewed as working hypotheses subject to verification, which is the same status accorded their analogs in the non-parametrized theory.

The principle of detailed balance is usually presented as a theorem based on the Hermitian property of transition probabilities calculated to lowest order in nonrelativistic, Galilean time-dependent perturbation theory. Blokhintsev [1964] has pointed out that nonrelativistic quantum mechanical (NRQM) transition probabilities are not, in general, Hermitian. They are unitary. A similar analysis applies to PRQT. Limitations on the validity of the NRQM master equation have also been noted. Kittel [1961], for example, remarked that superposition of accessible states could effect the validity of the master equation. Consequently, we restrict the status of Eqs. (17.3.1) and (17.3.2) to plausible, working hypotheses.

Combining Eqs. (17.3.1) and (17.3.2) yields the revised master equation

$$\frac{dP_\lambda}{ds} = \sum_{\lambda'} W_{\lambda\lambda'}(P_{\lambda'} - P_\lambda). \tag{17.3.3}$$

We now employ Boltzmann's principle expressed in Eq. (17.1.32) to relate quantum probabilities with thermodynamic quantities. Writing entropy as

$$S = -k_B \sum_\lambda P_\lambda \ln P_\lambda \tag{17.3.4}$$

and differentiating with respect to s gives

$$\begin{aligned} \frac{dS}{ds} &= -k_B \sum_\lambda \left[\frac{dP_\lambda}{ds} \ln P_\lambda + \frac{dP_\lambda}{ds} \right] \\ &= -k_B \sum_\lambda \left[\ln P_\lambda \frac{dP_\lambda}{ds} \right] - k_B \frac{d}{ds} \sum_\lambda P_\lambda . \end{aligned} \tag{17.3.5}$$

The last term on the right hand side vanishes because of probability conservation and we are left with

$$\frac{dS}{ds} = -k_B \sum_{\lambda} \left[\ell n P_\lambda \frac{dP_\lambda}{ds} \right]. \tag{17.3.6}$$

Substituting Eq. (17.3.3) into (17.3.6) and rearranging the double summation yields

$$\frac{dS}{ds} = -\frac{k_B}{2} \sum_{\lambda} \sum_{\lambda'} W_{\lambda\lambda'}(P_{\lambda'} - P_\lambda)(\ell n P_\lambda - \ell n P_{\lambda'}) \geq 0. \tag{17.3.7}$$

The inequality is valid because each term in the double summation is negative. Noting that Boltzmann's H is defined by

$$S = -k_B H \tag{17.3.8}$$

lets us rewrite Eq. (17.3.7) as

$$\frac{dH}{ds} = \frac{1}{2} \sum_{\lambda} \sum_{\lambda'} W_{\lambda\lambda'}(P_{\lambda'} - P_\lambda)(\ell n P_\lambda - \ell n P_{\lambda'}) \leq 0. \tag{17.3.9}$$

Equation (17.3.9) is Boltzmann's H-theorem.

Equation (17.3.7) shows that entropy can only be constant or increase with respect to the monotonically increasing evolution parameter s. A similar result has been obtained by Horwitz, Shashoua and Schieve [1989] using a parametrized Boltzmann equation to obtain a proof of Boltzmann's H-theorem.

By recognizing the existence of a unique coordinate time and a unique parametrized time, there is no longer a conflict between a universal direction of time and a time which may proceed as readily from future to past as from past to future. The two times are different: the former is a parameter, and the latter is a coordinate. These distinctions remove ambiguities in the properties associated with what should be, and apparently are, two entirely different temporal concepts.

PART IV:

INTRODUCTION TO TACHYON PHYSICS

CHAPTER 18
PARAMETER DEPENDENT PERTURBATION THEORY

Parameter independent perturbation calculations within the context of PRQT were presented in Chapter 14. It is a legitimate question to ask if parameter dependent perturbations exist in nature. We hypothesize that parameter dependent perturbations have physically measurable consequences. The purpose of the present chapter is to develop a parameter dependent perturbation formalism [Fanchi, 1990]. Given the formalism, we focus our attention on testing the hypothesis in Chapter 19 and Chapter 20.

§18.1 DISCRETE SCALAR SYSTEMS

The mathematical development of a parameter dependent perturbation theory closely parallels the development of Galilean time-dependent perturbation theory in nonrelativistic quantum mechanics. Important conceptual differences arise, however, and must be considered. Rather than immediately presenting a general formalism, it is instructive to begin with a relatively simple system consisting of scalar eigenfunctions and discrete eigenvalues.

Consider a scalar field equation of the form

$$i \frac{\partial \Phi(x,s)}{\partial s} = [K_0 + U_I] \Phi(x,s) \tag{18.1.1}$$

where $\mathbf{K}_0$ is independent of s and the s-dependent perturbation is U_I. We require U_I to be Hermitian so that subsequent matrix elements involving U_I are real and therefore physically meaningful. A general constraint on allowed forms of perturbations is derived from probability conservation in §18.2.

The state Φ may be represented in the usual manner as the expansion

$$\Phi(x,s) = \sum_n a_n(s)\,\phi_n(x,s) \tag{18.1.2}$$

where ϕ are solutions of the unperturbed field equation

$$i\frac{\partial \phi_n(x,s)}{\partial s} = \boldsymbol{K}_0\,\phi_n(x,s) \tag{18.1.3}$$

for all eigenvalues $\{n\}$. The normality condition on Φ

$$\int \Phi^*\,\Phi\, d^4x = 1 \tag{18.1.4}$$

implies

$$\sum_n a_n^*(s)\,a_n(s) = 1\,. \tag{18.1.5}$$

The term $a_n^*(s)a_n(s)$ is interpreted as the probability that the system will be in mass state ϕ_n at evolution parameter s. More detailed justification of this interpretation is presented in Chapter 19. We now wish to solve for the expansion coefficients.

Substituting Eq. (18.1.2) into (18.1.1), expanding, and using Eq. (18.1.3) gives

$$i\sum_n \phi_n \frac{\partial a_n}{\partial s} = \sum_n U_I a_n \phi_n\,. \tag{18.1.6}$$

Multiplying by ϕ_m^*, integrating over d^4x and using the orthonormality relationship

$$\int \phi_m^*\phi_n\, d^4x = \delta_{mn} \tag{18.1.7}$$

yields

$$i\frac{\partial a_m}{\partial s} = \sum_n V_{mn}\, a_n\,. \tag{18.1.8}$$

The matrix element V_{mn} of the perturbing potential is

$$V_{mn} = \int \phi_m^* U_I \phi_n d^4x . \tag{18.1.9}$$

By requiring U_I to be Hermitian, we have guaranteed that V_{mn} is real. Equation (18.1.8) is a set of differential equations for the expansion coefficients. Does Eq. (18.1.8) make sense physically? We begin to develop an answer to this question by evaluating the formalism as it is applied to some simple systems.

SINGLE-STATE SYSTEM

The simplest case is a single-state system. Equation (18.1.8) reduces to

$$i\frac{\partial a_1}{\partial s} = V_{11} a . \tag{18.1.10}$$

Equation (18.1.10) has the integral form

$$a_1(s) = a_1(s_0) \exp\left[-i \int_{s_0}^{s} V_{11}(s') ds' \right] \tag{18.1.11}$$

where s_0 is the initial value of the evolution parameter. The integral in Eq. (18.1.11) is real since the matrix element is real. Substituting Eq. (18.1.11) into (18.1.5) shows that

$$a_1^*(s) a_1(s) = a_1^*(s_0) a_1(s_0) = 1 . \tag{18.1.12}$$

We obtain the reasonable conclusion that a system with only one mass state will remain in that state after being perturbed by a Hermitian potential.

ADIABATIC TRANSITION

Another interesting -- albeit academic -- application is to consider the adiabatic case in which the perturbation matrix element in Eq. (18.1.8) changes slowly with respect to s. We consider the extreme case in which

$$V_{mn} = \begin{Bmatrix} u_{m,n}, & 0 \le s \le s_f \\ 0, & s > s_f \end{Bmatrix} \tag{18.1.13}$$

where u_{mn} is a constant and the perturbation acts only during the perturbation interval $0 \le s \le s_f$. Substituting (18.1.13) into (18.1.8) gives

$$i\frac{\partial a_m}{\partial s} = \begin{Bmatrix} \sum_n u_{mn} a_n, & 0 \le s \le s_f \\ 0, & s > s_f \end{Bmatrix}. \tag{18.1.14}$$

Integrating Eq. (18.1.14) with respect to s yields the formal solution

$$a_m(s) = \begin{Bmatrix} -i\sum_n u_{mn} \int_0^s a_n(s')\,ds', & 0 \le s \le s_f \\ a_m(s_f), & s > s_f \end{Bmatrix}. \tag{18.1.15}$$

The solution for values of $s > s_f$ is required to assure continuity of the transition amplitudes. As a first order approximation, we assume that the system is predominantly in the state ϕ_ℓ, thus

$$a_n \approx \delta_{\ell n} \tag{18.1.16}$$

and Eq. (18.1.15) becomes

$$a_m(s) = \begin{Bmatrix} -i u_{m\ell} s, & 0 \le s \le s_f \\ a_m(s_f), & s \ge s_f \end{Bmatrix}. \tag{18.1.17}$$

In this case the transition amplitudes during the interval $0 \le s \le s_f$ are proportional to s. The corresponding transition probability $a_m^* a_m$ is therefore proportional to s^2. This result appears again in §19.3 when we consider the small s behavior of the transition probability calculated for a realistic problem.

OSCILLATING MASS STATES

We saw in §13.3 that PRQT allows the possibility of oscillating mass states. The discussion in §13.3 focused on oscillating states from the perspective of a single state. Rather than viewing the system from the perspective of a single state, we consider a system with two mass states. Applying Eq. (18.1.8), we find that the two expansion coefficients must satisfy the following set of coupled differential equations:

$$i\frac{\partial a_1}{\partial s} = V_{11}a_1 + V_{12}a_2, \tag{18.1.18}$$

and

$$i\frac{\partial a_2}{\partial s} = V_{21}a_1 + V_{22}a_2. \tag{18.1.19}$$

It is not possible to analytically solve the above set of equations for an arbitrary perturbation. However, with a few simple assumptions we can obtain conditions under which mass states may oscillate.

Let us suppose that the perturbation forces the particle out of its incident state, i.e. we assume $V_{11}=V_{22}=0$. Equations (18.1.18) and (18.1.19) become

$$\begin{aligned} i\frac{\partial a_1}{\partial s} &= V_{12}a_2, \\ i\frac{\partial a_2}{\partial s} &= V_{21}a_1. \end{aligned} \tag{18.1.20}$$

Differentiating Eqs. (18.1.20) with respect to s gives

$$\begin{aligned} i\frac{\partial^2 a_1}{\partial s^2} &= V_{12}\frac{\partial a_2}{\partial s}, \\ i\frac{\partial^2 a_2}{\partial s^2} &= V_{21}\frac{\partial a_1}{\partial s}. \end{aligned} \tag{18.1.21}$$

Comparing Eqs. (18.1.21) with (18.1.20) lets us replace the single order s derivatives. The resulting equations are

$$\frac{\partial^2 a_1}{\partial s^2} = -V_{12} V_{21} a_1 ,$$
$$\frac{\partial^2 a_2}{\partial s^2} = -V_{21} V_{12} a_2 . \tag{18.1.22}$$

These equations have well known general solutions. We determine a physically reasonable solution by imposing the initial conditions

$$a_1(0) = 1 ,$$
$$a_2(0) = 0 . \tag{18.1.23}$$

In other words, we assume the state ϕ_1 is occupied initially and there is no mixing of states. The corresponding solutions are

$$a_1(s) = \cos\left(\sqrt{V_{12} V_{21}}\, s\right),$$
$$a_2(s) = \sin\left(\sqrt{V_{12} V_{21}}\, s\right). \tag{18.1.24}$$

Forming the sum

$$a_1^* a_1 + a_2^* a_2 = 1 \tag{18.1.25}$$

shows that the normalization condition is satisfied.

Equations (18.1.24) represent a particle oscillating between two mass states. A physical example may be neutrino oscillation -- as was discussed in §13.3 -- but this has not been established experimentally. We could continue to discuss simple models for arbitrary perturbations, but these mathematical exercises do not answer the fundamental question: do parameter dependent perturbations exist? To answer this question, we study a more realistic example in Chapter 19 and Chapter 20. Before proceeding, however, we must develop a more sophisticated mathematical framework. Let us turn now to this task.

§18.2 GENERAL FORMALISM

A perturbation solution for an equation of the form

$$i\partial_s \underline{\Psi} = K_0 \underline{\Psi} + \underline{K}_I \tag{18.2.1}$$

is sought. Solutions of Eq. (18.2.1) are denoted by the eigenfunctions $\underline{\Psi}(x,s)$. The operator $\mathbf{K}_0$ is an s-independent operator and $\underline{\mathbf{K}}_I$ is an s-dependent perturbation. The operator $\mathbf{K}_0$ must be Hermitian to assure physically realistic expectation values.

The eigenfunction $\underline{\Psi}(x,s)$ is represented in the usual manner as an expansion $\underline{\psi}_\beta(x,s)$ such that

$$\underline{\Psi}(x,s) = \Sigma_n a_n(s) \underline{\psi}_n(x,s) + \int a_\eta(s) \underline{\psi}_\eta(x,s)\, d\eta \tag{18.2.2}$$

where $\underline{\psi}_\beta(x,s)$ are solutions of the unperturbed equations

$$i\partial_s \underline{\psi}_\beta = K_0 \underline{\psi}_\beta = K_{0\beta} \underline{\psi}_\beta \tag{18.2.3}$$

and $K_{0\beta}$ are eigenvalues of the unperturbed operator $\mathbf{K}_0$. The set of unperturbed basis eigenfunctions satisfy the orthonormality condition

$$\begin{aligned} &\int \underline{\psi}_\alpha^+ \underline{\psi}_\beta\, d^4x = \delta_{\alpha\beta},\ \textit{discrete}; \\ &\int \underline{\psi}_\alpha^+ \underline{\psi}_\beta\, d^4x = \delta(\alpha - \beta),\ \textit{continuous}. \end{aligned} \tag{18.2.4}$$

The delta function is a Kronecker delta function if the eigenvalues are discrete, and a Dirac delta function if the eigenvalues are continuous. Subscripts α, β signify either discrete or continuous eigenvalues, and a_n, a_η are complex, scalar expansion coefficients for discrete and continuous eigenvalues, respectively.

The expansion in Eq. (18.2.2) applies to an eigenfunction basis with an eigenvalue spectrum having both discrete and continuous eigenvalues. The two cases in which the spectrum contains only discrete or continuous eigenvalues are special cases of the general formalism developed here for spinless particles. Our interest is in calculating the expansion coefficients.

MASS STATE TRANSITION AMPLITUDES

Substituting Eq. (18.2.2) into (18.2.1) and simplifying with (18.2.4) gives

$$i\left[\Sigma_n \underline{\psi}_n \mathring{a}_n + \int \underline{\psi}_\eta \mathring{a}_\eta \, d\eta\right] = \underline{K}_I . \tag{18.2.5}$$

Multiplying Eq. (18.2.5) by $\underline{\psi}^+_\alpha$ and integrating over d^4x results in

$$i\sum_n \left[\int \underline{\psi}^+_\alpha \underline{\psi}_n d^4x\right]\mathring{a}_n + i\int\left[\int \underline{\psi}^+_\alpha \underline{\psi}_\eta d^4x\right]\mathring{a}_\eta \, d\eta$$
$$= \int \underline{\psi}^+_\alpha \underline{K}_I d^4x. \tag{18.2.6}$$

Equation (18.2.6) is simplified by invoking the normality condition of $\underline{\Psi}$:

$$\int \underline{\Psi}^+ \underline{\Psi} d^4x = 1. \tag{18.2.7}$$

Substituting Eq. (18.2.2) into (18.2.7) gives the normalization condition for the expansion coefficients

$$\sum_n a_n^* a_n + \int a_\eta^* a_\eta \, d\eta = 1. \tag{18.2.8}$$

Combining Eq. (18.2.8) with (18.2.6) for $\underline{\psi}_\alpha$ belonging to either the set of eigenfunctions with discrete or continuous eigenvalues gives

$$i\mathring{a}_\alpha = \int \underline{\psi}^+_\alpha \underline{K}_I d^4x. \tag{18.2.9}$$

The corresponding integral equation for the transition probability amplitude is

$$a_\alpha = a_\alpha^0 - i\int_0^s \left[\int \underline{\psi}^+_\alpha \underline{K}_I d^4x\right] ds', \quad a_\alpha^0 \equiv a_\alpha(0) \tag{18.2.10}$$

where the superscript 0 denotes evaluation of the quantity at s=0.

We have identified the transition probability amplitudes as amplitudes for calculating the transition probability associated with transitions between mass states. Our interpretation is directly analogous to the nonrelativistic theory of transitions between energy states. The identification is further established in Chapter 19 within the context of a physically plausible application.

ALLOWED PERTURBATIONS

The perturbation term $\underline{K}_I$ is not arbitrary. To determine the constraints on $\underline{K}_I$, we first differentiate Eq. (18.2.7) with respect to s:

$$\int \left[(\partial_s \underline{\Psi}^+) \underline{\Psi} + \underline{\Psi}^+ (\partial_s \underline{\Psi}) \right] d^4x = 0. \tag{18.2.11}$$

Expressions for the s derivatives are obtained from Eq. (18.2.1) and its complex conjugate

$$-i \partial_s \underline{\Psi}^* = \underline{K}_0^* \underline{\Psi}^* + \underline{K}_I^* \tag{18.2.12}$$

Substituting Eqs. (18.2.1) and (18.2.12) into (18.2.11) gives

$$\int \left[i (\underline{K}_0^+ \underline{\Psi}^+ + \underline{K}_I^+) \underline{\Psi} - i \underline{\Psi}^+ (\underline{K}_0 \underline{\Psi} + \underline{K}_I) \right] d^4x = 0. \tag{18.2.13}$$

The terms involving $\mathbf{K}_0$ cancel because $\mathbf{K}_0$ is Hermitian. The resulting integral is

$$\int \left[\underline{K}_I^+ \underline{\Psi} - \underline{\Psi}^+ \underline{K}_I \right] d^4x = 0. \tag{18.2.14}$$

Equation (18.2.14) constrains the form of $\underline{K}_I$. Its applicability is demonstrated in Chapter 19. To complete the parameter dependent perturbation formalism, we need definitions of transition probabilities and transition rates.

§18.3 TRANSITION PROBABILITY AND TRANSITION RATE

The transition probability to state α is

$$P_\alpha = |\langle \underline{\psi}_\alpha | \underline{\Psi} \rangle|^2 \tag{18.3.1}$$

where the matrix element $<\underline{\psi}_\alpha | \underline{\Psi}>$ in coordinate representation has the form

$$\begin{aligned} \langle \underline{\psi}_\alpha | \underline{\Psi} \rangle &= \int \underline{\psi}_\alpha^+ \underline{\Psi} \, d^4x \\ &= \sum_n a_n \int \underline{\psi}_\alpha^+ \underline{\psi}_n \, d^4x + \int a_\eta \left[\int \underline{\psi}_\alpha^+ \underline{\psi}_\eta \, d^4x \right] d\eta \, . \end{aligned} \tag{18.3.2}$$

In the specific case of either a discrete eigenvalue spectrum or a continuous eigenvalue spectrum, Eq. (18.3.1) simplifies to

$$P_\alpha = a_\alpha^* a_\alpha \, . \tag{18.3.3}$$

The rate of transition between states is

$$R_\alpha = \mathring{P}_\alpha = \frac{\partial P_\alpha}{\partial s} \, . \tag{18.3.4}$$

To simplify terminology, transition probability density and transition rate density are referred to as transition probability and transition rate, respectively. It should be understood that density is implied when a calculation uses unperturbed functions having a continuous eigenvalue spectrum. Given the above definitions and mathematical framework, all we need now is a physically plausible parameter dependent perturbation. We postulate one in the next chapter.

CHAPTER 19
MASS STATE TRANSITIONS AT SCATTERING VERTICES

Incoming particles and outgoing particles in the scattering formalism of Chapter 15 are distinguished relative to the interaction region by the invariant evolution parameter s. The interaction region corresponds to the scattering vertex of Feynman diagrams. In many Feynman diagrams the identifying characteristics of incoming particles -- such as mass, charge, spin -- are not the same as the characteristics of outgoing particles. The state of the system before interaction with the scattering vertex, as determined by s, is the incoming particle state. The state of the system after interaction is the outgoing particle state. Perturbation of an incoming free particle state at a scattering vertex is treated below as a parameter dependent interaction that can induce a mass transition. To simplify the discussion, the perturbation application presented below is restricted to spin-0 particles.

§19.1 STATEMENT OF THE INTERACTION PROBLEM

As an application of the formalism presented in Chapter 18, we consider a system of scalar particles in which a projectile interacts with a target to form a product [Fanchi, 1990]. The system is sketched in Figure 19.1. This three-particle system is the simplest physically realizable system. Its Lagrangian density must include a description of the noninteracting target and projectile particles, as well as a term describing their interaction. The restriction to scalar particles does not preclude us from considering realistic experiments because spinless mesons are described by scalar fields. Given this physical picture, we can hypothesize a physically reasonable classical Lagrangian density and derive its corresponding classical field equations. Quantum field equations are then constructed from the classical field equations by requiring that the wave function satisfy the PRQT probability conservation constraint.

FIGURE 19.1

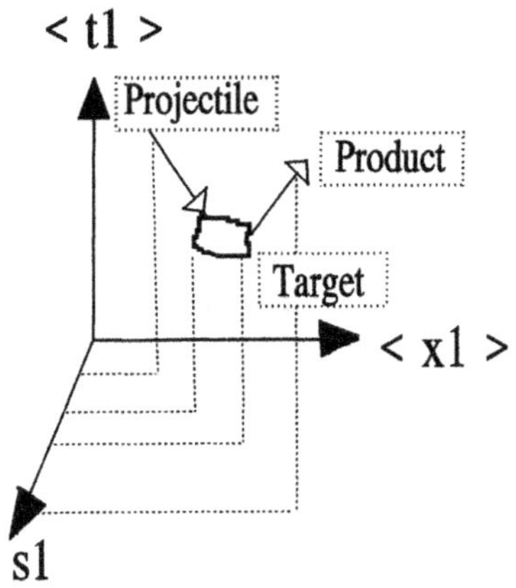

CLASSICAL LAGRANGIAN FORMULATION

The total Lagrangian density of an interacting system is the sum of free particle terms and an interaction term:

$$\mathcal{L} = \mathcal{L}_{0\Psi} + \mathcal{L}_{0\Phi} + \mathcal{L}_I . \tag{19.1.1}$$

A free particle Lagrangian density for a scalar field $\zeta = \Psi, \Phi$ and mass parameter m_ζ is

$$\mathcal{L}_{0\zeta} = \frac{1}{2i}[\zeta^*(\partial_s \zeta) - (\partial_s \zeta^*)\zeta] + \frac{1}{2m_\zeta}\frac{\partial \zeta^*}{\partial x_\mu}\frac{\partial \zeta}{\partial x^\mu} . \tag{19.1.2}$$

Specification of an interaction term is the point of greatest uncertainty. For simplicity, we study a perturbing interaction of the form

$$\mathcal{L}_I = g(\Phi_s \Psi^* \Psi + \Psi_s \Phi^* \Phi) \tag{19.1.3}$$

where g is a real coupling constant and

$$\begin{aligned} \Phi_s &\equiv \Phi + \Phi^* , \\ \Psi_s &\equiv \Psi + \Psi^* . \end{aligned} \tag{19.1.4}$$

This interaction term satisfies several constraints: it is Hermitian; it represents a three-particle interaction; it is symmetric with respect to the interchange of Φ and Ψ fields; and it does not contain derivatives.

CLASSICAL FIELD EQUATIONS

Given a Lagrangian density, it is straightforward to derive the corresponding field equations from the parametrized Euler-Lagrange equation derived in Chapter 5:

$$\frac{\partial \mathcal{L}}{\partial \zeta} - \frac{\partial}{\partial x^{\mu}}\left[\frac{\partial \mathcal{L}}{\partial(\partial \mathcal{L}/\partial x^{\mu})}\right] - \frac{d}{ds}\left[\frac{\partial \mathcal{L}}{\partial \mathring{\zeta}}\right] = 0. \tag{19.1.5}$$

Substituting the Lagrangian density into Eq. (19.1.5) gives field equations for the particles and their complex conjugate antiparticles. Varying Ψ^*, for example, gives the field equation

$$i\frac{\partial \Psi}{\partial s} = -\frac{1}{2m_{\Psi}}\frac{\partial^2 \Psi}{\partial x^{\mu}\partial x_{\mu}} + g(\Phi_s \Psi + \Phi^* \Phi). \tag{19.1.6}$$

Varying Ψ gives a field equation that is the complex conjugate of Eq. (19.1.6). Similar results apply to variations of Φ and Φ^*, as shown in Fanchi [1990]. The field equations exhibit an interchange symmetry as expected.

QUANTUM FIELD EQUATIONS

The classical field Ψ derived from the Euler-Lagrange equations cannot be directly used as a quantum amplitude unless it satisfies the probability conservation constraint (normalization condition). The appropriate constraint was found in §18.2 to be Eq. (18.2.14) for a field equation of the form of Eq. (18.2.1).

A field equation satisfying Eq. (18.2.14) and retaining most of the features of Eq. (19.1.6) is

$$i\frac{\partial \Psi}{\partial s} = -\frac{1}{2m_{\Psi}}\frac{\partial^2 \Psi}{\partial x^{\mu}\partial x_{\mu}} + g\Phi_s\Psi . \tag{19.1.7}$$

The free particle operator and the interaction term are

$$K_0 = -\frac{1}{2m_{\Psi}}\frac{\partial^2}{\partial x^{\mu}\partial x_{\mu}} ,$$
$$K_I = g\Phi_s\Psi . \tag{19.1.8}$$

Unperturbed basis functions for free, scalar particles are given by eigenfunctions of the form

$$\psi_{\eta}(x,s) = (2\pi)^{-2} \exp[-iq_{\eta}s + ik_{\eta}\cdot x] \tag{19.1.9}$$

where

$$q_{\eta} = \frac{k_{\eta}\cdot k_{\eta}}{2m_{\eta}} ,$$
$$k_{\eta}\cdot x = (k_{\eta})_0 x^0 - \vec{k}_{\eta}\cdot\vec{x} \tag{19.1.10}$$

and m_{η} is particle mass. Similar expressions hold for the basis functions of the target particle.

We hypothesize that Eq. (19.1.7) is the appropriate quantum field equation for describing our three-body system. Analogous equations apply to the other fields Ψ^*, Φ, Φ^*. By solving the field equation for one of the particles -- Eq. (19.1.7) in our case -- we obtain a set of kinematic constraints that applies to each of the particles in this system. The complex interdependence justifies our selection of perturbation theory as the most effective means of solving the problem. A second-order correct calculation procedure for our three-body system is now presented.

§19.2 ITERATIVE SOLUTION PROCEDURE

The problem of interest here is the study of transitions from an initial free particle state (actually a limited range of initial states) to other free particle states after interaction of the initial state with a perturbing potential. A procedure analogous to the standard nonrelativistic procedure is followed with one important exception. Standard discussions of the nonrelativistic perturbation problem usually specify the initial condition as a Dirac delta function for free particle scattering problems where basis states have a continuous eigenvalue spectrum. This simplifies the calculation, but the physical interpretation suffers because of the irregular mathematical properties of the Dirac delta function. Initial probability conservation, for example, contains an integral over the product of a Dirac delta function with itself. Other quantities of physical interest, such as transition probabilities, contain Dirac delta functions outside of integrals. Rather than retain these inconsistencies, a more rigorous procedure is presented below.

SECOND-ORDER CORRECT CALCULATIONS

An integral equation for the transition probability amplitude is

$$a_\alpha = a_\alpha^0 - ig\left[\int_0^s \left(\int a_\eta V_{\alpha\eta}\, d\eta\right) ds'\right] \tag{19.2.1}$$

where a_α^0 is the value of a_α at s=0. The term $V_{\alpha\eta}$ is defined in terms of free particle basis functions Eqs. (19.1.9) and (19.1.10) as

$$\begin{aligned} V_{\alpha\eta} &= \int b_\mu\left[\int \psi_\alpha^* \phi_\mu \psi_\eta\, d^4x\right] d\mu \\ &+ \int b_\mu^*\left[\int \psi_\alpha^* \phi_\mu^* \psi_\eta\, d^4x\right] d\mu . \end{aligned} \tag{19.2.2}$$

It is reasonable to assume that the probability amplitudes b_μ of the target particle are independent of a_α and the approximation $b_\mu \approx b_\mu^0$ holds for the duration of the perturbation. These assumptions make the problem tractable.

A rigorous solution of the problem would have to simultaneously solve all of the field equations for Ψ, Φ and their conjugates. This would significantly complicate the problem and an analytic perturbation theory approach would be questionable. On the other hand, the above assumptions let us examine the slight perturbation case. This case is a limiting special case of a more rigorous and considerably more complicated solution. Given these assumptions, the iterative expansion of Eq. (19.2.1) to second order in the coupling constant g is

$$\begin{aligned} a_\alpha = a_\alpha^0 &- ig\left\{\int_0^s \left[\int a_\eta^0 V_{\alpha\eta}\, d\eta\right] ds'\right\} \\ &- g^2 \left\{\int_0^s \left[\int \left(\int_0^{s'}\int a_{\eta'}^0 V_{\eta\eta'}\, d\eta'\, ds''\right) V_{\alpha\eta}\, d\eta\right] ds'\right\}. \end{aligned} \tag{19.2.3}$$

Calculation of the probability amplitude requires forming the product

$$P_\alpha = a_\alpha^* a_\alpha . \tag{19.2.4}$$

This could get quite complicated if all of the terms are needed. Most of the terms drop out, however, if we recognize that P_α should only be calculated up to terms that are second order in g so that P_α is consistent with a_α. To determine the appropriate terms, we write the amplitudes in the functional form

$$\begin{aligned} a_\alpha &= a_\alpha^0 + G_1 g + G_2 g^2 \\ a_\alpha^* &= a_\alpha^{0*} + G_1^* g + G_2^* g^2 \end{aligned} \tag{19.2.5}$$

where G_1 and G_2 are defined in Eq. (19.2.3). Keeping only terms to second order in g gives

$$\begin{aligned} P_\alpha &= a_\alpha^0 (a_\alpha^{0*} + G_1^* g + G_2^* g^2) \\ &\quad + a_\alpha^{0*} (G_1 g + G_2 g^2) + G_1 G_1^* g^2 . \end{aligned} \tag{19.2.6}$$

For our case, we assume a slight perturbation so that the initial state does not change significantly over the duration of the perturbation. We are also interested only in calculations of transitions away from the initial state. Consequently we can take $a_\alpha^0 \to 0$ for all values of α not equal to the initial value λ. The probability of a transition into the state α simplifies to the form

$$P_\alpha = a_\alpha^* a_\alpha = G_1 G_1^* g^2 \tag{19.2.7}$$

where

$$a_\alpha \approx G_1 g = -ig \left\{ \int_0^s \left[\int a_\eta^0 V_{\alpha\eta} d\eta \right] ds' \right\}. \tag{19.2.8}$$

Our problem is now one of evaluating the integrals in Eq. (19.2.8).

EVALUATING INTEGRALS

Unlike the discrete eigenvalue case in which the initial state λ could be specified with a Kronecker delta function, the initial state in this case has an eigenvalue in a range

$$\lambda - \delta/2 < \alpha < \lambda + \delta/2. \tag{19.2.9}$$

We assume the increment δ is much smaller than the total available eigenvalue range, thus

$$\delta \ll \alpha_{max} - \alpha_{min} \tag{19.2.10}$$

where the integration extends from a minimum eigenvalue α_{min} to a maximum eigenvalue α_{max}. Notice that the eigenvalue range is a mathematical convenience that does not necessarily bound the free particle spectrum. No constraint on the eigenvalue difference has been specified. The ensuing initial condition is

$$\begin{aligned} a_\alpha^0 &= N, \quad |\alpha - \lambda| \le \delta/2 \\ a_\alpha^0 &= 0, \quad |\alpha - \lambda| > \delta/2. \end{aligned} \tag{19.2.11}$$

The constant N is fixed by probability conservation:

$$\int a_\xi^* a_\xi d\xi = \int_{\xi_{min}}^{\lambda-\delta/2} a_\xi^* a_\xi d\xi + \int_{\lambda-\delta/2}^{\lambda+\delta/2} a_\xi^* a_\xi d\xi + \int_{\lambda+\delta/2}^{\xi_{max}} a_\xi^* a_\xi d\xi = \int_{\lambda-\delta/2}^{\lambda+\delta/2} a_\xi^* a_\xi d\xi = 1. \tag{19.2.12}$$

Solving Eq. (19.2.12) gives

$$N = \delta^{-1/2}. \tag{19.2.13}$$

Combining Eq. (19.2.13) with (19.2.11) yields the initial condition in the form

$$\begin{aligned} a_\alpha^0 &= \delta^{-1/2}, \ |\alpha - \lambda| \le \delta/2 \\ a_\alpha^0 &= 0, \ |\alpha - \lambda| > \delta/2. \end{aligned} \tag{19.2.14}$$

Equation (19.2.14) represents a Dirac delta function in the limit as δ goes to zero. A more rigorous and easily interpretable perturbation analysis is obtained if the limit is not taken, at least not until the very end.

Finding an explicit expression for a_α requires evaluating integrals of the form

$$I_\alpha = \int a_\eta^0 V_{\alpha\eta} d\eta. \tag{19.2.15}$$

Substituting Eq. (19.2.14) into (19.2.15) gives

$$I_\alpha = \int_{\eta_{min}}^{\lambda-\delta/2} 0 \cdot V_{\alpha\eta} d\eta + \int_{\lambda-\delta/2}^{\lambda+\delta/2} \delta^{-1/2} V_{\alpha\eta} d\eta + \int_{\lambda+\delta/2}^{\eta_{max}} 0 \cdot V_{\alpha\eta} d\eta = \delta^{-1/2} \int_{\lambda-\delta/2}^{\lambda+\delta/2} V_{\alpha\eta} d\eta . \tag{19.2.16}$$

The eigenvalue integral over the matrix element demands special attention. For $\delta \lll \eta_{max} - \eta_{min}$, we assume $V_{\alpha\eta} \approx V_{\alpha\lambda}$ and $V_{\alpha\eta}$ is independent of η in the range of integration. This is a good approximation for any function that is piecewise continuous in the neighborhood $|\eta - \lambda| \leq \delta / 2$. It is exact in the limit as δ goes to zero. Given these conditions, I_α becomes

$$I_\alpha = \delta^{-1/2} V_{\alpha\lambda} \int_{\lambda-\delta/2}^{\lambda+\delta/2} d\eta = V_{\alpha\lambda} \delta^{1/2}. \tag{19.2.17}$$

Substituting Eq. (19.2.17) into (19.2.8) yields the transition probability amplitude

$$a_\alpha = -ig \int_0^s \delta_a^{1/2} \delta_b^{1/2} \left[\int \psi_\alpha^* \phi_{\lambda_b} \psi_{\lambda_a} d^4x \right] ds' + -ig \int_0^s \delta_a^{1/2} \delta_b^{1/2} \left[\int \psi_\alpha^* \phi_{\lambda_b}^* \psi_{\lambda_a} d^4x \right] ds'. \tag{19.2.18}$$

The space-time integrals are evaluated to give

$$\int \psi_\alpha^* \phi_\mu \psi_\eta d^4x = \int_{-\epsilon}^{\epsilon} e^{-i(q_\eta + Q_\mu - q_\alpha)s + i(k_\eta + K_\mu - k_\alpha)\cdot x} \frac{d^4x}{(2\pi)^6} \equiv Re(\eta + \mu - \alpha) \exp[-i(\eta + \mu - \alpha)s] \tag{19.2.19}$$

where we are using the notation

$$\exp[-i(\eta + \mu - \alpha)s] \equiv \exp[-i(q_\eta + Q_\mu - q_\alpha)s] \tag{19.2.20}$$

and

$$Re(\eta + \mu - \alpha) \equiv (2\pi)^{-6} \prod_{\sigma=0}^{3} \frac{2\sin(k_\eta + K_\mu - k_\alpha)_\sigma \epsilon_\sigma}{(k_\eta + K_\mu - k_\alpha)_\sigma} . \tag{19.2.21}$$

The resulting transition probability amplitude is

$$\begin{aligned} a_\alpha = & -ig\,\delta_a^{1/2}\delta_b^{1/2} \int_0^s e^{-i(\lambda_a + \lambda_b - \alpha)s'} Re(\lambda_a + \lambda_b - \alpha)\, ds' \\ & -ig\,\delta_a^{1/2}\delta_b^{1/2} \int_0^s e^{-i(\lambda_a - \lambda_b - \alpha)s'} Re(\lambda_a - \lambda_b - \alpha)\, ds' . \end{aligned} \tag{19.2.22}$$

Evaluating the integral over s yields

$$\begin{aligned} a_\alpha = & -ig\,\delta_a^{1/2}\delta_b^{1/2} \left[\frac{e^{-i(\lambda_a + \lambda_b - \alpha)s} - 1}{-i(\lambda_a + \lambda_b - \alpha)} \right] Re(\lambda_a + \lambda_b - \alpha) \\ & -ig\,\delta_a^{1/2}\delta_b^{1/2} \left[\frac{e^{-i(\lambda_a - \lambda_b - \alpha)s} - 1}{-i(\lambda_a - \lambda_b - \alpha)} \right] Re(\lambda_a - \lambda_b - \alpha) . \end{aligned} \tag{19.2.23}$$

These results are used in Chapter 20 to develop a set of kinematic constraints governing mass state transitions for the three-particle system being studied.

§19.3 TRANSITION PROBABILITIES

We are interested in a system of scalar particles in which a projectile interacts with a target to form a product. Much of the mathematical preparation for the solution of this problem was presented in §19.2. Adopting the notation introduced there, we write the most probable initial states of the projectile and target particles as λ_a and λ_b, respectively. The increments δ_a and δ_b denote the range of eigenvalues associated with the most probable initial states.

The transition probability P_α from the initial projectile state to a state α is given by the product of the transition amplitude a_α with its complex conjugate. The transition amplitude for this problem is given by Eq. (16.4.23). Forming the product $P_\alpha = a_\alpha^* a_\alpha$ and simplifying gives

$$\begin{aligned}\frac{P_\alpha}{g^2\delta_a\delta_b} &= Re(a+b-\alpha)\frac{\left[\sin\frac{(q_{\lambda_a}+Q_{\lambda_b}-q_\alpha)}{2}s\right]^2}{\left[\frac{q_{\lambda_a}+Q_{\lambda_b}-q_\alpha}{2}\right]^2} \\ &+ Re(a-b-\alpha)\frac{\left[\sin\frac{(q_{\lambda_a}-Q_{\lambda_b}-q_\alpha)}{2}s\right]^2}{\left[\frac{q_{\lambda_a}-Q_{\lambda_b}-q_\alpha}{2}\right]^2} \\ &+ \frac{2Re(a+b-\alpha)Re(a-b-\alpha)}{(q_{\lambda_a}-q_\alpha)^2-Q_{\lambda_b}^2} \\ &\cdot\{1+\cos[(q_{\lambda_a}+Q_{\lambda_b})s]-\cos(q_\alpha s) \\ &-\cos[(q_{\lambda_a}-Q_{\lambda_b}-q_\alpha)s]\}\end{aligned} \tag{19.3.1}$$

where $Re(\xi)$ is defined by

$$Re(\xi) \equiv (2\pi)^{-6}\prod_{\sigma=0}^{3}\frac{2\sin(k_\xi)_\sigma\,\epsilon^\sigma}{(k_\xi)_\sigma}. \tag{19.3.2}$$

As a check on the validity of Eq. (19.3.1), we look at the small s behavior. In this case P_α becomes

$$(P_\alpha)_{s\to 0} = \frac{g^2s^2}{\hbar^2}\delta_a\delta_b[Re(a+b-\alpha)^2+Re(a-b-\alpha)^2] \tag{19.3.3}$$

which exhibits a dependence on the square of s. This is analogous to the nonrelativistic case. A "golden rule" in which the transition probability is proportional to s could be derived by summing over eigenvalues as in the nonrelativistic case [Landé, pp. 171-172, 1951].

Before discussing the kinematics of the system, we should note that Eq. (19.3.1) satisfies the principle of detailed balance. According to this principle, the transition probability from state λ_a to state α, and the inverse transition from state α to state λ_a, must be equal. This can be shown by repeating the preceding analysis with the state α (rather than λ_a) as the dominant initial state for the inverse transition.

Returning to our study of the kinematics of the system, we observe that the most probable product particle mass states of the interaction between a projectile and a target particle are the extrema of the transition probability P_α. When $s \to 0$, Eq. (19.3.3) shows that P_α does not exhibit well-defined extrema. If we let Q represent the nonzero mass state part of the argument of the trigonometric functions in Eq. (19.3.1), the parameter s must satisfy a criterion of the form

$$Qs \ll 1 . \tag{19.3.4}$$

For an electron at rest -- one of the smallest, nonzero, physically realistic values possible -- Q is on the order of 10^{19}/sec ($Q \sim m_e c^2/2\hbar$). Using this value in Eq. (19.3.4), we can show that the perturbation duration s for an interaction that does not exhibit sharp peaks must be on the order of 10^{-19} sec or less. By contrast, for larger values of s, P_α has two sharply defined maxima which define the system kinematics, which are summarized in Table 19.1. Subscripts $\{\lambda_a, \lambda_b\}$ are replaced by {a,b} to simplify the table. Lower case k,q refer to Ψ particles, and upper case K,Q refer to Φ particles.

The kinematic constraints listed in Table 19.1 include a mass constraint in addition to four-momentum conservation for sufficiently long perturbation intervals. This is not surprising since a new conservation law can be expected as a consequence of expanding the symmetry group in Part II. The rest of this chapter looks at the feasibility of using the enlarged set of conservation laws to make experimental predictions which could test the validity of PRQT.

TABLE 19.1 SUMMARY OF KINEMATIC CONSTRAINTS

VARIED FIELD	FOUR-MOMENTUM	MASS
Ψ or Ψ^*	$k_\alpha = k_a + K_b$ $k_\alpha = k_a - K_b$	$q_\alpha = q_a + Q_b$ $q_\alpha = q_a - Q_b$
Φ or Φ^*	$K_\alpha = K_b + k_a$ $K_\alpha = K_b - k_a$	$Q_\alpha = Q_b + q_a$ $Q_\alpha = Q_b - q_a$

§19.4 KINEMATICS OF MASS STATE TRANSITIONS

We have implicitly assumed that q_α, $q_{\lambda a}$, and $Q_{\lambda b}$ are nonzero in our analysis. Each of these quantities represents the mass of a particle, where mass has the meaning presented in §9.2. The mass constraint is

$$q_\alpha = q_{\lambda_a} \pm Q_{\lambda_b} \tag{19.4.1}$$

where

$$q_\zeta = \frac{\hbar}{2m_\zeta}(k \cdot k)_\zeta . \tag{19.4.2}$$

Bradyon states have a positive $(k \cdot k)_\zeta$, while $(k \cdot k)_\zeta$ is negative for tachyon states. If expectation values are sharply peaked, as they would be for observable free particles, we can use the following form of Eq. (9.2.12):

$$m_\zeta^2 = \frac{\hbar^2}{c^2}|(k \cdot k)_\zeta| . \tag{19.4.3}$$

Both positive and negative values of m_ζ are allowed by Eq. (19.4.3). In the case of bradyons, negative "rest masses" do not have any experimental meaning. To preserve agreement between theory and observations, we assume that only positive values of m_ζ are acceptable. At this point there is no compelling theoretical reason for doing so. Given these assumptions and substituting Eq. (19.4.3) into (19.4.2) gives

$$q_\zeta = \frac{c}{2}\,|(k\cdot k)_\zeta|^{1/2}\,sgn(k\cdot k)_\zeta \tag{19.4.4}$$

where sgn(x) denotes the sign of x.

Kinematic calculations require the specification of all pertinent constraints (see Table 19.1). Combining the constraints yields

$$\begin{aligned} |(k\cdot k)_\alpha|^{1/2}\,sgn(k\cdot k)_\alpha &= |(k\cdot k)_{\lambda_a}|^{1/2}\,sgn(k\cdot k)_{\lambda_a} \\ &\quad \pm |(K\cdot K)_{\lambda_b}|^{1/2}\,sgn(K\cdot K)_{\lambda_b} \end{aligned} \tag{19.4.5}$$

subject to four-momentum conservation. Equation (19.4.5) may be used to determine what transitions are possible without regard to four-momentum conservation. The possibilities are summarized in Table 19.2. Subscripts $\{\lambda_a, \lambda_b\}$ are again replaced by {a,b} to simplify the table. Upper case letters B and T refer to bradyon and tachyon respectively. The first two columns represent reacting particles, and the last two columns denote product particles. Among the possibilities listed in Table 19.2 is faster-than-light (FTL) travel.

TABLE 19.2 MASS STATE TRANSITIONS

TARGET Q_b	PROJECTILE q_a	$q_\alpha = q_a + Q_b$	$q_\alpha = q_a - Q_b$
B	B	B	B T
B	T	B T	T
T	B	B T	B
T	T	T	B T

Many physicists have given serious consideration to FTL travel ever since Bilaniuk et al. [1962] showed that the concept was not necessarily precluded by special relativity. Bilaniuk et al. introduced the term "luxon" to denote a particle traveling at the speed of light. To distinguish between FTL particles and everything else, Feinberg [1967] called FTL particles "tachyons," which he coined from the Greek word for swift. Bilaniuk and Sudarshan [1969] used the term "tardyons" to denote slower-than-light particles. "Tardyon" has since been replaced by "bradyon," which was coined from the Greek word for slow [Recami, 1987]. We use the terms bradyon, luxon, and tachyon to denote subluminal, luminal, and superluminal particles, respectively.

We have implicitly assumed that the masses of all interacting projectile and target states are nonzero. A separate treatment of luxon (mass=0) states would require a careful perturbation analysis of massless states. Rather than unnecessarily complicating the present analysis, we continue to neglect luxon states. Despite this limitation, the theory presented thus far does make significant physical predictions.

According to Table 19.2, tachyons may be created when two particles interact, even if the two interacting particles are bradyons. Furthermore, an existing tachyon may be annihilated in interactions with bradyons or tachyons. Are these observations still valid when four-momentum conservation is added to the kinematics of mass state transitions? Specific examples will show that the answer is yes.

Equation (19.4.5) is the one feature of PRQT which makes the PRQT formulation different from other tachyon kinematic systems [Fanchi, 1990]. It serves to constrain the types of physically allowable processes. The additional constraint arises from an additional conservation law that is fundamental to PRQT. A legitimate question to ask is whether the new mass constraint violates existing experimental data.

Several examples are presented in Chapter 20 which show that the addition of a new constraint does not contradict corresponding experimental results. There are mass relations in the literature which are used to compute the mass of a product particle as the sum of the masses of its constituents. These relations are usually accompanied by corrections for binding energy and other interactions.

Examples include the quark model (§12.5) and Weizacker's semiempirical mass formula [e.g. Williams, 1991] . We must keep in mind that the system we are considering is a relatively simple system. At this level of complexity, the new mass constraint is not in violation of known experimental data.

Perhaps the most significant physical point of this presentation is that tachyon creation within the context of PRQT does not require smooth acceleration of a classical timelike entity through the light cone. Instead, tachyons are created or annihilated by quantum transitions. This is an important conceptual advantage of PRQT. The true test of the theory is in the realm of experiment, to which we now turn.

CHAPTER 20
TACHYON KINEMATICS

We saw in Chapter 19 that mass state transitions are possible within the context of PRQT. The analysis led to a kinematic description that includes the possibility of tachyon creation and annihilation. In this chapter we apply the kinematics introduced in Chapter 19 to some specific experiments [Fanchi, 1990].

Two types of experiments have been frequently performed in high-energy physics: scattering of a particle beam by a thin foil; and counter-circulating beam experiments. If tachyons exist, why have they not been observed in these experiments? The answer to this question is provided below by looking at particular values of the projectile and target particle four-momenta and mass.

§20.1 ANALYSIS OF A PION SYSTEM

PRQT kinematics is applied here to pion interactions. This system is selected because it consists of spinless particles with nonzero masses. Each of these features complies with the physical picture underlying our mathematical development.

PION INTERACTIONS

We consider three physically distinct interactions:

$$\textit{Case BBB}\text{: } \pi^0 + \pi^\pm \rightarrow B^\pm \tag{20.1.1}$$

$$Case\ BBT\colon\ \pi^0 + \pi^{\pm} \rightarrow T^{\pm} \tag{20.1.2}$$

$$Case\ TBB\colon\ T^{\pm} + \pi^0 \rightarrow B^{\pm} \tag{20.1.3}$$

The case acronym is designed to symbolically describe the system being studied. The first, second, and third letters denote the projectile, target, and product particles, respectively. For example, the first letter in Case BBT indicates a bradyon is the projectile, the second letter means a bradyon is the target, and the third letter shows a tachyon was produced. Cases BBB and BBT represent the production of a bradyon and a tachyon, respectively, from a bradyon-bradyon interaction. The annihilation of a tachyon by interaction with a bradyon is represented by Case TBB to avoid introducing an electromagnetic interaction with the product tachyon from Case BBT. Finding a consistent set of particle properties -- four-momenta and masses -- to satisfy Eqs. (20.1.1) through (20.1.3) will demonstrate the theoretical viability of the mechanism described in §19.1.

CASE BBB: BRADYON PRODUCTION

This is the simplest of the three cases to analyze. Referring to Table 19.2, we see that either $q_\alpha = q_{\lambda a} + Q_{\lambda b}$ or $q_\alpha = q_{\lambda a} - Q_{\lambda b}$ could be used. For simplicity, we take the $q_\alpha = q_{\lambda a} + Q_{\lambda b}$ case. Equation (19.4.5) becomes

$$(k{\cdot}k)_\alpha^{1/2} = (k{\cdot}k)_{\lambda_a}^{1/2} + (K{\cdot}K)_{\lambda_b}^{1/2}. \tag{20.1.4}$$

Replacing k_α by four-momentum conservation, squaring both sides of the resulting equation, and simplifying gives

$$k_{\lambda_a}{\cdot}K_{\lambda_b} = (k{\cdot}k)_{\lambda_a}^{1/2}(K{\cdot}K)_{\lambda_b}^{1/2}. \tag{20.1.5}$$

Several theoretically acceptable solutions of Eq. (20.1.5) are possible. Taking, for example, the case $k_\alpha = k_{\lambda a} + K_{\lambda b}$, we obtain a solution of Eq. (20.1.5); namely

$$k_{\lambda_a} = \kappa K_{\lambda_b} \tag{20.1.6}$$

where κ is a positive proportionality constant. Substituting Eq. (20.1.6) into the kinematic constraints yields

$$k_\alpha = (\kappa + 1) K_{\lambda_b} \tag{20.1.7}$$

and

$$q_\alpha = \frac{c}{2}(\kappa + 1)(K \cdot K)^{1/2}_{\lambda_b} > 0. \tag{20.1.8}$$

Thus we obtain a consistent set of particle properties when the four-momentum of the target bradyon is proportional to the four-momentum of the projectile bradyon.

CASE BBT: TACHYON CREATION

Finding a consistent set of particle properties for tachyon production from a bradyon-bradyon interaction is much more involved than in Case BBB. For Case BBT, Eq. (19.4.5) becomes

$$|(k \cdot k)_\alpha|^{1/2} = |(K \cdot K)_{\lambda_b}|^{1/2} - |(k \cdot k)_{\lambda_a}|^{1/2} \tag{20.1.9}$$

where $q_\alpha = q_{\lambda a} - Q_{\lambda b}$.

Solutions of Eq. (20.1.9) in conjunction with four-momentum conservation may be found by considering the simpler problem in 1+1 dimensions. The mass constraints of the interacting bradyons and resulting tachyon may be written in 1+1 dimensions as

$$a_0^2 - a_1^2 = m_a^2 > 0, \; m_a > 0, \tag{20.1.10}$$

$$b_0^2 - b_1^2 = m_b^2 > 0, \; m_b > 0, \tag{20.1.11}$$

$$\alpha_1^2 - \alpha_0^2 = m_\alpha^2 > 0, \; m_\alpha > 0, \tag{20.1.12}$$

and

$$m_{\alpha} = m_{b} - m_{a} > 0 \tag{20.1.13}$$

where $\{a_0,a_1\}$, $\{b_0,b_1\}$, and $\{\alpha_0,\alpha_1\}$ are the nonzero components of the four-momenta for the projectile, target, and product particles, respectively. Units have been chosen so that $\hbar=c=1$. Equation (20.1.13) is the constraint $q_{\alpha} = q_{\lambda a} - Q_{\lambda b}$.

Two nontrivial relations are derived from four-momentum conservation:

$$\alpha_0 = a_0 - b_0 \tag{20.1.14}$$

and

$$\alpha_1 = a_1 - b_1 . \tag{20.1.15}$$

Three more relations stem from Eqs. (20.1.10) through (20.1.12):

$$b_0 = \rho b_1, \ \rho > 1 \tag{20.1.16}$$

$$a_0 = \beta a_1, \ \beta > 1 \tag{20.1.17}$$

$$\alpha_1 = \gamma \alpha_0, \ \gamma > 1 . \tag{20.1.18}$$

The quantities ρ and β are the ratios of energy to momentum of the target and incident bradyons; γ is the ratio of momentum to energy for the product tachyon.

A solution of the kinematic equations is obtained by using the above set of relations. We assume the bradyon masses m_a, and m_b are known, as well as the proportionality constant ρ. Equations (20.1.10) through (20.1.18) yield two equations for the proportionality constants β and γ. The equations are

$$\begin{aligned}&\gamma^2\{2[\rho^2\beta^2 - (\rho^2 + \beta^2) + \rho\beta(1 - \sqrt{\rho^2 - 1}\sqrt{\beta^2 - 1})]\} \\ &+ \gamma\{2[(\rho + \beta)(1 + \sqrt{\rho^2 - 1}\sqrt{\beta^2 - 1} - \rho\beta)]\} \\ &+ 2(\rho^2 + \beta^2 - \rho\beta - 1 - \sqrt{\rho^2 - 1}\sqrt{\beta^2 - 1}) = 0\end{aligned} \tag{20.1.19}$$

and

$$\gamma = \frac{\left[m_b - m_a \frac{\sqrt{\rho^2 - 1}}{\sqrt{\beta^2 - 1}}\right]}{\left[\rho m_b - \beta m_a \frac{\sqrt{\rho^2 - 1}}{\sqrt{\beta^2 - 1}}\right]}. \tag{20.1.20}$$

A simple computer program with interactive interpolation was used to calculate the proportionality constants presented below. Once the set of proportionality constants is known, it is straightforward to calculate all particle four-vectors and masses. The resulting solution set is then confirmed by substitution into the kinematic equations.

As an example, we consider an incident π^0 interacting with a target $\pi^{\pm}$. The rest masses of π^0 and $\pi^{\pm}$ are 134.9642 MeV and 139.5685 MeV respectively [Particle Data Group, 1992]. Arbitrarily assuming $\rho = 2$, we calculate the values of the remaining two proportionality constants to be $\beta = 2.1571$ and $\delta = 1.1275$ where the last digit is uncertain because of numerical round off. The calculated particle properties are given in Table 20.1.

TABLE 20.1 CASE BBT

$a_0 = 152.3208$	$a_1 = 70.6137$	$m_a = 134.9642$
$b_0 = 161.1602$	$b_1 = 80.5801$	$m_b = 139.5688$
$\alpha_0 = 8.8394$	$\alpha_1 = 9.9664$	$m_\alpha = 4.6037$

The values in Table 20.1 are at least 99.9% accurate. Many other solutions are possible depending on the choice of ρ. Table (20.1) shows the viability of tachyon creation within the context of PRQT.

CASE TBB: TACHYON ANNIHILATION

The process of interest here is depicted by Eq. (20.1.3). Beginning with the tachyon produced in Case BBT, we determine the conditions under which it can be annihilated. The calculation is performed in $1+1$ dimensions as in Case BBT. In this case, we know the energy-momentum and mass of the tachyon projectile, and we specify the mass of the target particle. The problem then becomes one of determining the energy-momenta of the target and product particles, and the product particle mass which satisfy the kinematic constraints.

Appropriate mass and four-momentum constraints are

$$\begin{aligned} q_\alpha &= q_{\lambda_a} + Q_{\lambda_b}, \\ k_\alpha &= k_{\lambda_a} + K_{\lambda_b} \end{aligned} \tag{20.1.21}$$

which correspond to the relations

$$m_\alpha = m_b - m_a \tag{20.1.22}$$

and

$$\begin{aligned} \alpha_0 &= a_0 + b_0, \\ \alpha_1 &= a_1 + b_1 \end{aligned} \tag{20.1.23}$$

in $1+1$ dimensions. Additional relationships include

$$\begin{aligned} a_1^2 - a_0^2 &= m_a^2 \\ b_0^2 - b_1^2 &= m_b^2 \end{aligned} \tag{20.1.24}$$

and

$$\alpha_0^2 - \alpha_1^2 = m_\alpha^2 . \tag{20.1.25}$$

These relations may be used to obtain a quadratic equation for the magnitude of b_1:

$$\begin{aligned} &[-4m_a^2]\, b_1^2 - [4a_1(2m_\alpha m_a)]\, b_1 \\ &+ [4a_0^2 m_b^2 - (2m_\alpha m_a)^2] = 0 . \end{aligned} \tag{20.1.26}$$

Once we have the magnitude of b_1, we obtain the magnitude of b_0 from Eq. (20.1.24). A trial-and-error procedure is then used to solve the problem. We assign signs to b_0, b_1 and use Eqs. (20.1.23) to calculate the corresponding values of α_0, α_1. We accept as solutions those energy-momentum components which simultaneously satisfy Eq. (20.1.22) and (20.1.24). Assuming a neutral pion target and using the tachyon properties of the antiparticle tachyon produced in Case BBT, we obtain the solution in Table 20.2.

TABLE 20.2 CASE TBB - SOLUTION 1

a_0= -8.8394	a_1=-9.9664	m_a= 4.6037
b_0=155.9174	b_1=78.0699	m_b=134.9642
α_0=147.0780	α_1=68.1035	m_α=130.3605

Another acceptable solution is given in Table 20.3.

TABLE 20.3 CASE TBB - SOLUTION 2

a_0= -8.8394	a_1= -9.9664	m_a= 4.6037
b_0=-656.5187	b_1=-642.4963	m_b=134.9642
α_0=-665.3582	α_1=-652.4628	m_α=130.3605

Notice that the product particle of Table 20.2 is traveling at a speed of 0.463 c, while the product particle of Table 20.3 is traveling at 0.981 c. The particular product particle obtained depends on the energy-momentum of the target particle. Had we used a charged pion instead of a neutral pion, the kinematics would have yielded a bradyon with the mass of a neutral pion. In this case, however, there would also be an electromagnetic interaction that has been neglected in our relatively simple physical model.

Tables 20.2 and 20.3 show that tachyon annihilation is theoretically possible. The mechanism of bradyon production by tachyon annihilation could be used as the basis for designing tachyon detectors. More discussion about tachyon detectors is given below.

§20.2 THIN-FOIL EXPERIMENTS

Many particle experiments involve the interaction of a projectile beam with a thin foil. Except for thermal vibrations, the target particles in the foil may be viewed as stationary targets relative to an energetic incident beam. According to PRQT, it is kinematically possible to create tachyons in this system. The conditions for tachyon creation are derived here. Calculations for realistic systems are then performed.

PROJECTILE-STATIONARY TARGET KINEMATICS

Of the two mass operator constraints in Table 19.2, only $q_\alpha = q_{\lambda a} - Q_{\lambda b}$ is valid for Case BBT. Proceeding in 1+1 dimensions, we reduce the four-momentum conservation constraint to

$$\begin{aligned} \alpha_1 &= a_1, \\ \alpha_0 &= a_0 - b_0 \end{aligned} \tag{20.2.1}$$

where we have recognized that a stationary target has zero momentum ($b_1=0$). Mass relations for the interacting particles are

$$\begin{aligned} a_0^2 - a_1^2 &= m_a^2, \\ b_0^2 = m_b^2 &\Rightarrow |b_0| = m_b, \\ \alpha_1^2 - \alpha_0^2 &= m_\alpha^2. \end{aligned} \tag{20.2.2}$$

and

$$m_\alpha = m_b - m_a. \tag{20.2.3}$$

We want to know if a physically meaningful set of tachyon values can be calculated from Eqs. (20.2.1) through (20.2.3) given m_a and m_b. To find out, we must establish constraints, if any, on the allowable properties of our incident beam. Solving the above equations for a_0 gives

$$a_0 = \frac{1}{2m_b}[m_a^2 + m_b^2 + (m_b - m_a)^2]\,sgn(b_0). \tag{20.2.4}$$

If m_a is much smaller than m_b, the value of a_0 becomes $a_0 \approx m_b\ sgn(b_0) = b_0$. Equation (20.2.3) does not allow m_a to be greater than m_b as long as we require m_α to be positive. Another point worth noting about Eq. (20.2.4) is the case when $m_a = m_b$. The only allowed value of a_0 is m_a, which implies by Eq. (20.2.2) that $a_1 = 0$. This case is unrealistic because it asserts that the projectile must be stationary relative to the target. Consequently tachyons cannot be created when $m_a = m_b$ for three-body kinematics in 1+1 dimensions. Extensions of the analysis to more particles and more spacial dimensions are needed.

CALCULATIONS

To show that the above equations do predict tachyon production for a realistic system, we again consider the interaction of a $\pi^\pm$ target with a π^0 projectile. The resulting properties are presented in Table 20.4.

TABLE 20.4 $\pi^\pm$-π^0 INTERACTION

$a_0 = \pm 135.1161$	$a_1 = \pm 6.4051$	$m_a = 134.9642$
$b_0 = \pm 139.5685$	$b_1 = 0.0000$	$m_b = 139.5685$
$\alpha_0 = \mp 4.4524$	$\alpha_1 = \pm 6.4051$	$m_\alpha = 4.6045$

Alternatively, the inverse interaction of a stationary π^0 target with a $\pi^\pm$ projectile cannot produce a tachyon because Eq. (20.2.3) would yield a negative value of m_α.

We conclude from the above analysis that a tachyon can be produced in a collision with a stationary target particle if our formulation of PRQT is correct. The four-momentum of a given projectile beam is constrained by Eqs. (20.2.2) and

(20.2.4). Consequently, the product tachyon has very specific kinematic properties. Thus, even though PRQT predicts tachyons can be produced in experiments involving bradyon projectiles with stationary bradyon targets, the bradyon properties must be finely tuned and tachyon detectors must look for product particles with equally well-defined four-momenta and masses.

Some obvious questions arise. Have any experiments been performed which satisfy the above constraints? If so, why were tachyons not observed? If not, what design parameters would lead to tachyon creation? To answer these questions, we consider a broader range of particle interactions than those between pions only. If we consider nonscalar particles by extrapolating our scalar particle model, we can evaluate a larger class of experiments. Results for some examples are shown in Table 20.5. We have avoided the electromagnetic interaction by requiring that at least one of the incident particles be electrically neutral. The interactions are also restricted to the most experimentally accessible particles.

TABLE 20.5 CASE BBT SAMPLE CALCULATIONS

PROCESS	TARGET	PROJECTILE	PRODUCT
$e^{\pm}+\pi^0$	a_0=134.4551	b_0=134.9642	α_0= -0.5091
	a_1=134.4542	b_1= 0.0000	α_1= 134.4542
	m_a= 0.5110	m_b=134.9642	m_α= 134.4532
$\pi^0+p^{\pm}$	a_0=822.7289	b_0=938.2796	α_0=-115.5507
	a_1=811.5834	b_1= 0.0000	α_1= 811.5834
	m_a=134.9642	m_b=938.2796	m_α= 803.3154
$\pi^{\pm}+n^0$	a_0=820.7365	b_0=939.5731	α_0=-118.8363
	a_1=808.7828	b_1= 0.0000	α_1= 808.7828
	m_a=139.5685	m_b=939.5731	m_α= 800.0046
$p^{\pm}+n^0$	a_0=938.2814	b_0=939.5731	α_0= -1.2917
	a_1= 1.8200	b_1= 0.0000	α_1= 1.8200
	m_a=938.2796	m_b=939.5731	m_α= 1.2935

Table 20.5 shows that experimentally attainable energies and readily available bradyons are theoretically suitable for producing tachyons. If our model is physically realistic, it is possible that tachyons may have already been created. The question of detectability becomes paramount. In each example of Table 20.5, a detector must be tuned to either record a Case BBT superluminal antiparticle trajectory or a Case TBB bradyon particle trajectory. An example of a set of properties that might be expected for a Case TBB bradyon particle can be obtained by applying the Case TBB techniques of §20.1 to the neutral pion-proton interaction shown in Table 20.5. Assuming the Case BBT product tachyon is the Case TBB tachyon projectile, and taking the Case TBB target particle to be a proton, we find the two unique solutions shown in Tables 20.6 and 20.7.

TABLE 20.6 π^0-p: SOLUTION 1

a_0=-115.5507	a_1=811.5834	m_a=803.3154
b_0= 938.2796	b_1= 0.0000	m_b=938.2796
α_0= 822.7289	α_1=811.5834	m_α=134.9642

TABLE 20.7 π^0-p: SOLUTION 2

a_0= -115.5507	a_1= 811.5834	m_a=803.3154
b_0= -977.1067	b_1= 272.7066	m_b=938.2796
α_0=-1092.6573	α_1=1084.2900	m_α=134.9642

Table (20.6) is just the reverse relation in which a neutral pion is produced with the same properties as the projectile pion of Case BBT. A totally different particle is produced in Table (20.7), but the target has to be an antiproton.

Case BBT provides a method for predicting the properties of a product tachyon (or antitachyon). Case TBB shows how the produced tachyon can be converted to a detectable bradyon. For example, Table (20.6) represents an antitachyon projectile impinging on a stationary proton target. The problem is that the resulting bradyon has the same properties as the original incident bradyon beam. This could explain why tachyons have not been observed in projectile beam-thin foil scattering experiments. By contrast, Table (20.7) would require use of an antiproton beam as the detector target, but the resulting product antibradyon has significantly different momentum from either the incident projectile beam or target beam. Alternative target particles could and should be considered. The important point of this discussion is that the above techniques provide a means for designing an experiment capable of producing and detecting tachyons.

§20.3 COUNTER-CIRCULATING BEAMS

As our last application, we consider the possibility of creating tachyons by letting two beams of particles with opposite momenta interact. This type of counter-circulating beam experiment includes many particle-antiparticle experiments including positron-electron scattering and nucleon-antinucleon scattering. One way to study this system is to work in the center of mass (CM) frame. After casting the kinematic constraints in a CM format, we show by example that tachyon creation by means of Case BBT is possible using counter-circulating beams of bradyons with different rest masses. The analysis uses the stationary target results of §20.2 and a Lorentz transformation.

CONSTRAINTS IN THE CENTER OF MASS FRAME

Four-momenta in the CM frame are defined by

$$P_\sigma = a_\sigma + b_\sigma , \qquad p_\sigma = \frac{m_b a_\sigma - m_a b_\sigma}{M} \tag{20.3.1}$$

where

$$M \equiv m_a + m_b . \tag{20.3.2}$$

In terms of CM four-momenta, the projectile and target four-momenta are

$$a_\sigma = \frac{m_a}{M} P_\sigma + p_\sigma ,$$
$$b_\sigma = \frac{m_b}{M} P_\sigma - p_\sigma . \tag{20.3.3}$$

The PRQT four-momentum constraints become

$$\alpha_\sigma = a_\sigma + b_\sigma = P_\sigma , \tag{20.3.4}$$

$$\alpha_\sigma = a_\sigma - b_\sigma = \left[\frac{m_a - m_b}{M} \right] P_\sigma + 2 p_\sigma . \tag{20.3.5}$$

Analysis of Counter-Circulating Beams

A counter-circulating beam experiment is characterized by the interaction of two particles having opposite momenta, namely

$$a_1 = b_1 . \tag{20.3.6}$$

Equation (20.3.6) must be used in the two cases represented by Eqs. (20.3.4) and (20.3.5).

Substituting Eq. (20.3.6) into Eq. (20.3.4) gives

$$\alpha_1 = P_1 = 0 . \tag{20.3.7}$$

Equation (20.3.7) says that the spacial momentum component of a product particle must be zero. Noting that a tachyon must satisfy

$$\alpha_1^2 - \alpha_0^2 > 0 , \tag{20.3.8}$$

we see that the product particle cannot be a tachyon when Eq. (20.3.4) applies. We already knew this from Table 19.2, but the above analysis independently verifies our discussion in §19.4 for a special case. The analysis is not as simple when Eq. (20.3.5) is considered.

The energy, momentum, and mass constraints for a tachyon produced by Case BBT are

$$\begin{aligned}\alpha_0 &= a_0 - b_0,\\ \alpha_1 &= a_1 - b_1,\\ \alpha_1^2 - \alpha_0^2 &= m_\alpha^2,\end{aligned} \tag{20.3.9}$$

and

$$m_\alpha = m_b - m_a. \tag{20.3.10}$$

One way to find tachyon values satisfying Eqs. (20.3.7) through (20.3.10) is to Lorentz-transform the results of the projectile-stationary target case from its reference frame K' to the counter-circulating reference frame K. The appropriate reference frame notation is

Frame K	Frame K'
m_a, a_0, a_1	m_a, a_0', a_1'
m_b, b_0, b_1=-a_1	m_b, b_0', b_1'=0
m_α, α_0, α_1	m_α, α_0', α_1'

where the target particle momentum requirements in each frame have been made explicit. The Lorentz transformation which maps K' into K is

$$\begin{aligned}a_0 &= \gamma_L[a_0' + \beta_L a_1'],\\ a_1 &= \gamma_L[a_1' + \beta_L a_0'],\end{aligned} \tag{20.3.11}$$

where

$$\gamma_L = (1-\beta_L^2)^{-1/2} \tag{20.3.12}$$

and β_L is speed/c of the K' frame relative to the K frame. Similar expressions apply to b_0, b_1 and α_0, α_1. To find β_L, we substitute values of a_1 and b_1 from the above table into Eq. (20.3.11) to get

$$a_1 = \gamma_L [a_1' + \beta_L a_0'] = -\gamma_L \beta_L b_0' \tag{20.3.13}$$

or

$$\beta_L = -\frac{a_1'}{(a_0' + b_0')}. \tag{20.3.14}$$

If we know all of the values in the stationary target frame K', we can calculate β_L from Eq. (20.3.14) and then use the Lorentz transformation to calculate the remaining unknowns. Doing so for the neutral pion-proton example of §20.2 gives β_L=-0.4609 and the results in Table 20.8. Numerical round-off error accounts for slight differences between calculated values and theoretically expected values.

TABLE 20.8 π^0-p: SOLUTION 3

a_0= 505.5166	a_1= 487.1712	m_a=134.9642
b_0=1057.1596	b_1=-487.2449	m_b=938.2796
α_0=-551.6430	α_1= 974.3424	m_α=803.1395

Once again we see that tachyon production is possible, this time using counter-circulating beams. It should be noted, however, that the particles comprising the beams must have different masses. If they do not, Eq. (20.3.10) does not yield a positive, nonzero value of m_α. A similar result was obtained in §20.2. The discussion at the end of §20.2 for projectile-thin foil scattering applies to this case as well.

The experimental failure to detect free tachyons has split the physics community. Corben [1978] identified three points of view: those who believe tachyons are different from other particles; those who believe tachyons are like other particles; and those who believe tachyons are a waste of time. Part of the problem is that people do not know what types of properties tachyons should have, especially properties of tachyon-bradyon interactions [for example, see Hanni and Hugentobler, 1978]. There is considerable uncertainty about tachyon properties. Without this knowledge, definitive experiments cannot be designed. It is at this point that PRQT connects with the experimental study of tachyons.

§20.4 COMPARISON WITH EXPERIMENT

We have shown that PRQT kinematics is consistent for a particular physical (pion) system, and that tachyon production is theoretically possible. Our analysis assumes that parametrization of a relativistic particle description is meaningful, and that parameter-dependent perturbation interactions between particles do occur. Furthermore, we have assumed that all theoretically allowed interactions should occur in nature. This has been a longstanding bias of the physics community. An example of the bias is the historical case in which the Klein-Gordon equation was originally rejected because its negative energy solutions were not understood. The Stückelberg-Feynman interpretation justified resurrecting the Klein-Gordon equation.

The constraints imposed on the allowed physical properties of tachyons by our simple model are quite restrictive. This implies that an experimental search must be very specific. Some possible experiments are suggested by the above calculations. The calculation in §20.2, for example, shows that a projectile bradyon interacting with a thin foil of target bradyons should produce tachyons if the projectile energy and momentum are properly set and a nonzero coupling constant for a parameter-dependent potential exists between the projectile and target. In this regard the question of detectability becomes relevant since we have seen that tachyons may be annihilated as well as created. Thus, the observable signature of the produced tachyon may not be tachyonic as in Case BBT, but bradyonic as in Case TBB. Specific predictions will depend on the particular physical experiment being considered. Meson factories such as LAMPF (U.S.A.), TRIUMF (Canada), and SIN (Switzerland) would be appropriate sites for such experiments.

Recent experimental work with pion interactions has shown that neutrinos may be tachyons [Particle Data Group, 1992]. Kostelecky [1993] has reviewed results from several different laboratories. His findings are summarized below.

ELECTRON NEUTRINO MASS FROM TRITIUM BETA DECAY

Group	Mass Squared eV^2/c^4
Tokyo	- 65 ± 85 ± 65 [a]
Los Alamos	- 147 ± 68 ± 41
Zürich	- 24 ± 48 ± 61
Livermore	- 72 ± 41 ± 30
Mainz	- 39 ± 34 ± 15
AVERAGE	- 69 ± 33 [b]

MUON NEUTRINO MASS FROM PION DECAY

Method	Mass Squared MeV^2/c^4
Pion decay in flight	- 0.14 ± 0.20
Rest-frame decay	- 0.163 ± 0.080
Improved pion mass	- 0.097 ± 0.072
AVERAGE	- 0.154 ± 0.045 [b]

[a] Includes statistical and systematic error.

[b] Statistical and systematic error combined.

The mass squared in the above tables is negative because it is calculated as the difference between the square of energy and momentum. Current experiments have found that neutrino mass is spacelike to within experimental error. The experimental status of neutrino mass is presently viewed as not definitive, however, because of experimental uncertainty. Many conventional theorists would like to see a null value for the neutrino mass, or at least a small, positive value. Additional experiments should reduce the uncertainty associated with the neutrino mass.

PART V:

INTRODUCTION TO PARAMETRIZED RELATIVISTIC QUANTUM FIELD THEORY

CHAPTER 21
FOUNDATIONS OF FIELD THEORY

The most widely used formulation of Lagrangian Quantum Field Theory (LQFT) is that by Schwinger [1951]. By generalizing Schwinger's formulation of LQFT [Fanchi, 1979], we obtain a field theoretic framework including a mass operator. Other generalizations have been proposed [e.g. Kyprianidis, 1987; Pavsic, 1991b], but all build on the framework presented below. The resulting formalism can be used to describe particles with arbitrary spin undergoing fundamental interactions. This claim is substantiated by showing that LQFT is a special case of the general formalism. Broad conclusions, particularly with regard to conservation laws, and an application of the mass operator concept are presented in Chapter 22.

§21.1 VARIATIONAL PRINCIPLE

Schwinger's formulation began with the definition of an action operator W_S in terms of a function $\mathcal{L}_S$ of field operators $\phi_\alpha(x)$ and their first derivatives such that

$$W_S = \int_{R_S} \mathcal{L}_S(\phi_\alpha(x), \partial_\mu \phi_\alpha(x))\, dx. \tag{21.1.1}$$

The quantity R_S denotes an infinite four-volume in space-time that is bounded by the spacelike surfaces σ_1, σ_2 in Figure 21.1. Performing the variation δW_S and then postulating that δW_S is equal to the difference between the generators of canonical transformations F_S at σ_1 and σ_2, i.e.

$$\delta W_S = F_S[\sigma_2] - F_S[\sigma_1], \tag{21.1.2}$$

Schwinger developed LQFT. A correspondence between LQFT and classical field theory may be drawn as follows.

FIGURE 21.1

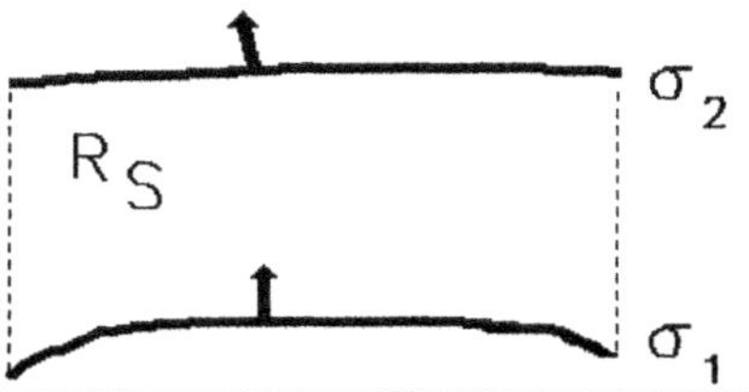

Classically the invariant parameter used to trace the evolution of a system point in configuration space is the proper time τ_{cl}. Furthermore, the Lagrangian L_{cl} used in a covariant formulation of Hamilton's principle must satisfy certain specified transformation properties, e.g. Lorentz invariance. The subsequent classical action integral has the form

$$I_{cl} = \int_{\tau_{cl1}}^{\tau_{cl2}} L_{cl} d\tau, \tag{21.1.3}$$

and Hamilton's principle requires that

$$\delta I_{cl} = 0 \tag{21.1.4}$$

where both I_{cl} and L_{cl} are invariant scalars. The proper times $\tau_{cl\,1}$ and $\tau_{cl\,2}$ are kept fixed. In a relativistic classical field theory, the quantity L_{cl} becomes an integral of a classical Lagrangian density $\mathscr{L}_{cl}$ such that

$$L_{cl} = \int_{R_{cl}} \mathscr{L}_{cl} dx \tag{21.1.5}$$

where R_{cl} is the appropriate four-volume.

The point of interest here is the similarity of Eqs. (21.1.1) and (21.1.5). Schwinger's procedure is classically analogous to varying L_{cl} rather than I_{cl}. This raises the question: What would quantum field theory look like if the action operator was redefined to more closely parallel classical field theory, such as the formalism presented in §2?

An answer to this question is obtained by defining an action operator A analogous to I_{cl} and then evaluating δA. The resulting formalism is consistent with the treatment presented throughout this text, and is aesthetically appealing for two reasons: it closely parallels classical field theory [e.g. Barut, 1980; Bogoliubov and Shirkov, 1980]; and it retains all of LQFT as a special case.

FORMULATION OF THE VARIATIONAL INTEGRAL

Following Schwinger, we work in the Heisenberg picture, however its meaning in the present concept needs to be made explicit. In particular, a state vector is specified as the simultaneous eigenket of a complete set of commuting observables in space-time at a fixed value of an independent scalar s. The scalar s is an invariant which is used to parametrize the evolution of a physical system.

By choosing the Heisenberg picture it is possible to describe the dynamical development of the physical system with respect to s. This is achieved by determining the transformation law that connects the observables at s_1, s_2 separated by the interval

$$\Delta s = s_2 - s_1 > 0. \tag{21.1.6}$$

The quantity Δs can be thought of as the invariant interval during which a physical system is observed. We want to describe the evolution of the physical system in a space-time region R during the interval Δs.

Let ζ_1, ζ_2 denote complete sets of commuting operators at parameter values s_1, s_2 respectively, and denote the corresponding state vectors $\xi(\zeta_1)$, $\xi(\zeta_2)$. The transformation law that connects these state vectors and operators also expresses the dynamical properties -- the evolution -- of the physical system. This law may be written in terms of the transformation matrix which transforms the family of state vectors $\xi(\zeta_1)$ into the family $\xi(\zeta_2)$. The transformation matrix is just the inner product

$$\langle \zeta_1 | \zeta_2 \rangle = \langle \xi(\zeta_1) | \xi(\zeta_2) \rangle. \tag{21.1.7}$$

The transformation matrix $< \zeta_1 | \zeta_2 >$ represents one path of evolution of the system from s_1 to s_2. Alternative evolution paths are obtained by canonically varying $<\zeta_1|\zeta_2>$. Canonical transformations preserve eigenvalues of operators, which implies that the physically interesting results -- the eigenvalues -- are unchanged. A canonical variation is induced by the infinitesimal canonical transformation discussed in §6.3:

$$\delta\psi_\alpha = i[\boldsymbol{F},\psi_\alpha] \tag{21.1.8}$$

where ψ_α is an operator and **F** is the generator of the infinitesimal canonical transformation. The new operators at s_2 can be written as $\zeta_2+\delta\zeta_2$, where

$$\delta\zeta_2 = i[\boldsymbol{F},\zeta_2]. \tag{21.1.9}$$

The eigenvalues of the new operators are the same as the eigenvalues of the old operators since the variation is canonical, but the corresponding state vector is changed by an amount

$$\delta\xi(\zeta_2) = i\boldsymbol{F}(s_2)\,\xi(\zeta_2). \tag{21.1.10}$$

A similar variation of the operators at s_1 changes the eigenvectors at s_1 by a similar amount:

$$\delta\xi(\zeta_1) = i\boldsymbol{F}(s_1)\,\xi(\zeta_1). \tag{21.1.11}$$

The new transformation matrix corresponding to the inner product of the new eigenvectors can be written to first order in the variations as $<\zeta_1|\zeta_2>$ + $\delta<\zeta_1|\zeta_2>$, where

$$\delta\langle\zeta_1|\zeta_2\rangle = i\langle\zeta_1|\delta\boldsymbol{A}|\zeta_2\rangle \tag{21.1.12}$$

and the operator δ**A** has the form

$$\delta\boldsymbol{A} = \boldsymbol{F}(s_2) - \boldsymbol{F}(s_1). \tag{21.1.13}$$

Equations (21.1.12) and (21.1.13) show the effect of the canonical transformation on the transformation matrix and thus on the evolution path. This effect is determined by evaluating δ**A**.

ACTION INTEGRAL

We assume that an invariant scalar function L -- subsequently called the Lagrangian -- can be found such that $\delta\mathbf{A}$ can be obtained by varying the action integral

$$A = \int_{s_1}^{s_2} L ds . \tag{21.1.14}$$

The points s_1, s_2 are fixed and **A** is the action operator. Substituting Eq. (21.1.14) into (21.1.13) yields the variational integral

$$\delta A = \delta \int_{s_1}^{s_2} L ds = F(s_2) - F(s_1) . \tag{21.1.15}$$

In words, the variational integral asserts that the variation of the action operator is equal to the difference between the generators of infinitesimal canonical transformations at s_1, s_2. An explicit evaluation of $\delta\mathbf{A}$ and the generators depends on the functional form of L which, in turn, depends on the symmetry properties of the physical system.

The field theoretic specification of the functional form of L is in terms of fields $\psi_\alpha(x,s)$ where the subscript $\{\alpha=1,2,\ldots,N\}$ represents the N degrees of freedom of the physical system. We may write the Lagrangian L as a four-volume integral of a Lagrangian density $\mathcal{L}$ such that

$$L \equiv \int_R \mathcal{L}(\psi_\alpha, \partial_\mu \psi_\alpha, \partial_s \psi_\alpha) dx , \tag{21.1.16}$$

where R is an arbitrary four-volume that contains the entire physical system and

$$\partial_s \psi_\alpha(x,s) \equiv \frac{\partial \psi_\alpha(x,s)}{\partial s} . \tag{21.1.17}$$

The fields ψ_α must be defined such that $\mathcal{L}$ is integrable. Furthermore, the neglect of higher-order derivatives of ς_α in $\mathring{\alpha}$ is an arbitrary assumption designed to simplify the subsequent mathematical manipulations. Additional restrictions are imposed by the symmetry properties of the physical system.

§21.2 PARAMETRIZED RELATIVISTIC QUANTUM FIELD THEORY

We assumed in Chapter 6 that the symmetry properties of the parametrized theory are those associated with the JARS transformation. An equivalent description of the physical system presently described in the coordinate frame $\{x^\mu\}$ at parameter s is achieved by an infinitesimal transformation to a new frame $\{x'^\mu\}$ at s' given by

$$x'^\mu = x^\mu + \delta x^\mu, \tag{21.2.1}$$

$$s' = s + \delta s, \tag{21.2.2}$$

and by making an infinitesimal transformation of the fields such that

$$\psi'_\alpha(x,s) = \psi_\alpha(x,s) + \delta_0 \psi_\alpha(x,s). \tag{21.2.3}$$

Equation (21.2.3) defines the total variation $\delta_0\psi_\alpha(x,s)$. Observe that the argument of ψ'_α has the same numerical value as the argument of ψ_α, therefore ψ'_α and ψ_α refer to different physical values.

The total variation of the field $\psi_\alpha(x,s)$ due to $\delta\mathbf{A}$ is given by Eq. (6.3.12), namely

$$\delta_0\psi_\alpha(x,s) = i[\delta A, \psi_\alpha(x,s)] \tag{21.2.4}$$

where x must be a point in the region R since the generator $\delta\mathbf{A}$ depends on R. We show below that $\delta\mathbf{A}$ is the generator of the infinitesimal canonical transformation Eqs. (21.2.1) through (21.2.3). With the symmetry properties of the physical system specified, the explicit evaluation of $\delta\mathbf{A}$ can be made.

The form of the initial action operator $\mathbf{A}_i$ is

$$A_i = \int_{s_1}^{s_2}\int_R \mathcal{L}(x,s)\,dx\,ds \tag{21.2.5}$$

where

$$\mathcal{L}(x,s) \equiv \mathcal{L}(\psi_\alpha(x,s), \partial_\mu\psi_\alpha(x,s), \partial_s\psi_\alpha(x,s)) . \tag{21.2.6}$$

The canonical transformation induces a change in $\mathbf{A}_i$. The form of the final, canonically transformed action operator $\mathbf{A}_f$ is

$$A_f = \int_{s_1}^{s_2}\int_{R'} \mathcal{L}'(x',s')\,dx'ds' \tag{21.2.7}$$

where

$$\mathcal{L}'(x',s') \equiv \mathcal{L}(\psi'_\alpha(x',s'), \partial_\mu\psi'_\alpha(x',s'), \psi'_\alpha(x',s')) . \tag{21.2.8}$$

The change in the boundaries of R due to Eq. (21.2.1) is denoted by R'. Notice that ds is unchanged because the s end points are fixed so that ds'=ds. The variation of $\delta\mathbf{A}$ is just the difference

$$\delta A = A_f - A_i \tag{21.2.9}$$

or

$$\delta A = \int_{s_1}^{s_2}\int_{R'} \mathcal{L}'(x',s')\,dx'ds' - \int_{s_1}^{s_2}\int_R \mathcal{L}(x,s)\,dx\,ds . \tag{21.2.10}$$

Equation (21.2.10) is evaluated by first expressing $\mathbf{A}_f$ in the unprimed system.

The four-volume element dx' is related to dx by the Jacobian $\partial(x')/\partial(x)$ of the coordinate transformation, thus

$$dx' = \frac{\partial(x')}{\partial(x)}dx \approx \left[1 + \frac{\partial(\delta x^\mu)}{\partial x^\mu}\right]dx \tag{21.2.11}$$

where only terms up to first order in the variations are kept. The quantity $\mathcal{L}'(x',s')$ is written in the unprimed system by using Eqs. (21.2.1) through

(21.2.3) and Taylor's series to get

$$\begin{aligned}\mathcal{L}'(x',s') &= \mathcal{L}'(x,s) + \frac{\partial \mathcal{L}'(x,s)}{\partial x^{\mu}}\,\delta x^{\mu} + \frac{\partial \mathcal{L}'(x,s)}{\partial s}\,\delta s \\ &= \mathcal{L}(x,s) + \bar{\delta}\mathcal{L}(x,s) \\ &\quad + \left\{\frac{\partial}{\partial x^{\mu}}[\mathcal{L}(x,s) + \bar{\delta}\mathcal{L}(x,s)]\right\}\delta x^{\mu} \\ &\quad + \left\{\frac{\partial}{\partial s}[\mathcal{L}(x,s) + \bar{\delta}\mathcal{L}(x,s)]\right\}\delta s\end{aligned} \tag{21.2.12}$$

where

$$\mathcal{L}'(x,s) \equiv \mathcal{L}(\psi'_{\alpha}(x,s), \partial_{\mu}\psi'_{\alpha}(x,s), \partial_{s}\psi'_{\alpha}(x,s)) \tag{21.2.13}$$

and

$$\begin{aligned}\bar{\delta}\mathcal{L}(x,s) &= \frac{\partial \mathcal{L}(x,s)}{\partial \psi_{\alpha}}\delta_0\psi_{\alpha} + \frac{\partial \mathcal{L}(x,s)}{\partial(\partial_{\mu}\psi_{\alpha})}\delta_0(\partial_{\mu}\psi_{\alpha}) \\ &\quad + \frac{\partial \mathcal{L}(x,s)}{\partial(\partial_{s}\psi_{\alpha})}\delta_0(\partial_{s}\psi_{\alpha}).\end{aligned} \tag{21.2.14}$$

Again dropping terms that are second order in the variations lets us simplify Eq. (21.2.12):

$$\begin{aligned}\mathcal{L}'(x',s') &= \mathcal{L}(x,s) + \bar{\delta}\mathcal{L}(x,s) \\ &\quad + \frac{\partial \mathcal{L}(x,s)}{\partial x^{\mu}}\delta x^{\mu} + \frac{\partial \mathcal{L}(x,s)}{\partial s}\delta s.\end{aligned} \tag{21.2.15}$$

Substituting Eqs. (21.2.11) and (21.2.15) into (21.2.10) yields

$$\delta A = \int_{s_1}^{s_2}\int_R \left[\mathcal{L}(x,s) + \bar{\delta}\mathcal{L}(x,s) + \frac{\partial \mathcal{L}(x,s)}{\partial x^\mu}\delta x^\mu + \frac{\partial \mathcal{L}(x,s)}{\partial s}\delta s \right] \cdot \left[1 + \frac{\partial(\delta x^\mu)}{\partial x^\mu} \right] dx\, ds - \int_{s_1}^{s_2}\int_R \mathcal{L}(x,s)\, dx\, ds. \tag{21.2.16}$$

Expanding and dropping terms that are second order in the variation gives

$$\delta A = \int_{s_1}^{s_2}\int_R \left[\bar{\delta}\mathcal{L}(x,s) + \frac{\partial \mathcal{L}(x,s)}{\partial x^\mu}\delta x^\mu \right] dx\, ds + \int_{s_1}^{s_2}\int_R \left[\frac{\partial \mathcal{L}(x,s)}{\partial s}\delta s + \mathcal{L}(x,s)\frac{\partial(\delta x^\mu)}{\partial x^\mu} \right] dx\ ds. \tag{21.2.17}$$

Integration by parts shows that the integral

$$\int_{s_1}^{s_2}\int_R \frac{\partial \mathcal{L}(x,s)}{\partial s}(\delta s)\, dx\ ds$$

vanishes since the s end points are fixed. Thus Eq. (21.2.17) becomes

$$\delta A = \int_{s_1}^{s_2}\int_R dx\, ds \cdot \left[\bar{\delta}\mathcal{L}(x,s) + \frac{\partial \mathcal{L}(x,s)}{\partial x^\mu}\delta x^\mu + \mathcal{L}(x,s)\frac{\partial(\delta x^\mu)}{\partial x^\mu} \right] \tag{21.2.18}$$

or

$$\delta A = \int_{s_1}^{s_2}\int_R \left[\frac{\partial \mathcal{L}(x,s)}{\partial \psi_\alpha}\delta_0\psi_\alpha + \frac{\partial \mathcal{L}(x,s)}{\partial(\partial_\mu \psi_\alpha)}\partial_\mu(\delta_0\psi_\alpha)\right] dx\, ds$$
$$+ \int_{s_1}^{s_2}\int_R \left[\frac{\partial \mathcal{L}(x,s)}{\partial(\partial_s \psi_\alpha)}\partial_s(\delta_0\psi_\alpha) + \frac{\partial}{\partial x^\mu}[\mathcal{L}(x,s)\,\delta x^\mu]\right] dx\; ds \tag{21.2.19}$$

where the relations

$$\delta_0(\partial_\mu \psi_\alpha) = \partial_\mu(\delta_0\psi_\alpha),$$
$$\delta_0(\partial_s \psi_\alpha) = \partial_s(\delta_0\psi_\alpha) \tag{21.2.20}$$

have been used. We now simplify Eq. (21.2.19).

Integrating the term

$$\int_{s_1}^{s_2}\int_R \frac{\partial \mathcal{L}}{\partial(\partial_s \psi_\alpha)}\delta_0(\partial_s\psi_\alpha)\, dx\, ds$$

by parts, using Eq. (21.2.20), and postulating the constraint that $\delta_0\psi_\alpha$ vanishes at the end points s_1, s_2 gives

$$\int_{s_1}^{s_2}\int_R \frac{\partial \mathcal{L}}{\partial(\partial_s \psi_\alpha)}\delta_0(\partial_s\psi_\alpha)\, dx\, ds = -\int_{s_1}^{s_2}\int_R \left[\frac{d}{ds}\left(\frac{\partial \mathcal{L}}{\partial(\partial_s \psi_\alpha)}\right)\right]\delta_0\psi_\alpha\, dx\, ds. \tag{21.2.21}$$

The rule for differentiating a product is used to rearrange the second term in Eq. (21.2.19) such that

$$\frac{\partial \mathcal{L}}{\partial(\partial_\mu \psi_\alpha)}\partial_\mu(\delta_0\psi_\alpha) = \partial_\mu\left[\left(\frac{\partial \mathcal{L}}{\partial(\partial_\mu \psi_\alpha)}\right)\delta_0\psi_\alpha\right]$$
$$- \left[\partial_\mu \frac{\partial \mathcal{L}}{\partial(\partial_\mu \psi_\alpha)}\right]\delta_0\psi_\alpha . \tag{21.2.22}$$

Substituting Eqs. (21.2.21) and (21.2.22) into (21.2.19) and then rearranging gives

$$\delta A = \int_{s_1}^{s_2}\int_R \left\{ \left[\frac{\partial \mathcal{L}}{\partial \psi_\alpha} - \partial_\mu \pi^{\alpha\mu} - \frac{d}{ds}\frac{\partial \mathcal{L}}{\partial(\partial_s \psi_\alpha)} \right] \delta_0 \psi_\alpha \right\} dx\, ds$$
$$+ \int_{s_1}^{s_2}\int_R \{\partial_\mu [\pi^{\alpha\mu}\delta_0\psi_\alpha + \mathcal{L}\delta x^\mu]\} dx\, ds \tag{21.2.23}$$

where

$$\pi^{\alpha\mu}(x,s) \equiv \frac{\partial \mathcal{L}(x,s)}{\partial(\partial_\mu \psi_\alpha(x,s))}. \tag{21.2.24}$$

Equation (21.2.24) associates a four-vector $\pi^{\alpha\mu}$ with every field ψ_α.

The integrand of the second integral in Eq. (21.2.23) is the divergence of a four-vector defined by

$$J^\mu \equiv \pi^{\alpha\mu}\delta_0\psi_\alpha + \mathcal{L}\delta x^\mu. \tag{21.2.25}$$

The quantity J^μ is put into a physically more interesting form by first introducing the local variation

$$\delta\psi_\alpha \equiv \psi'_\alpha(x',s') - \psi_\alpha(x,s) \tag{21.2.26}$$

where x,s and x',s' refer to the same physical point as viewed from canonically transformed reference frames. Performing a Taylor series expansion of ψ'_α(x',s') and keeping only first-order terms gives

$$\delta\psi_\alpha = \psi'_\alpha(x,s) + \partial^\nu \psi'_\alpha(x,s)\,\delta x_\nu$$
$$+ \partial_s \psi'_\alpha(x,s)\,\delta s - \psi_\alpha(x,s). \tag{21.2.27}$$

Equation (2.21.3) is used to express $\delta\psi_\alpha$ in terms of the original fields ψ_α. The result is

$$\delta\psi_\alpha = \delta_0\psi_\alpha + \partial^\nu \psi_\alpha(x,s)\,\delta x_\nu + \partial_s \psi_\alpha(x,s)\,\delta s \tag{21.2.28}$$

where second order terms such as $[\partial^\nu(\delta_0\psi_\alpha(x,s))]\delta x_\nu$ and $[\partial(\delta_0\psi_\alpha(x,s))/\partial s]\delta s$ have again been neglected. As a point of interest, Kubo [1985] included the former term in his analysis, while Pavsic [1991b] kept both terms in his analysis. Neither term is essential to our discussion, and the interested reader should consult the references for details.

Combining Eqs. (21.2.25) and (21.2.28) lets us write the four-vector J^μ in the form

$$J^\mu = \pi^{\alpha\mu}\delta\psi_\alpha - (\pi^{\alpha\mu}\partial^\nu\psi_\alpha - \delta^{\nu\mu}\mathcal{L})\,\delta x_\nu - \pi^{\alpha\mu}\partial_s\psi_\alpha\,\delta s \tag{21.2.29}$$

where $\delta^{\nu\mu}$ is the Kronecker delta. Substituting the above results into $\delta\mathbf{A}$ yields

$$\begin{aligned}\delta A &= F(s_2) - F(s_1)\\ &+ \int_{s_1}^{s_2}\int_R \left(\frac{\partial \mathcal{L}}{\partial \psi_\alpha} - \partial_\mu \pi^{\alpha\mu} - \frac{d}{ds}\frac{\partial \mathcal{L}}{\partial(\partial_s \psi_\alpha)} \right) \delta_0 \psi_\alpha \, dx\, ds\,.\end{aligned} \tag{21.2.30}$$

The quantities $\pi_\alpha{}^\mu$ and J^μ are defined by Eqs. (21.2.25) and (21.2.29) respectively, and the notation

$$F(s_2) - F(s_1) \equiv \int_{s_1}^{s_2}\int_R \partial_\mu J^\mu \, dx\, ds \tag{21.2.31}$$

has been used.

At this point the mathematical formalism is essentially complete. Readers familiar with Schwinger's work will undoubtedly notice the similarity of the above formalism to LQFT. The principal differences are due to the incorporation of the evolution parameter and this, in turn, arises because the symmetry properties of the JARS group have been imposed on the physical system.

§21.3 PARAMETRIZED FIELD EQUATIONS

The notation used in Eq. (21.2.30) is intended to suggest the identification of $\mathbf{F}(s)$. Equation (21.2.30) is consistent with the postulated variational integral in Eq. (21.1.15) only if the integral

$$I_2 = \int_{s_1}^{s_2} \int_R \left(\frac{\partial \mathcal{L}}{\partial \psi_\alpha} - \partial_\mu \pi^{\alpha\mu} - \frac{d}{ds} \frac{\partial \mathcal{L}}{\partial(\partial_s \psi_\alpha)} \right) \delta_0 \psi_\alpha \, dx \, ds \tag{21.3.1}$$

is always zero. Observing that $\delta_0 \psi_\alpha$ and R are arbitrary implies that I_2 vanishes identically only if the bracketed quantity vanishes. Therefore Eq. (21.2.31) is valid only if

$$\frac{\partial \mathcal{L}}{\partial \psi_\alpha} - \partial_\mu \left(\frac{\partial \mathcal{L}}{\partial(\partial_\mu \psi_\alpha)} \right) - \frac{d}{ds} \frac{\partial \mathcal{L}}{\partial(\partial_s \psi_\alpha)} = 0 \tag{21.3.2}$$

is true for all values of α. This relation is the Euler-Lagrange equation. It is the field equation for the field $\psi_\alpha(x,s)$ and differs from the corresponding field equation of LQFT in that the field can be a function of s. If the fields are independent of s, then $\mathcal{L}$ is independent of $\partial\psi_\alpha/\partial s$ and Eq. (21.3.2) reduces to the LQFT result.

The above derivation of field theory from Hamilton's principle yields equations for the fields ψ_α. If the Lagrangian density $\mathcal{L}$ is invariant under some symmetry group, such as the JARS group, then the field equations are covariant with respect to the same symmetry group.

Attempts to second quantize PRQFT have been made by Kubo [1985], Saad, Horwitz and Arshansky [1989], and Pavsic [1991b]. Kubo worked with a specific model that postulated a parametrized Dirac equation. We saw in §10.1 that the Dirac equation is not consistent with the probabilistic basis at the heart of the theory presented here. Nevertheless, Kubo's work shows how we might approach second quantization using the 4x4 representation of the spin-½ Stückelberg equation introduced in §10.1. By contrast, Pavsic and Saad, et al worked only with scalar field equations. It is interesting to note that the commutators proposed by Pavsic differ from those proposed by Kubo. Saad, et al proposed equal s commutation relations for the pre-Maxwell fields discussed briefly in §9.3. Clearly much work remains to be done in this area.

§21.4 LAGRANGIAN QUANTUM FIELD THEORY

To be a valid physical theory, parametrized relativistic quantum field theory (PRQFT) must describe all of the phenomena described by LQFT. By showing that LQFT is a special case of PRQFT, we verify that the results of LQFT may be obtained from PRQFT. We also find the conditions under which LQFT applies.

Equation (21.2.30) can be written as

$$\delta A = \int_{s_1}^{s_2} \delta L \, ds = \int_{s_1}^{s_2} \int_R \partial_\mu J^\mu \, dx \, ds \tag{21.4.1}$$

where Eqs. (21.1.15), (21.2.31) and (21.3.2) have been used. If the fields are independent of s, as all fields of LQFT are, then Eq. (21.4.1) becomes

$$\delta A_S = \Delta s \, \delta L_s = \Delta s \int_R \partial_\mu J^\mu dx. \tag{21.4.2}$$

The subscript S indicates the operator is independent of s.

Dividing Eq. (21.4.2) by Δs yields

$$\delta L_S = \frac{\delta A_S}{\Delta s} = \int_R \partial_\mu J^\mu \, dx. \tag{21.4.3}$$

Applying the divergence theorem to Eq. (21.4.3) gives

$$\delta L_S = \int_{\sigma_2} J^\mu d\sigma_\mu - \int_{\sigma_1} J^\mu d\sigma_\mu \tag{21.4.4}$$

where $d\sigma_\mu$ is the μth surface element defined by

$$\begin{aligned} d\sigma_0 &= dx^1 dx^2 dx^3, \\ d\sigma_1 &= dx^0 dx^2 dx^3, \\ d\sigma_2 &= dx^0 dx^1 dx^3, \\ d\sigma_3 &= dx^0 dx^1 dx^2. \end{aligned} \tag{21.4.5}$$

We have assumed that the fields vanish at spacial infinity. The variation δL_S can be written as the difference of generators of observables on the spacelike surfaces σ_1, σ_2 such that

$$\delta L_S = F_S[\sigma_2] - F_S[\sigma_1] \tag{21.4.6}$$

where

$$F_S[\sigma] \equiv \int_\sigma J^\mu d\sigma_\mu \tag{21.4.7}$$

which reproduce exactly the relations used by Schwinger [1951b]. We obtain the familiar fields of LQFT by altering the weight distribution of the volume elements of four-momentum space.

LQFT FIELDS

All volume elements of four-momentum space are weighted equally in the PRQT definition of the scalar product (§7.3). This is illustrated by the general solution of the noninteracting, spin-0 particle Stückelberg equation

$$\begin{aligned}\psi(\boldsymbol{x},t,s) = \int\int & A(\boldsymbol{k},\omega)\exp(i\omega t - i\boldsymbol{k}\cdot\boldsymbol{x}) \\ & \cdot\exp\left[\frac{-is}{2m}(\omega^2 - \boldsymbol{k}\cdot\boldsymbol{k})\right] d^3k\, d\omega\end{aligned} \tag{21.4.8}$$

where $A(\mathbf{k},\omega)$ is an arbitrary expansion coefficient and $\mathbf{x}$, $\mathbf{k}$ denote 3-space vectors with spacial components only. By contrast, LQFT assumes that any given elementary particle has a unique rest mass which can be measured to an arbitrary degree of accuracy.

The assumption of a unique mass is equivalent to considering only the on mass shell state. It corresponds to asserting that only a single mass state contributed to the PRQT wave function. For our noninteracting particle example, the expansion coefficient would have the form

$$A(\boldsymbol{k},\omega) = b(\boldsymbol{k},\omega)\,\delta(\omega^2 - \boldsymbol{k}\cdot\boldsymbol{k} - m_0^2) \tag{21.4.9}$$

where $b(\mathbf{k},\omega)$ is arbitrary and m_0 is the particle rest mass. We rewrite the delta function as [Dennery and Krzywicki, 1967]

$$\delta(\omega^2 - \boldsymbol{k}\cdot\boldsymbol{k} - m_0^2) = \frac{\delta(\omega - \omega_0) + \delta(\omega + \omega_0)}{2\omega_0} \tag{21.4.10}$$

where

$$\omega_0(\boldsymbol{k}) \equiv (m_0^2 + \boldsymbol{k}\cdot\boldsymbol{k})^{1/2} > 0. \tag{21.4.11}$$

Substituting Eqs. (21.4.9) and (21.4.10) into (21.4.8) and evaluating the ω integral yields

$$\begin{aligned}\psi(\boldsymbol{x},t,s) = \exp\left(\frac{-im_0^2 s}{2m}\right)\int \frac{d^3k}{2\omega_0}\\ \cdot\{[b(\boldsymbol{k},\omega_0)\exp[i(\omega_0 t - \boldsymbol{k}\cdot\boldsymbol{x})]\\ + b(\boldsymbol{k},-\omega_0)\exp[-i(\omega_0 t + \boldsymbol{k}\cdot\boldsymbol{x})]\}.\end{aligned} \tag{21.4.12}$$

The corresponding result for the antiparticle is

$$\begin{aligned}\psi^*(\boldsymbol{x},t,s) = \exp\left(\frac{im_0^2 s}{2m}\right)\int \frac{d^3k}{2\omega_0}\\ \cdot\{[b^*(\boldsymbol{k},\omega_0)\exp[-i(\omega_0 t - \boldsymbol{k}\cdot\boldsymbol{x})]\\ + b^*(\boldsymbol{k},-\omega_0)\exp[i(\omega_0 t + \boldsymbol{k}\cdot\boldsymbol{x})]\}.\end{aligned} \tag{21.4.13}$$

The s dependence of ψ and ψ^* is now just a numerical factor which has no effect on the probability density ($\psi^*\psi$) and can therefore be removed by a gauge transformation or, equivalently, by setting $s=0$ in Eqs. (21.4.12) and (21.4.13).

If we compare $\psi(\mathbf{x},t,0)$ and $\psi^*(\mathbf{x},t,0)$ to the ψ operators of LQFT, we observe that these functions are precisely the same when the coefficients $b(\mathbf{k},\omega_0)$

and $b^*(\mathbf{k},\omega_0)$ are interpreted as particle creation and annihilation operators, respectively, and the coefficients $b(\mathbf{k},-\omega_0)$ and $b^*(\mathbf{k},-\omega_0)$ are interpreted as antiparticle creation and annihilation operators, respectively. Thus the field operators of LQFT can be derived from PRQFT by assuming that only a single mass state contributes to the wave function and then replacing coefficients according to the techniques of conventional second quantization [Schwinger, 1951b; Bjorken and Drell, 1965; Bogoliubov and Shirkov, 1980; Wu and Hwang, 1991].

CHAPTER 22
PROPERTIES OF THE FIELD THEORY

We showed in Chapter 21 that Hamilton's principle yields the field equations for the fields ψ_α. Furthermore, if the Lagrangian density $\mathcal{L}$ is invariant under some symmetry group, such as the JARS group, then the field equations are covariant with respect to the same symmetry group. The purpose of this chapter is to develop some of the properties of parametrized relativistic quantum field theory (PRQFT). These properties include identifying generators of the JARS group and obtaining conservation laws associated with the generators [Fanchi, 1979]. Among the more novel properties of PRQFT is the mass operator concept. As an illustration of its use, a phenomenological model of meson mass spectra and mixing angles is developed.

§22.1 GENERATORS OF THE JARS GROUP

A physical interpretation can be assigned to specific mathematical constructs of PRQFT by obtaining the field theoretic form of the JARS group generators. These forms are developed from the equation

$$\delta A = \int_{s_1}^{s_2} \int_R \partial_\mu J^\mu \, dx \, ds, \tag{22.1.1}$$

which was obtained by combining Eqs. (21.2.30), (21.2.31) and (21.3.2). The four-volume integral in Eq. (22.1.1) can be transformed into a hypersurface integral by using the divergence theorem. The result is

$$\delta A = \int_{s_1}^{s_2} \oint J^{\mu} dS_{\mu} ds \tag{22.1.2}$$

where dS_μ is an infinitesimal, covariant four-vector characterizing the hypersurface elements of the closed boundary of R. If the particular four-volume R_S is used as the region of integration, then dS_μ becomes $d\sigma_\mu$ on σ_2 and $-d\sigma_\mu$ on σ_1.

Equation (22.1.2) is used to determine the total variation of ψ_α due to δ**A** as in Eq. (21.2.4), thus

$$\delta_0 \psi_\alpha(x',s') = i \left[\int_{s_1}^{s_2} \oint J^{\mu} dS_{\mu} ds, \psi_\alpha(x',s') \right]. \tag{22.1.3}$$

To evaluate $\delta_0\psi_\alpha$ for an infinitesimal JARS transformation, it is necessary to determine δx^μ, δs, and $\delta\psi_\alpha$. These increments are developed in §6.4 and may be written as

$$\begin{aligned} \delta x_\nu &= \epsilon_{\nu\mu} x^{\mu} + \epsilon_{p\nu} + \epsilon_{x\nu} s, \\ \delta s &= \epsilon, \\ \delta \psi &= \frac{1}{2} \Sigma_\alpha^{\beta\nu\rho} \psi_\beta \epsilon_{\nu\rho}. \end{aligned} \tag{22.1.4}$$

Equation (22.1.4) may be used to write J^μ in the form

$$J^{\mu} = \frac{1}{2} M^{\mu\nu\rho} \epsilon_{\nu\rho} + t^{\mu\nu}(\epsilon_{p\nu} + \epsilon_{x\nu} s) + K^{\mu}\epsilon \tag{22.1.5}$$

where $M^{\mu\nu\rho}$ is the covariant angular momentum tensor density defined by

$$M^{\mu\nu\rho} \equiv t^{\mu\rho} x^{\nu} - t^{\mu\nu} x^{\rho} + \pi^{\alpha\mu} \Sigma_\alpha^{\beta\nu\rho} \psi_\beta, \tag{22.1.6}$$

$t^{\mu\nu}$ is the canonical energy-momentum tensor defined by

$$t^{\mu\nu} \equiv \delta^{\mu\nu} \mathcal{L} - \pi^{\alpha\mu} \partial^{\nu} \psi_\alpha, \tag{22.1.7}$$

and the notation

$$K^{\mu} \equiv -\pi^{\alpha\mu}\partial_s\psi_{\alpha} \tag{22.1.8}$$

has been used. Observe that $M^{\mu\nu\rho}$ is antisymmetric, i.e.

$$M^{\mu\nu\rho} = -M^{\mu\rho\nu} \tag{22.1.9}$$

Substituting Eq. (22.1.5) into (22.1.3) yields

$$\begin{aligned}
\delta_0\psi_{\alpha}(x',s') = {} & \frac{i}{2}\left[\int_{s_1}^{s_2}\oint M^{\mu\nu\rho}dS_{\mu}ds, \psi_{\alpha}(x',s')\right]\epsilon_{\nu\rho} \\
& + i\left[\int_{s_1}^{s_2}\oint t^{\mu\nu}dS_{\mu}ds, \psi_{\alpha}(x',s')\right]\epsilon_{p\nu} \\
& + i\left[\int_{s_1}^{s_2}\oint s t^{\mu\nu}dS_{\mu}ds, \psi_{\alpha}(x',s')\right]\epsilon_{x\nu} \\
& + i\left[\int_{s_1}^{s_2}\oint K^{\mu}dS_{\mu}ds, \psi_{\alpha}(x',s')\right]\epsilon .
\end{aligned} \tag{22.1.10}$$

A comparison of Eq. (22.1.10) with (6.4.14) and (6.3.12) implies that the generators of the JARS transformation are given by

$$J^{\mu\nu} = \int_{s_1}^{s_2}\oint M^{\mu\nu\rho}dS_{\mu}ds, \tag{22.1.11}$$

$$P^{\mu} = \int_{s_1}^{s_2}\oint t^{\mu\nu}dS_{\mu}ds, \tag{22.1.12}$$

$$X^{\mu} = \int_{s_1}^{s_2} \oint s t^{\mu\nu} dS_{\mu}\, ds, \tag{22.1.13}$$

and

$$K = \int_{s_1}^{s_2} \oint K^{\mu} dS_{\mu}\, ds, \tag{22.1.14}$$

since the coefficients of the arbitrary infinitesimals must be equal.

We have previously assumed that ψ_{α} vanishes at spacial infinity. If we choose the four-volume R_S as the region of integration. Eqs. (22.1.11) through (22.1.14) become

$$J^{\mu\nu} = J^{\mu\nu}(\sigma_2) - J^{\mu\nu}(\sigma_1), \tag{22.1.15}$$

$$P^{\mu} = P^{\mu}(\sigma_2) - P^{\mu}(\sigma_1), \tag{22.1.16}$$

$$X^{\mu} = X^{\mu}(\sigma_2) - X^{\mu}(\sigma_1), \tag{22.1.17}$$

and

$$K = K(\sigma_2) - K(\sigma_1). \tag{22.1.18}$$

The hypersurface dependent operators on the right hand side of Eqs. (22.1.15) through (22.1.18) are

$$J^{\mu\nu}(\sigma) = \int_{s_1}^{s_2} \int_{R_S} M^{\mu\nu\rho} d\sigma_{\mu}\, ds, \tag{22.1.19}$$

$$P^{\mu}(\sigma) = \int_{s_1}^{s_2} \int_{R_S} t^{\mu\nu} d\sigma_{\mu}\, ds, \tag{22.1.20}$$

$$X^{\mu}(\sigma) = \int_{s_1}^{s_2}\int_{R_S} s t^{\mu\nu} d\sigma_{\mu}\, ds, \tag{22.1.21}$$

and

$$K(\sigma) = \int_{s_1}^{s_2}\int_{R_S} K^{\mu} d\sigma_{\mu}\, ds. \tag{22.1.22}$$

Notice that if the eigenvalues of $\mathbf{J}^{\mu\nu}$, $\mathbf{P}^{\mu}$, $\mathbf{X}^{\mu}$, $\mathbf{K}$ are zero, the respective eigenvalues of these generators evaluated at σ_2 equal those evaluated at σ_1, i.e. the eigenvalues are conserved. Conservation conditions are discussed in more detail in §22.2.

§22.2 CONSERVATION LAWS

The scope of PRQFT is illustrated by using the theory to derive conservation laws with broad physical validity. Conservation laws are a result of canonical transformations which do not change the action operator **A**, i.e.

$$\delta A = 0 = \int_{s_1}^{s_2}\int_{R} \partial_{\mu} J^{\mu}\, dx\, ds \tag{22.2.1}$$

where the four-volume R is again arbitrary. The necessary and sufficient condition for Eq. (22.2.1) to hold is the continuity equation

$$\partial_{\mu} J^{\mu} = 0 \tag{22.2.2}$$

which is the differential form of a conservation law. The form of J^{μ} depends on the type of continuous symmetry transformation that leaves **A** invariant. For the case of the JARS transformation, Eq. (22.2.2) becomes

$$\partial_{\mu} J^{\mu} = 0 = \partial_{\mu}[\frac{1}{2} M^{\mu\nu\rho} \epsilon_{\nu\rho} + t^{\mu\nu} \epsilon_{p\nu} + s t^{\mu\nu} \epsilon_{x\nu} + K^{\mu} \epsilon] \tag{22.2.3}$$

where Eq. (22.1.5) has been used. By expressing $\partial_\mu J^\mu$ as in Eq. (22.2.3), it is easy to determine the form of the conservation laws for four particular interesting transformations.

(i) For a coordinate translation only:

$$\epsilon_{\nu\rho} = \epsilon_{x\nu} = \epsilon = 0 \tag{22.2.4}$$

and Eq. (22.2.3) gives

$$\partial_\mu t^{\mu\nu} = 0. \tag{22.2.5}$$

(ii) For an evolution parameter translation only:

$$\epsilon_{\nu\rho} = \epsilon_{p\nu} = \epsilon_{x\nu} = 0 \tag{22.2.6}$$

and Eq. (22.2.3) gives

$$\partial_\mu K^\mu = 0. \tag{22.2.7}$$

(iii) For a proper homogeneous Lorentz transformation only:

$$\epsilon_{p\nu} = \epsilon_{x\nu} = \epsilon = 0 \tag{22.2.8}$$

and Eq. (22.2.3) gives

$$\partial_\mu M^{\mu\nu\rho} = 0. \tag{22.2.9}$$

(iv) For a relativistic Galilean boost only:

$$\epsilon_{\nu\rho} = \epsilon_{p\nu} = \epsilon = 0 \tag{22.2.10}$$

and Eq. (22.2.3) gives

$$\partial_\mu (s t^{\mu\nu}) = 0 = s \partial_\mu t^{\mu\nu} \tag{22.2.11}$$

or

$$\partial_\mu t^{\mu\nu} = 0 \tag{22.2.12}$$

since s is an independent variable.

These relations are independent of all spacelike surfaces because the space-time four-volume R is arbitrary. Equations (22.2.5) and (22.2.9) represent conservation of energy-momentum and angular momentum respectively. Equation (22.2.12) reduces to (22.2.5) and does not represent a new conservation law. Only Eq. (22.2.7) is a new conservation law. It has appeared because the symmetry properties of the JARS group have been imposed on the physical system. We have seen some of its consequences in our study of tachyon kinematics in Chapter 18 through Chapter 20.

Equations (22.2.5), (22.2.7), (22.2.9) and (22.2.12) are differential forms of the conservation laws. Another form is obtained by applying the divergence theorem to Eqs. (22.1.19) through (22.1.22) to get

$$J^{\mu\nu} = \int_{s_1}^{s_2}\int_R \partial_\mu M^{\mu\nu\rho}\, dx\, ds, \tag{22.2.13}$$

$$P^{\mu} = \int_{s_1}^{s_2}\int_R \partial_\mu t^{\mu\nu}\, dx\, ds, \tag{22.2.14}$$

$$X^{\mu} = \int_{s_1}^{s_2}\int_R \partial_\mu (s t^{\mu\nu})\, dx\, ds, \tag{22.2.15}$$

and

$$K = \int_{s_1}^{s_2}\int_R \partial_\mu K^{\mu}\, dx\, ds. \tag{22.2.16}$$

Substituting Eqs. (22.2.5), (22.2.7), (22.2.9) and (22.2.12) into (22.2.13) through (22.2.16) yields

$$J^{\mu\nu} = 0 = J^{\mu\nu}(\sigma_2) - J^{\mu\nu}(\sigma_1), \tag{22.2.17}$$

$$P^{\mu} = 0 = P^{\mu}(\sigma_2) - P^{\mu}(\sigma_1), \tag{22.2.18}$$

$$X^{\mu} = 0 = X^{\mu}(\sigma_2) - X^{\mu}(\sigma_1), \tag{22.2.19}$$

and

$$K = 0 = K(\sigma_2) - K(\sigma_1) \tag{22.2.20}$$

where Eqs. (22.1.15) through (22.1.18) have been used. Equations (22.2.17) through (22.2.20) are the alternative expression for conservation laws that was alluded to in §22.1.

§22.3 INTERPRETATION OF THE MASS OPERATOR

The interpretation of **K** as a mass operator is verified by evaluating the eigenvalues of **K** for a free particle. This exercise also illustrates an application of the field theoretic formalism.

We begin by defining a vacuum state $|0>$ such that

$$K(\sigma)|0\rangle = 0, \tag{22.3.1}$$

$$X^{\mu}(\sigma)|0\rangle = 0, \tag{22.3.2}$$

$$P^{\mu}(\sigma)|0\rangle = 0, \tag{22.3.3}$$

$$J^{\mu\nu}(\sigma)|0\rangle = 0, \tag{22.3.4}$$

and

$$\langle 0|0\rangle = 1. \tag{22.3.5}$$

The specification of σ is intended to indicate that the eigenvalues of the respective operators belong to a state vector at the spacelike surface σ.

The field equation for a free, spinless particle is

$$i\frac{\partial \psi}{\partial s} = -\frac{1}{2m}\frac{\partial^2 \psi}{\partial x^{\mu}\partial x_{\mu}}. \tag{22.3.6}$$

Equation (22.3.6) can be derived from a Lagrangian density having the form

$$\mathcal{L}_0 = \frac{i}{2}[(\partial_s \psi^*)\psi - \psi^*(\partial_s \psi)] + \frac{1}{2m}\partial^\mu \psi \partial_\mu \psi^* . \tag{22.3.7}$$

Natural units ($\hbar = c = 1$) are being used, and m is the mass of the particle with the meaning given in §9.2. Equation (22.3.6) has the general solution

$$\psi(x,s) = \int A(k) \exp(-iq(k)s + ik \cdot x)\, dk \tag{22.3.8}$$

where

$$q(k) = \frac{k \cdot k}{2m} . \tag{22.3.9}$$

If Eq. (22.3.8) is used in Eq. (6.4.18) and the result applied to the vacuum state, the result is

$$\mathbf{K}(\sigma) A(k) |0\rangle - A(k) \mathbf{K}(\sigma) |0\rangle = q(k) A(k) |0\rangle . \tag{22.3.10}$$

Equation (22.3.10) is simplified by using Eq. (22.3.1) to get

$$\mathbf{K}(\sigma) |1\rangle = q(k) |1\rangle \tag{22.3.11}$$

where

$$|1\rangle \equiv A(k) |0\rangle . \tag{22.3.12}$$

In words, the eigenvalues of **K** are dimensionally proportional to the mass of state $|1>$. Repetition of the above procedure can be used to build up a multiparticle state.

Correspondence to LQFT Field

The s-dependent free particle field given in Eq. (22.3.8) can be related to an LQFT field by recognizing that LQFT assumes that a given elementary particle has a unique mass which can be measured to an arbitrary degree of accuracy. This assumption corresponds to asserting that only a single mass state contributes to the wave function; thus

$$A(k) = b(k)\,\delta(\omega^2 - \vec{k}\cdot\vec{k} - m^2) \tag{22.3.13}$$

where $k^0 = \omega$ in natural units and b(k) is arbitrary. The delta function can be written as

$$\delta(\omega^2 - \vec{k}\cdot\vec{k} - m^2) = \frac{\delta(\omega - \omega_0) + \delta(\omega + \omega_0)}{2\omega_0} \tag{22.3.14}$$

where

$$\omega_0(\vec{k}) = (m^2 + \vec{k}\cdot\vec{k})^{1/2} > 0. \tag{22.3.15}$$

Substituting Eqs. (22.3.14) and (22.3.15) into (22.3.8) and evaluating the ω integral yields

$$\begin{aligned}\psi(x,s) &= \exp\left(\frac{-ims}{2}\right)\\ &\cdot\int [b(\vec{k},\omega_0)\exp(i\omega_0 t - i\vec{k}\cdot\vec{x})\\ &+ b(\vec{k},-\omega_0)\exp(-i\omega_0 t - i\vec{k}\cdot\vec{x})]\frac{d^3k}{2\omega_0}.\end{aligned} \tag{22.3.16}$$

The corresponding result for the antiparticle is

$$\begin{aligned}\psi^*(x,s) &= \exp\left(\frac{ims}{2}\right)\\ &\cdot\int \{b^*(\vec{k},\omega_0)\exp[-i(\omega_0 t - \vec{k}\cdot\vec{x})]\\ &+ b^*(\vec{k},-\omega_0)\exp[i(\omega_0 t + \vec{k}\cdot\vec{x})]\}\frac{d^3k}{2\omega_0}.\end{aligned} \tag{22.3.17}$$

The s dependence of ψ and ψ^* is now just a numerical factor which has no effect on the probability density $<\psi|\psi>$ and can therefore be removed by a gauge transformation or, equivalently, by setting s=0. If we compare $\psi(x,0)$ and $\psi^*(x,0)$ to the ψ fields of LQFT, we see that these functions are the same when the coefficients b(k) and $b^*(k)$ with $\omega = \omega_0$ are interpreted as particle creation and annihilation operators, respectively. Similarly, the coefficients $b^*(k)$ and $b(k)$ with $\omega=-\omega_0$ are interpreted as antiparticle creation and annihilation operators, respectively. Thus the fields of LQFT can be derived from the parametrized fields by assuming that only a single mass state contributes to the wave function.

ADDITIVITY OF MASS OPERATORS

Previous applications of the mass operator concept in PRQT have shown that the mass operator can be written as a sum of contributions. The ability to add and subtract terms from a mass operator is a valuable feature of PRQT. Our purpose here is to show that the additivity of mass operators in PRQT also applies in PRQFT.

Suppose the Lagrangian density can be written as the following sum of terms:

$$\mathcal{L} = \sum_j \mathcal{L}_j. \tag{22.3.18}$$

We obtain the mass operator **K** corresponding to $\mathcal{L}$ by first noting that

$$\begin{aligned} \pi^\mu_\alpha &= \frac{\partial \mathcal{L}}{\partial \partial_\mu \psi_\alpha} \\ &= \sum_j \left(\frac{\partial \mathcal{L}_j}{\partial \partial_\mu \psi_\alpha} \right) \equiv \sum_j (\pi^\mu_\alpha)_j. \end{aligned} \tag{22.3.19}$$

Substituting Eq. (22.3.19) in the mass flux operator gives

$$\boldsymbol{K}^\mu = -\pi^\mu_\alpha \partial_s \psi_\alpha = \sum_j (\boldsymbol{K}^\mu)_j \tag{22.3.20}$$

where

$$(\boldsymbol{K}^\mu)_j = (\pi^\mu_\alpha)_j \partial_s \psi_\alpha. \tag{22.3.21}$$

Performing the appropriate integrations over the mass flux operator yields

$$\boldsymbol{K} = \int_{s_1}^{s_2}\int \boldsymbol{K}^{\mu} dS_{\mu} ds = \sum_j \boldsymbol{K}_j \tag{22.3.22}$$

where

$$\boldsymbol{K}_j \equiv \int_{s_1}^{s_2}\int (\boldsymbol{K}^{\mu})_j dS_{\mu} ds. \tag{22.3.23}$$

Equation (22.3.22) confirms that mass operator contributions are additive.

§22.4 A MASS OPERATOR APPLICATION

An unsolved riddle of modern particle physics is the problem of explaining hadron mass spectra [Particle Data Group, 1992]. The standard model -- quantum chromodynamics (QCD) -- is based on the idea that hadrons are systems of bound quarks. Quarks were first introduced by Gell-Mann [1964] and Zweig [1964]. Numerous references contain detailed discussions of the standard quark model. Some examples include Kokkedee [1969], Lichtenberg [1970], Scadron [1979], Hughes [1985], Kobzarev and Manin [1989], and Williams [1991].

Quark models treat mesons as bound states of quark-antiquark pairs and baryons as bound states of three quarks. The baryon calculations relate combinations of linear quark masses and quark interactions to observed baryon masses. A similar approach applied to vector mesons yields satisfactory results but fails for pseudoscalar mesons. To obtain the meson mass spectra, quark models postulate that the squares of meson and quark masses must be directly related, but the use of squared quark masses is not *a priori* justifiable. An alternative analysis is possible within the context of PRQT.

We saw in Chapter 12 that PRQT provides a mathematically tractable formalism for bound state models of the strong interaction. These ideas are developed further here. In particular, a model relating meson masses to linear combinations of effective quark masses is presented. The model combines simple quark concepts for a triplet of quarks {u, d, s} and the mass operator concepts presented in §22.3. We show that this relatively simple model is equivalent to the Gell-Mann-Okubo mass formula, but now provided with a consistent theoretical foundation.

PHYSICAL MODEL OF MESON MASS SPECTRA

Noninteracting mesons can be viewed as either free particles (meson picture) or as bound quark-antiquark pairs (quark picture). In the meson picture, each noninteracting meson has a mass

$$\langle \boldsymbol{K}_M \rangle = \frac{m_{obs}^2}{2m(J^P)} \tag{22.4.1}$$

where $<\mathbf{K}_M>$ is the expectation value of the meson mass operator, m_{obs} is the observed meson mass, and $m(J^P)$ is identified as the average free particle rest mass of the meson nonet having spin J and spacial parity P. The model identifications are made plausible by the discussion of spin-0 particles presented in Chapter 9, and by phenomenological results presented below.

The quark picture views the same meson as a system composed of bound quark-antiquark pairs. The starting point for most modern treatments of hadron physics is the work of DeRujula, Georgi and Glashow (1975). Details of the QCD Lagrangian in the standard paradigm are presented by the Particle Data Group [1992]. An alternative, and somewhat simpler analysis, may be presented as follows.

We assume the Lagrangian density describing the two-particle system has the form

$$\mathscr{L}_Q = \mathscr{L}_S + \mathscr{L}_3 + \mathscr{L}_6 + \mathscr{L}_{EM} \tag{22.4.2}$$

where $\mathscr{L}_S$ represents the strong interaction, and the terms $\mathscr{L}_3$ and $\mathscr{L}_6$ denote interactions that break SU(3) and SU(6) symmetry respectively. The term $\mathscr{L}_{EM}$ denotes the relatively small electromagnetic interaction which we henceforth neglect. The Lagrangian densities $\mathscr{L}_S$, $\mathscr{L}_3$ and $\mathscr{L}_6$ correspond to mass operators $\mathbf{K}_S$, $\mathbf{K}_3$ and $\mathbf{K}_6$ respectively. By §22.3, we know the mass operators contribute additively to the mass operator $\mathbf{K}_Q$ corresponding to the Lagrangian density $\mathscr{L}_Q$.

The simplest description of the bound system is to assume that $\mathscr{L}_S$ has the form

$$\mathscr{L}_S = \mathscr{L}_A + \mathscr{L}_{B'} + \mathscr{L}_{AB'} \tag{22.4.3}$$

where $\mathscr{L}_A$ represents a free quark, $\mathscr{L}_{B'}$ a free antiquark, and $\mathscr{L}_{AB'}$ their interaction. Corresponding mass operators are $\mathbf{K}_A$, $\mathbf{K}_{B'}$, $\mathbf{K}_{AB'}$ respectively. An example of a PRQT model of the strong interaction is the covariant harmonic oscillator treatment described in Chapter 12. At this level of approximation we assume the free quark masses are degenerate, i.e. $<\mathbf{K}_A> = <\mathbf{K}_{B'}>$. Thus the contribution of $\mathscr{L}_S$ to the total mass of the bound system is

$$\langle \boldsymbol{K}_S \rangle = 2m^* \tag{22.4.4}$$

where the effective mass is defined as

$$m^* \equiv \langle \boldsymbol{K}_A \rangle + \frac{1}{2} \langle \boldsymbol{K}_{AB'} \rangle . \tag{22.4.5}$$

Let us now consider the symmetry breaking perturbations $\mathscr{L}_3$ and $\mathscr{L}_6$. We assume that $\mathscr{L}_3$ splits the quark triplet such that

$$\begin{aligned} m_u &\approx m_d \approx m^*, \\ m_s &= m^* + \Delta \end{aligned} \tag{22.4.6}$$

where Δ is a scalar constant and $\{m_u, m_d, m_s\}$ are effective masses of the quarks $\{u, d, s\}$ respectively. Effective $\{u, d, s\}$ quark masses are poorly known [Particle Data Group, 1992]. Kokkedee [1969] estimated values of Δ ranging from 147 MeV for the $3^+/2$ decuplet and 113 MeV for vector mesons. An additional contribution to the mass operator arises from the $\mathscr{L}_6$ term. We write its contribution as $-U(\alpha)$ where U denotes the SU(6) breaking quark-antiquark potential [Kokkedee, 1969] and α denotes the SU(3) multiplet. Combining these effects yields a total mass

$$\langle \boldsymbol{K}_Q \rangle = \langle A \,|\, \sum_\ell m_\ell - U(\alpha) \,|A\rangle. \tag{22.4.7}$$

The sum includes the effective quark masses of the quark-antiquark pairs composing meson A in the state labeled | A>.

Physically the meson and quark pictures are equivalent. Therefore we should have the relation

$$\langle \boldsymbol{K}_M \rangle = \langle \boldsymbol{K}_Q \rangle. \tag{22.4.8}$$

Employing Eqs. (22.4.1) and (22.4.7) in (22.4.8) yields

$$\frac{m_{obs}^2}{2m(J^P)} = \langle A \,|\, \sum_\ell m_\ell - U(\alpha) \,|A\rangle. \tag{22.4.9}$$

Equation (22.4.9) provides a link between observed meson masses and the parameters of the simple quark model outlined above. It contains several parameters which must be determined from experimental data.

CALCULATION OF MODEL PARAMETERS

We are interested in applying our triplet quark model to three meson nonets: pseudoscalar, vector and tensor. Postulated quark content and observed masses are presented in Tables 22.1 and 22.2. The symbol ^ used in d^, u^, s^ and K^ denotes the respective antiparticle. A detailed calculation of model parameters is given in this section for pseudoscalar mesons. The calculation procedure for vector and tensor mesons is the same as that given here. Therefore only the results for vector and tensor mesons are presented.

TABLE 22.1 QUARK CONTENT OF PSEUDOSCALAR MESONS

SU(3) MULTIPLET	QUARK CONTENT	PSEUDOSCALAR $J^P=0^-$ (MeV)
{8}	ud^	π^+ (140)
	(uu^-dd^)/$\sqrt{2}$	π^0 (135)
	du^	π^- (140)
	ds^	K^0 (498)
	us^	K^+ (494)
	d^s	$K^{\hat{}0}$ (498)
	u^s	K^- (494)
	(u^u+d^d-2s^s)/$\sqrt{6}$	η_8 (?)
{1}	(u^u+d^d+s^s)/$\sqrt{3}$	η_1 (?)
{1} AND {8} MIXTURE		η (549) η' (958)

TABLE 22.2 QUARK CONTENT OF VECTOR AND TENSOR MESONS

SU(3) MULTIPLET	QUARK CONTENT	VECTOR $J^P=1^-$ (MeV)	TENSOR $J^P=2^+$ (MeV)
{8}	ud^	ρ^+ (770)	a_2^+ (1320)
	(uu^-dd^)/$\sqrt{2}$	ρ^0 (770)	a_2^0 (1320)
	du^	ρ^- (770)	a_2^- (1320)
	ds^	K^{*0} (892)	K^{*0} (1430)
	us^	K^{*+} (892)	K^{*+} (1430)
	d^s	$K^{\hat{}*0}$ (892)	$K^{\hat{}*0}$ (1430)
	u^s	K^{*-} (892)	K^{*-} (1430)
	(u^u+d^d-ds^s)/$\sqrt{6}$	ω_8 (?)	f_8 (?)
{1}	(u^u+d^d+s^s)/$\sqrt{3}$	ω_1 (?)	f_1 (?)
{1} AND {8} MIXTURE		ω (783) ϕ' (1020)	f_2 (1270) f_2' (1525)

The calculation of model parameters begins by rewriting Eq. (22.4.9) in the form

$$\frac{m_{obs}^2}{2m_{J^P}} = 2m^* - B(J^P,\alpha) \tag{22.4.10}$$

where the perturbation matrix element $B(J^P,\alpha)$ is defined by

$$B(J^P,\alpha) \equiv \langle A | U(\alpha) | A \rangle . \tag{22.4.11}$$

Applying Eqs. (22.4.10) and (22.4.11) to pseudoscalar mesons gives the set of equations

$$\frac{m_\pi^2}{2m_0} = 2m^* - B(0^-,8) \equiv m_{08}, \tag{22.4.12}$$

$$\frac{m_K^2}{2m_0} = 2m^* - B(0^-,8) + \Delta = m_{08} + \Delta , \tag{22.4.13}$$

$$\frac{m_{\eta_8}^2}{2m_0} = 2m^* - B(0^-,8) + \frac{4}{3}\Delta = m_{08} + \frac{4}{3}\Delta , \tag{22.4.14}$$

and

$$\frac{m_{\eta_1}^2}{2m_0} = 2m^* - B(0^-,1) + \frac{2}{3}\Delta \equiv m_{01} + \frac{2}{3}\Delta \tag{22.4.15}$$

where $m(J^P)$ for pseudoscalar mesons is rewritten as

$$m(J^P) = m(0^-) \equiv m_0 . \tag{22.4.16}$$

The subscripts of the meson masses refer to the respective particles, e.g. m_π is the pion mass. The observed masses m_η and $m_{\eta'}$ are obtained by diagonalizing the matrix

$$\underline{\underline{M}} = \begin{bmatrix} \dfrac{m_{\eta_8}^2}{2m_0} & \delta(0^-) \\ \delta(0^-) & \dfrac{m_{\eta_1}^2}{2m_0} \end{bmatrix} \tag{22.4.17}$$

where $\delta(J^P) = \delta(0^-)$ is a negative mass effect due to mixing of η_1 and η_8 states. Notice that $\delta(J^P)$ is different for each meson nonet. Diagonalization yields

$$\frac{m_{\eta'}^2}{2m_0} = \frac{1}{2}\left(\frac{m_{\eta_8}^2}{2m_0} + \frac{m_{\eta_1}^2}{2m_0}\right) + R \tag{22.4.18}$$

and

$$\frac{m_{\eta}^2}{2m_0} = \frac{1}{2}\left(\frac{m_{\eta_8}^2}{2m_0} + \frac{m_{\eta_1}^2}{2m_0}\right) - R \tag{22.4.19}$$

where

$$R = \frac{1}{2}\left[\left(\frac{m_{\eta_8}^2}{2m_0} + \frac{m_{\eta_1}^2}{2m_0}\right)^2 - 4\left(\frac{m_{\eta_8}^2 m_{\eta_1}^2}{4m_0^2} - \delta(0^-)^2\right)\right]^{1/2} \tag{22.4.20}$$

Additional algebraic manipulations of the above equations provide the following simplified results:

$$m_{01} = \frac{\Delta}{m_K^2 - m_\pi^2}\left(m_{\eta'}^2 + m_\eta^2 - 2m_K^2 + m_\pi^2\right), \tag{22.4.21}$$

$$m_{08} = \frac{\Delta}{m_K^2 - m_\pi^2} m_\pi^2, \tag{22.4.22}$$

$$m_{\eta_8}^2 = \frac{1}{3}\left(4m_K^2 - m_\pi^2\right), \tag{22.4.23}$$

$$m_{\eta_1}^2 = m_{\eta'}^2 + m_\eta^2 + \frac{1}{3}m_\pi^2 - \frac{4}{3}m_K^2, \tag{22.4.24}$$

and

$$\delta(0^-) = -\frac{\Delta}{m_K^2 - m_\pi^2}\left(m_{\eta_8}^2 m_{\eta_1}^2 - m_\eta^2 m_{\eta'}^2\right)^{\frac{1}{2}}. \tag{22.4.25}$$

Notice that the masses of η_1 and η_8 depend only on observable masses. The poorly known quantity Δ appears only in m_{01}, m_{08} and $\delta(0_-)$. Estimates of model parameters are obtained by taking $\Delta = 147$ MeV. Results are reported in Table 22.3.

TABLE 22.3 MESON MODEL PARAMETERS

PARAMETER	PSEUDOSCALAR (MeV)	VECTOR (MeV)	TENSOR (MeV)
$m(J^P)$	780	658	1140
m_{J1}	475	505	665
m_{J8}	12	458	755
$m(\eta,\omega,f)_1$	946	891	1319
$m(\eta,\omega,f)_8$	569	927	1472
$\delta(J^P)$	-75	-160	-117

CALCULATION OF OCTET-SINGLET MIXING ANGLES

The SU(3) breaking interaction $\mathscr{L}_3$ induces a transition between the states η_1 and η_8 that is represented by $\delta(0^-)$ in the mass matrix $\underline{\underline{M}}$. The observable mass eigenvalues of the physical state η and η' are obtained by diagonalizing the mass matrix as discussed above. We make the usual assumption that the states η_1 and η_8 are related to the states η and η' by a rotation through an angle $\theta(J^P)$ where $\theta(J^P)$ corresponds to $J^P=0^-$. As a consequence, we obtain the transformation matrix $\underline{\underline{R}}(\theta)$ defined by

$$\begin{bmatrix} \eta_8 \\ \eta_1 \end{bmatrix} = \begin{bmatrix} \cos\theta(J^P) & \sin\theta(J^P) \\ -\sin\theta(J^P) & \cos\theta(J^P) \end{bmatrix} \begin{bmatrix} \eta \\ \eta' \end{bmatrix} \equiv \underline{\underline{R}}(\theta) \begin{bmatrix} \eta \\ \eta' \end{bmatrix}. \tag{22.4.26}$$

The mass matrix $\underline{\underline{M}}$ is diagonalized by first writing the total meson mass as

$$K_{mass} = [\eta_8 \ \eta_1] \ \underline{\underline{M}} \begin{bmatrix} \eta_8 \\ \eta_1 \end{bmatrix}. \tag{22.4.27}$$

An expression for the angle $\theta(J^P)$ is found by expanding Eq. (22.4.26), substituting the result into Eq. (22.4.27), collecting all terms having $\eta\eta'$ as a factor, and then defining $\theta(J^P)$ such that the factor multiplying $\eta\eta'$ vanishes. The result for the angle $\theta(0_-)$ is

$$\tan 2\theta(0^-) = -\frac{2\left(m_{\eta_8}^2 m_{\eta_1}^2 - m_\eta^2 m_{\eta'}^2\right)^{1/2}}{m_{\eta'}^2 + m_\eta^2 - \frac{8}{3}m_K^2 + \frac{2}{3}m_\pi^2}. \tag{22.4.28}$$

Similar relations exist for vector and tensor mesons. Model calculated mixing angles are presented in Table 22.4 for pseudoscalar, vector and tensor mesons.

TABLE 22.4 MESON MODEL MIXING ANGLES

MESON NONET	MIXING ANGLE (degrees)
PSEUDOSCALAR ($J^P=0^-$)	-11
VECTOR ($J^P=1^-$)	40
TENSOR ($J^P=2^+$)	26

Meson mixing angles in Table 22.4 are equivalent to the mixing angles calculated from the Gell-Mann-Okubo SU(3) mass formula

$$m^2 = M_0^2 + M_1^2\left[I(I+1) - \frac{1}{4}Y^2\right] \tag{22.4.29}$$

where Y is hypercharge, I is isospin, and $\{M_0, M_1\}$ are model parameters. To see the equivalence, we use Eq. (22.4.29) to write the mass formulae for pseudoscalar mesons as

$$\begin{aligned} m_\pi^2 &= M_{08}^2 + 2M_1^2, \\ m_K^2 &= M_{08}^2 + \frac{1}{2}M_1^2, \\ m_{\eta_8}^2 &= M_{08}^2, \\ m_{\eta_1}^2 &= M_{01}^2. \end{aligned} \tag{22.4.30}$$

For comparison, Eqs. (22.4.12) through (22.4.15) may be written in the form

$$\begin{aligned} m_\pi^2 &= 2m_0 m_{08}, \\ m_K^2 &= 2m_0\,[m_{08} + \Delta], \\ m_{\eta_8}^2 &= 2m_0\,[m_{08} + (4\Delta/3)], \\ m_{\eta_1}^2 &= 2m_0\,[m_{01} + (2\Delta/3)]. \end{aligned} \tag{22.4.31}$$

Making the replacements

$$\begin{aligned} M_{08}^2 &= 2m_0[m_{08} + (4\Delta/3)], \\ M_{01}^2 &= 2m_0[m_{01} + (2\Delta/3)], \\ M_1^2 &= 2m_0[-2\Delta/3] \end{aligned} \tag{22.4.32}$$

shows the relationship between the two models. The difference between the two models is not in their results, but in their respective theoretical bases.

CHAPTER 23
EPILOGUE -- A QUIXOTIC QUEST?

The task we set for ourselves in Chapter 1 is daunting: to provide a paradigm of relativistic quantum physics that is superior to the conventional paradigm. After much effort, it is fair to ask what has been achieved. We summarize results to date, then provide a few closing remarks.

SUMMARY OF THE TEXT

We originally introduced the objectives of this work in two contexts: philosophical and scientific. The adoption of a philosophical position made it easier to make consistent and meaningful decisions at various junctures in the construction of the scientific theory. One consequence of this approach is that tests of the scientific theory also test the underlying philosophy.

A classical basis for PRQT was provided in Part I. The classical theory is closely related to material that is familiar to most physicists. Presentation of the classical theory shows that the subsequent quantum theory is a natural extension of already familiar ideas, but it also begins to show some of the subtle changes in thinking that are required to understand PRQT.

The PRQT paradigm was presented in Part II. Among its tenets are the JARS transformation, probability conservation, the statistical interpretation and a postulated physical correspondence. One of the physically most novel results of Part II was the construction of an evolution parameter clock.

Most of the material in Part III was included to demonstrate that PRQT is capable of obtaining familiar results. Relatively simple techniques for studying bound state problems, including systems bound by the strong interaction, were presented. Although the solutions are model dependent, they show that the principles of PRQT are capable of supporting models that agree with experimental results. Beyond that, however, we showed in Chapter 13 that PRQT provides a consistent theoretical framework for describing particle stability in a single particle theory. In addition, we found from statistical PRQT that Fermi energy shifts relative to the conventional theory are possible. This provides a potentially fertile testing ground for experimentally evaluating the validity of PRQT. Perhaps the most conceptually significant result coming out of Part III is the relativistic Boltzmann's H-theorem. This theorem is the physical basis for establishing a metaphysical "arrow of time" within a scientific framework.

The chapters in Part IV take the most daring steps away from conventional science. They highlight the ability of PRQT to describe mass state transitions and to predict the behavior of particles as they make quantum transitions across the light cone. What is more startling than their predictions is the scientifically documented possibility that some of their unexpected experimental consequences may have already been observed. A detailed application of PRQT to neutrino physics -- notably neutrino production experiments and observations of neutrino oscillations -- may serve as a "crucial" test of PRQT.

Parametrized relativistic quantum field theory (PRQFT) was introduced in Part V. The most important result of Chapter 21 was the demonstration that conventional Lagrangian quantum field theory is a subset of PRQFT. Practitioners of conventional science who avoid the anomalies of conventional relativistic wave mechanics by retreating to field theory are in fact using a restricted version of PRQFT. Results in Chapter 22 demonstrated the potential utility of the mass conservation law. Despite these advances, the development of PRQFT is at an elementary stage. Among the outstanding problems are determining acceptable commutators and extending PRQFT to include interactions.

Concluding Remarks

Although we have just begun to scratch the surface of PRQT, we have shown that the persistent anomalies of conventional physics that were sketched in Chapter 1 can be resolved, or are at least amenable to resolution, within the context of PRQT. We have also shown that PRQT contains conventional physics as a subset. Most important of all, we have shown that PRQT is not a theory devoid of a future. Much work remains to be done!

We have already noted a few areas in need of attention. PRQT needs to be extended to include the electroweak and gravitational interactions. A few clues on how to proceed are available. For example, the electroweak problem may be solved by first formulating a Dirac-like 4x4 representation of fermions and then studying the local gauge invariance of the resulting theory. These problems represent opportunities for continuing our journey into the unknown, and PRQT appears to be a promising guide.

At last we can ask: should PRQT become the new paradigm? In my opinion, PRQT satisfies all criteria except one. There is no compelling experimental reason for switching.

When I first started working on parametrized theories in 1975, I did not believe the use of an invariant evolution parameter was a radical alteration of the status quo. The intervening years have changed my mind. It is now clear that a competing paradigm such as PRQT cannot be established without significant experimental support. An important objective of this book was to lay the foundation for designing "crucial" experiments. Hopefully this book will help motivate efforts to design and perform experimental tests.

APPENDIX

§A.1 NOTATION AND UNITS

The following notation is used throughout the book unless noted otherwise. Nonzero elements of the fundamental metric tensor are

$$g_{00}=1=-g_{11}=-g_{22}=-g_{33} \tag{A.1}$$

Greek indices run from 0 to 3 and Latin indices run from 1 to 3 unless indicated otherwise. Index 0 denotes the temporal component, while indices 1 to 3 denote spacial components.

The scalar product is defined by

$$\begin{aligned} A_\mu B^\mu &= A^0 B^0 - \vec{A}\cdot\vec{B} \\ &= A^\mu B_\mu \end{aligned} \tag{A.2}$$

where the four-vector A^μ has the elements

$$\{A^\mu\} = (A^0, \vec{A}) \ , \ \{A_\mu\} = (A^0, -\vec{A}) \ . \tag{A.3}$$

I adopt the notation

$$\partial^\mu = (\frac{\partial}{\partial t}, -\nabla) = \frac{\partial}{\partial x_\mu} \ , \ \partial_\mu = (\frac{\partial}{\partial t}, \nabla) = \frac{\partial}{\partial x^\mu} \tag{A.4}$$

and

$$\Box = \partial_\mu \partial^\mu = \frac{\partial^2}{\partial x^{0^2}} - \nabla^2 \tag{A.5}$$

for the usual differential operators. As a rule, operators will be bold characters, for example Ω.

The derivative of a quantity q with respect to the evolution parameter s may be written in several forms:

$$\frac{\partial q}{\partial s} = \partial_s q = \dot{q} . \tag{A.6}$$

The antisymmetric field strength tensor is defined as

$$F_{\mu\nu} = \partial_\mu A_\nu - \partial_\nu A_\mu , \tag{A.7}$$

and the antisymmetric spin tensor is defined in terms of Dirac gamma matrices such that

$$\sigma_{\mu\nu} = \frac{i}{2} [\gamma_\mu, \gamma_\nu] = i (\gamma_\mu \gamma_\nu - g_{\mu\nu}) . \tag{A.8}$$

For an explicit representation of the gamma matrices, see §10.1 or a text such as Scadron [1979] or Bjorken and Drell [1964].

In several cases a sum will need to be performed over an eigenvalue range that may include discrete and/or continuous values. We denote such a summation by the symbol $\uplus$. It should be obvious from the context whether $\uplus$ denotes a discrete sum, an integral, or a combination.

UNITS

Natural units $\hbar$=c=1 and Einstein's summation convention are used unless otherwise noted. Again, the choice of units should be obvious from the context or from dimensional analysis. As a rule, $\hbar$ and c are made explicit when it is deemed either instructive or clarifying to do so.

§A.2 ACRONYMS

ACRONYM	MEANING
CRQT	Conventional Relativistic Quantum Theory
FTL	Faster-Than-Light
HPR	Horwitz-Piron-Reuse
JARS	Johnson-Aghassi-Roman-Santilli
LQFT	Lagrangian Quantum Field Theory
NRQM	Nonrelativistic Quantum Mechanics
PRCM	Parametrized Relativistic Classical Mechanics
PRQFT	Parametrized Relativistic Quantum Field Theory
PRQM	Parametrized Relativistic Quantum Mechanics
PRQT	Parametrized Relativistic Quantum Theory
RCM	Relativistic Classical Mechanics
UR	Ultrarelativistic

§A.3 PRIMARY REFERENCES

CHAPTER	PRIMARY REFERENCE(S) FOR PART I
1	Fanchi [1981a]; Kyprianidis [1987]; Kuhn [1970]
2	Fanchi and Wilson [1983]; Horwitz and Piron [1973]; Reuse [1979]
3	Feynman [1948]; Fock [1937]; Morette [1951]; Pavsic [1991]
4	Horwitz, Schieve and Piron [1981]
5	Fanchi [1988]

CHAPTER	PRIMARY REFERENCE(S) FOR PART II
6	Aghassi, Roman and Santilli [1970]; Johnson [1970]; Fanchi [1979]
7	Fanchi and Wilson [1983]; Kyprianidis [1987]
8	Fanchi and Collins [1978]; Fanchi and Wilson [1983]
9	Fanchi and Collins [1978]; Fanchi and Wilson [1983]; Kyprianidis [1987]
10	Fanchi [1981]; Davidon [1955]; Horwitz, Piron and Reuse [1975]; Piron and Reuse [1977]; Reuse [1978]
11	Fanchi [1986]

CHAPTER	PRIMARY REFERENCE(S) FOR PART III
12	Arshansky and Horwitz [1989]; Fanchi [1977]; Kim and Noz [1986]; Reuse [1981]
13	Collins and Fanchi [1978]; Fanchi and Collins [1978]
14	Fanchi and Collins [1978]; Grelland [1981]
15-16	Horwitz and Lavie [1982]; Scadron [1979]
17	Fanchi [1987]; Horwitz, Schieve and Piron [1981]; Horwitz, Shashoua and Schieve [1989]

CHAPTER	PRIMARY REFERENCE(S) FOR PARTS IV AND V
18-20	Fanchi [1990]
21	Fanchi [1979]; Kyprianidis [1987]; Pavsic [1991]
22	Fanchi [1979]

BIBLIOGRAPHY

Abramowitz, M., and I.A. Stegun (1972): *Handbook of Mathematical Functions* (Dover, New York).

Aghassi, J.J., P. Roman, and R.M. Santilli (1970): *Phys. Rev.* **D 1**, 2753.

Aitchison, I.J.R., and A.J.G. Hey (1982): *Gauge Theories in Particle Physics* (Adam Hilger, Bristol).

Arfken, G. (1970): *Mathematical Methods for Physicists* (Academic, New York).

Arshansky, R.I., and L.P. Horwitz (1989): *J. Math. Phys.* **30**, 213.

Ballentine, L.E. (1970): *Rev. Mod. Phys.* **42**, 358.

Ballentine, L.E. (1990): *Quantum Mechanics* (Prentice-Hall, Old Tappan, New Jersey).

Bargmann, V. (1947): *Ann. Math.* **48**, 568

Bargmann, V., and E.P. Wigner (1948): *Proc. Natl. Acad. Sci. (US)* **34**, 211.

Barrow, J.D., and F.J. Tipler (1986): *The Anthropic Cosmological Principle* (Oxford University Press, Oxford).

Barut, A.O. (1980): *Electrodynamics and Classical Theory of Fields and Particles* (Dover, New York).

Bhabha, H.J. (1945): *Rev. Mod. Phys.* **17**, 200.

Bilaniuk, O.M.P., V.K. Deshpande, and E.C.G. Sudarshan (1962): *Am. J. Phys.* **30**, 718.

Bilaniuk, O.M.P., and E.C.G. Sudarshan (1969): *Phys. Today* **22**, 43.

Bjorken, J.D., and S.D. Drell (1964): *Relativistic Quantum Mechanics* (McGraw-Hill, New York).

Bjorken, J.D., and S.D. Drell (1965): *Relativistic Quantum Fields* (McGraw-Hill, New York).

Blokhintsev, D.I. (1964): *Quantum Mechanics*, 2nd edn. (Reidel, Dordrecht).

Bogoliubov, N.N., and D.V. Shirkov (1980): *Introduction to the Theory of Quantized Fields*, 3rd edn. (Wiley, New York).

Bohm, D. (1979): *Quantum Theory* (Dover, New York).

Born, M. (1926): *Z. Phys.* **37**, 863.

Borowitz, S. [1967]: *Fundamentals of Quantum Mechanics* (Benjamin, New York).

Brandenberger, R.H. (1985): *Rev. Mod. Phys.* **57**, 1.

Burke, W.L. (1985): *Applied Differential Geometry* (Cambridge University Press, Cambridge).

Carnahan, B., H.A. Luther, and J.O. Wilkes (1969): *Applied Numerical Methods* (Wiley, New York).

Chaichian, M., and N.F. Nelipa (1984): *Introduction to Gauge Field Theories* (Springer, Berlin).

Chandler, D. (1987): *Introduction to Modern Statistical Mechanics* (Oxford University Press, New York).

Chodos, A., V.A. Kostelecky, R. Potting, and E. Gates (1992): *Mod. Phys. Lett.* **A 7**, 467.

Collins, R.E. (1977a): *Found. Phys.* **7**, 475.

Collins, R.E. (1977b): *Lett. Nuovo Cimento* **18**, 581.

Collins, R.E. (1979a): *Lett. Nuovo Cimento* **25**, 473.

Collins, R.E. (1979b): Private communication.

Collins, R.E., and J.R. Fanchi (1978): *Nuovo Cimento* **48A**, 314.

Cook, J.L. (1972): *Aust. J. Phys.* **25**, 141.

Cooke, J.H. (1968): *Phys. Rev.* **166**, 1293.

Cooper, N.C., and G.B. West (1988): *Particle Physics: A Los Alamos Primer* (Cambridge University Press, Cambridge).

Corben, H.C. (1978): *Tachyons, Monopoles, and Related Topics*, ed. by E. Recami (North-Holland, Amsterdam), pp. 31-41.

Cramer, J.G. (1986): *Rev. Mod. Phys.* **58**, 647.

d'Abro, A. (1951): *The Rise of the NEW PHYSICS*, Vol. I (Dover, New York).

Davidon, W.C. (1955): *Phys. Rev.* **97**, 1131.

Dennery, P., and A. Krzywicki (1967): *Mathematics for Physicists* (Harper & Row, New York).

DeRujula, A., H. Georgi, and S. Glashow (1975): *Phys. Rev.* **D 12**, 147.

DeWitt, B.S. (1975): *Phys. Rep.* **19**, 295.

Dirac, P.A.M. (1928): *Proc. R. Soc. London* **A 117**, 610.

Dirac, P.A.M. (1930): *Proc. R. Soc. London* **A 126**, 360.

Dirac, P.A.M. (1945): *Proc. R. Soc. London* **A 183,** 284.

Dirac, P.A.M. (1949): *Rev. Mod. Phys.* **21**, 392.

Enatsu, H. (1963): *Prog. Theor. Phys.* **30**, 526.

Enatsu, H., and S. Kawaguchi (1975): *Nuovo Cimento* **27**, 458.

Evans, A.B. (1991): *Found. Phys.* **20**, 309; and ibid. **21**, 633.

Fanchi, J.R. (1977): Ph.D. Dissertation, University of Houston.

Fanchi, J.R. (1979): *Phys. Rev.* **D 20**, 3108.

Fanchi, J.R. (1981a): *Am. J. Phys.* **49**, 850.

Fanchi, J.R. (1981b): *J. Math. Phys.* **22**, 794.

Fanchi, J.R. (1981c): *Found. Phys.* **11**, 493.

Fanchi, J.R. (1986): *Phys. Rev.* **A 34**, 1677.

Fanchi, J.R. (1987): *Phys. Rev.* **A 35**, 4859.

Fanchi, J.R. (1988): *Phys. Rev.* **A 37**, 3956.

Fanchi, J.R. (1990): *Found. Phys.* **20**, 189.

Fanchi, J.R. (1993): *Found. Phys.* **23**, 487.

Fanchi, J.R., and R.E. Collins (1978): *Found. Phys.* **8**, 851.

Fanchi, J.R., and W.J. Wilson (1983): *Found. Phys.* **13**, 581.

Feinberg, G. (1967): *Phys. Rev.* **159**, 1089.

Feller, W. (1957): *An Introduction to Probability Theory and its Applications*, 2nd edn. (Wiley, New York).

Felten, J.E., and R. Isaacman (1986): *Rev. Mod. Phys.* **58**, 689.

Feshbach, H., and F. Villars (1958): *Rev. Mod. Phys.* **30**, 24.

Feynman, R.P. (1948): *Rev. Mod. Phys.* **20**, 367.

Feynman, R.P. (1949): *Phys. Rev.* **76**, 769.

Feynman, R.P. (1950): *Phys. Rev.* **80**, 440.

Feynman, R.P. (1951): *Phys. Rev.* **84**, 108.

Feynman, R.P. (1972): *Statistical Mechanics* (Benjamin, Reading, Massachusetts).

Feynman, R.P., M. Kislinger and F. Ravndal (1971): *Phys. Rev.* **D 3**, 2706.

Fock, V.A. (1937): *Phys. Z. Sowjetunion* **12**, 404.

Foldy, L.L., and S.A. Wouthuysen (1950): *Phys. Rev.* **78**, 29.

Flanders, H. (1989): *Differential Forms with Applications to the Physical Sciences* (Dover, New York)

Frauenfelder, H., and E.M. Henley (1974): *Subatomic Physics* (Prentice-Hall, Englewood Cliffs, New Jersey).

Gell-Mann, M. (1964): *Phys. Lett.* **8**, 214.

Goldstein, H. (1980): *Classical Mechanics*, 2nd edn. (Addison-Wesley, Reading, Massachusetts).

Gordon, W. (1926): *Z. Phys.* **40**, 117.

Greenberger, D.M. (1963): *Ann. Phys. (NY)* **25**, 290.

Greenberger, D.M. (1970): *J. Math. Phys.* **11**, 2329 and 2341.

Greenberger, D.M. (1974): *J. Math. Phys.* **15**, 395 and 406.

Grelland, H.H. (1981): *Int. J. Quant. Chem.* **XIX**, 873.

Hanni, E., and E. Hugentobler (1978): *Tachyons, Monopoles, and Related Topics*, ed. by E. Recami (North-Holland, Amsterdam), pp. 61-66.

Heisenberg, W. (1976): *Phys. Today*, 32.

Hiscock, W.A., and L. Lindblom (1985): *Phys. Rev.* **D 31**, 725.

Horwitz, L.P., R.I. Arshansky, and A.C. Elitzur (1988): *Found. Phys.* **18**, 1159.

Horwitz, L.P., and Y. Lavie (1982): *Phys. Rev.* **D 26**, 819.

Horwitz, L.P., and C. Piron (1973): *Helv. Phys. Acta* **46**, 316.

Horwitz, L.P., C. Piron, and F. Reuse (1975): *Helv. Phys. Acta* **48**, 546.

Horwitz, L.P., S. Shashoua, and W.C. Schieve (1989): *Physica* **A 161**, 300.

Horwitz, L.P., W.C. Schieve, and C. Piron (1981): *Ann. Phys. (NY)* **137**, 306.

Hostler, L. (1980): *J. Math. Phys.* **21**, 2461.

Hostler, L. (1981): *J. Math. Phys.* **22**, 2307.

Hughes, I.S. (1985): *Elementary Particles*, 2nd edn. (Cambridge University Press, Cambridge).

Iyer, B.R., N. Mukunda, and C.V. Vishveshwara (1989): *Gravitation, Gauge Theories and the Early Universe* (Kluwer, Dordrecht).

Jackson, J.D. (1975): *Classical Electrodynamics* (Wiley, New York).

Jauch, J.M., and F. Rohrlich (1976): *The Theory of Photons and Electrons*, 2nd edn. (Springer, New York).

Johnson, J.E. (1969): *Phys. Rev.* **181**, 1755.

Johnson, J.E., and K.K. Chang (1974): *Phys. Rev.* **10**, 2421.

Jüttner, F. (1911): *Ann. Phys. (Leipzig)* **34**, 856.

Kim, Y.S., and M.E. Noz (1986): *Theory and Applications of the Poincaré Group* (Reidel, Dordrecht).

Kittel, C. (1961): *Elementary Statistical Physics* (Wiley, New York).

Klein, O. (1927): *Z. Phys.* **41**, 407.

Klein, O. (1929): *Z. Phys.* **53**, 157.

Kobzarev, I. Yu., and Yu. I. Manin (1989): *Elementary Particles* (Kluwer, Dordrecht).

Kokkedee, J.J.J. (1969): *The Quark Model* (Benjamin, New York).

Kolb, E.W., and M.S. Turner (1990): *The Early Universe* (Addison-Wesley, Redwood City, California).

Konigl, A. (1980): *Phys. Fluids* **23**, 1083.

Kostelecky, V.A. (1993): "Mass Bounds for Spacelike Neutrinos," *Topics on Quantum Gravity and Beyond*, ed. by F. Mansouri and J.J. Scanio (World Scientific, Singapore).

Kreyszig, E. (1979): *Advanced Engineering Mathematics*, 4th edn. (Wiley, New York).

Kubo, R. (1985): *Nuovo Cimmento* **85 A**, 293.

Kuester, J.L., and J.H. Mize (1973): *Optimization Techniques with Fortran* (McGraw-Hill, New York).

Kuhn, T. (1970): *The Structure of Scientific Revolutions*, 2nd edn. (University of Chicago Press, Chicago).

Kyprianidis, A. (1987): *Phys. Rep.* **155**, 1.

Lakatos, I. (1970): *Criticism and the Growth of Knowledge*, edited by I. Lakatos and A. Musgrave (Cambridge University Press, Cambridge).

Lam, L. (1971): *J. Math. Phys.* **12**, 299.

Land, M.C., and L.P. Horwitz (1991): *Found. Phys.* **21**, 299.

Landau, L.D., and E.M. Lifshitz (1962): *The Classical Theory of Fields* (Addison-Wesley, Reading, Massachusetts).

Landé, A. (1951): *Quantum Mechanics* (Pitman, London).

Landsberg, P.T. (1978): *Thermodynamics and Statistical Mechanics* (Dover, New York).

Lattes, C.M.G., M. Muirhead, G.P.S. Occhialine, and C.F. Powell (1947): *Nature* **159**, 694.

Lebedev, N.N. (1965): *Special Functions and Their Applications* (Prentice-Hall, Englewood Cliffs, New Jersey).

Lichtenberg, D.B. (1970): *Unitary Symmetry and Elementary Particles* (Academic, New York).

Mayants, L. (1973): *Found. Phys.* **3**, 413.

Mayants, L. (1974): *Found. Phys.* **4**, 335.

Mayants, L. (1984): *The Enigma of Probability and Physics* (Kluwer, Dordrecht).

Mayants, L. (1993): Private communication.

Misner, C.W., K.S. Thorne, and J.A. Wheeler (1973): *Gravitation* (Freeman, San Francisco).

Morette, C. (1951): *Phys. Rev.* **81**, 848.

Nambu, Y. (1950): *Prog. Theor. Phys.* **5**, 82.

Newton, T.D., and E.P. Wigner (1949): *Rev. Mod. Phys.* **21**, 400.

Particle Data Group (1992): *Phys. Rev.* **D 45**, Part II.

Pauli, W. (1958): *Theory of Relativity* (Dover, New York).

Pauli, W., and V. Weisskoph (1934): *Helv. Phys. Acta* **7**, 709.

Pavsic, M. (1991a): *Found. Phys.* **21**, 1005.

Pavsic, M. (1991b): *Nuovo Cimento* **A104**, 1337.

Pearle, P.M. (1968): *Phys. Rev.* **168**, 1429.

Peebles, P.J.E. (1992): *Quantum Mechanics* (Princeton University Press, Princeton, New Jersey).

Piron, C., and F. Reuse (1978): *Helv. Phys. Acta* **51**, 146.

Recami, E. (1987): *Found. Phys.* **17**, 239.

Reif, F. (1965): *Fundamentals of Statistical and Thermal Physics* (McGraw-Hill, New York).

Reuse, F. (1978): *Helv. Phys. Acta* **51**, 157.

Reuse, F. (1979): *Found. Phys.* **9**, 865.

Reuse, F. (1980): *Helv. Phys. Acta* **53**, 416 and 552.

Saad, D., L.P. Horwitz, and R.I. Arshansky (1989): *Found. Phys.* **19**, 1125.

Scadron, M.D. (1979): *Advanced Quantum Theory and Its Applications Through Feynman Diagrams* (Springer, New York).

Schiff, L.I. (1968): *Quantum Mechanics* (McGraw-Hill, New York).

Schrödinger, E. (1926a): *Ann. Phys.* **79**, 361.

Schrödinger, E. (1926b): *Ann. Phys.* **81**, 109.

Schweber, S.S. (1962): *An Introduction to Relativistic Quantum Field Theory* (Harper & Row, New York).

Schwinger, J. (1951a): *Phys. Rev.* **82**, 664.

Schwinger, J. (1951b): *Phys. Rev.* **82**, 914.

Steeb, W.-H., and D.E. Miller (1982): *Found. Phys.* **12**, 531.

Stückelberg, E.C.G. (1941a): *Helv. Phys. Acta* **14**, 322.

Stückelberg, E.C.G. (1941b): *Helv. Phys. Acta* **14**, 588.

Stückelberg, E.C.G. (1942): *Helv. Phys. Acta* **14**, 23.

Urban, P. (1970): *Topics in Applied Quantumelectrodynamics* (Springer, Vienna).

Wigner, E.P. (1939): *Ann. Math.* **40**, 149.

Williams, W.S.C. (1991): *Nuclear and Particle Physics* (Clarendon, Oxford).

Wu, T-Y and W-Y Pauchy Hwang (1991): *Relativistic Quantum Mechanics and Quantum Fields* (World Scientific, Singapore).

Zweig, G. (1964): CERN preprint 8409/Th., 412.

INDEX

Fundamental Theories of Physics

Series Editor: Alwyn van der Merwe, *University of Denver, USA*

1. M. Sachs: *General Relativity and Matter.* A Spinor Field Theory from Fermis to Light-Years. With a Foreword by C. Kilmister. 1982 ISBN 90-277-1381-2
2. G.H. Duffey: *A Development of Quantum Mechanics.* Based on Symmetry Considerations. 1985 ISBN 90-277-1587-4
3. S. Diner, D. Fargue, G. Lochak and F. Selleri (eds.): *The Wave-Particle Dualism.* A Tribute to Louis de Broglie on his 90th Birthday. 1984 ISBN 90-277-1664-1
4. E. Prugovečki: *Stochastic Quantum Mechanics and Quantum Spacetime.* A Consistent Unification of Relativity and Quantum Theory based on Stochastic Spaces. 1984; 2nd printing 1986 ISBN 90-277-1617-X
5. D. Hestenes and G. Sobczyk: *Clifford Algebra to Geometric Calculus.* A Unified Language for Mathematics and Physics. 1984 ISBN 90-277-1673-0; Pb (1987) 90-277-2561-6
6. P. Exner: *Open Quantum Systems and Feynman Integrals.* 1985 ISBN 90-277-1678-1
7. L. Mayants: *The Enigma of Probability and Physics.* 1984 ISBN 90-277-1674-9
8. E. Tocaci: *Relativistic Mechanics, Time and Inertia.* Translated from Romanian. Edited and with a Foreword by C.W. Kilmister. 1985 ISBN 90-277-1769-9
9. B. Bertotti, F. de Felice and A. Pascolini (eds.): *General Relativity and Gravitation.* Proceedings of the 10th International Conference (Padova, Italy, 1983). 1984 ISBN 90-277-1819-9
10. G. Tarozzi and A. van der Merwe (eds.): *Open Questions in Quantum Physics.* 1985 ISBN 90-277-1853-9
11. J.V. Narlikar and T. Padmanabhan: *Gravity, Gauge Theories and Quantum Cosmology.* 1986 ISBN 90-277-1948-9
12. G.S. Asanov: *Finsler Geometry, Relativity and Gauge Theories.* 1985 ISBN 90-277-1960-8
13. K. Namsrai: *Nonlocal Quantum Field Theory and Stochastic Quantum Mechanics.* 1986 ISBN 90-277-2001-0
14. C. Ray Smith and W.T. Grandy, Jr. (eds.): *Maximum-Entropy and Bayesian Methods in Inverse Problems.* Proceedings of the 1st and 2nd International Workshop (Laramie, Wyoming, USA). 1985 ISBN 90-277-2074-6
15. D. Hestenes: *New Foundations for Classical Mechanics.* 1986 ISBN 90-277-2090-8; Pb (1987) 90-277-2526-8
16. S.J. Prokhovnik: *Light in Einstein's Universe.* The Role of Energy in Cosmology and Relativity. 1985 ISBN 90-277-2093-2
17. Y.S. Kim and M.E. Noz: *Theory and Applications of the Poincaré Group.* 1986 ISBN 90-277-2141-6
18. M. Sachs: *Quantum Mechanics from General Relativity.* An Approximation for a Theory of Inertia. 1986 ISBN 90-277-2247-1
19. W.T. Grandy, Jr.: *Foundations of Statistical Mechanics.*
    Vol. I: *Equilibrium Theory.* 1987 ISBN 90-277-2489-X
20. H.-H von Borzeszkowski and H.-J. Treder: *The Meaning of Quantum Gravity.* 1988 ISBN 90-277-2518-7
21. C. Ray Smith and G.J. Erickson (eds.): *Maximum-Entropy and Bayesian Spectral Analysis and Estimation Problems.* Proceedings of the 3rd International Workshop (Laramie, Wyoming, USA, 1983). 1987 ISBN 90-277-2579-9

22. A.O. Barut and A. van der Merwe (eds.): *Selected Scientific Papers of Alfred Landé. [1888-1975]*. 1988 ISBN 90-277-2594-2
23. W.T. Grandy, Jr.: *Foundations of Statistical Mechanics.* Vol. II: *Nonequilibrium Phenomena.* 1988 ISBN 90-277-2649-3
24. E.I. Bitsakis and C.A. Nicolaides (eds.): *The Concept of Probability.* Proceedings of the Delphi Conference (Delphi, Greece, 1987). 1989 ISBN 90-277-2679-5
25. A. van der Merwe, F. Selleri and G. Tarozzi (eds.): *Microphysical Reality and Quantum Formalism, Vol. 1.* Proceedings of the International Conference (Urbino, Italy, 1985). 1988 ISBN 90-277-2683-3
26. A. van der Merwe, F. Selleri and G. Tarozzi (eds.): *Microphysical Reality and Quantum Formalism, Vol. 2.* Proceedings of the International Conference (Urbino, Italy, 1985). 1988 ISBN 90-277-2684-1
27. I.D. Novikov and V.P. Frolov: *Physics of Black Holes.* 1989 ISBN 90-277-2685-X
28. G. Tarozzi and A. van der Merwe (eds.): *The Nature of Quantum Paradoxes.* Italian Studies in the Foundations and Philosophy of Modern Physics. 1988 ISBN 90-277-2703-1
29. B.R. Iyer, N. Mukunda and C.V. Vishveshwara (eds.): *Gravitation, Gauge Theories and the Early Universe.* 1989 ISBN 90-277-2710-4
30. H. Mark and L. Wood (eds.): *Energy in Physics, War and Peace.* A Festschrift celebrating Edward Teller's 80th Birthday. 1988 ISBN 90-277-2775-9
31. G.J. Erickson and C.R. Smith (eds.): *Maximum-Entropy and Bayesian Methods in Science and Engineering.* Vol. I: *Foundations.* 1988 ISBN 90-277-2793-7
32. G.J. Erickson and C.R. Smith (eds.): *Maximum-Entropy and Bayesian Methods in Science and Engineering.* Vol. II: *Applications.* 1988 ISBN 90-277-2794-5
33. M.E. Noz and Y.S. Kim (eds.): *Special Relativity and Quantum Theory.* A Collection of Papers on the Poincaré Group. 1988 ISBN 90-277-2799-6
34. I.Yu. Kobzarev and Yu.I. Manin: *Elementary Particles. Mathematics, Physics and Philosophy.* 1989 ISBN 0-7923-0098-X
35. F. Selleri: *Quantum Paradoxes and Physical Reality.* 1990 ISBN 0-7923-0253-2
36. J. Skilling (ed.): *Maximum-Entropy and Bayesian Methods.* Proceedings of the 8th International Workshop (Cambridge, UK, 1988). 1989 ISBN 0-7923-0224-9
37. M. Kafatos (ed.): *Bell's Theorem, Quantum Theory and Conceptions of the Universe.* 1989 ISBN 0-7923-0496-9
38. Yu.A. Izyumov and V.N. Syromyatnikov: *Phase Transitions and Crystal Symmetry.* 1990 ISBN 0-7923-0542-6
39. P.F. Fougère (ed.): *Maximum-Entropy and Bayesian Methods.* Proceedings of the 9th International Workshop (Dartmouth, Massachusetts, USA, 1989). 1990 ISBN 0-7923-0928-6
40. L. de Broglie: *Heisenberg's Uncertainties and the Probabilistic Interpretation of Wave Mechanics.* With Critical Notes of the Author. 1990 ISBN 0-7923-0929-4
41. W.T. Grandy, Jr.: *Relativistic Quantum Mechanics of Leptons and Fields.* 1991 ISBN 0-7923-1049-7
42. Yu.L. Klimontovich: *Turbulent Motion and the Structure of Chaos.* A New Approach to the Statistical Theory of Open Systems. 1991 ISBN 0-7923-1114-0

Fundamental Theories of Physics

43. W.T. Grandy, Jr. and L.H. Schick (eds.): *Maximum-Entropy and Bayesian Methods.* Proceedings of the 10th International Workshop (Laramie, Wyoming, USA, 1990). 1991 ISBN 0-7923-1140-X
44. P.Pták and S. Pulmannová: *Orthomodular Structures as Quantum Logics.* Intrinsic Properties, State Space and Probabilistic Topics. 1991 ISBN 0-7923-1207-4
45. D. Hestenes and A. Weingartshofer (eds.): *The Electron.* New Theory and Experiment. 1991 ISBN 0-7923-1356-9
46. P.P.J.M. Schram: *Kinetic Theory of Gases and Plasmas.* 1991 ISBN 0-7923-1392-5
47. A. Micali, R. Boudet and J. Helmstetter (eds.): *Clifford Algebras and their Applications in Mathematical Physics.* 1992 ISBN 0-7923-1623-1
48. E. Prugovečki: *Quantum Geometry.* A Framework for Quantum General Relativity. 1992 ISBN 0-7923-1640-1
49. M.H. Mac Gregor: *The Enigmatic Electron.* 1992 ISBN 0-7923-1982-6
50. C.R. Smith, G.J. Erickson and P.O. Neudorfer (eds.): *Maximum Entropy and Bayesian Methods.* Proceedings of the 11th International Workshop (Seattle, 1991). 1993 ISBN 0-7923-2031-X
51. D.J. Hoekzema: *The Quantum Labyrinth.* 1993 ISBN 0-7923-2066-2
52. Z. Oziewicz, B. Jancewicz and A. Borowiec (eds.): *Spinors, Twistors, Clifford Algebras and Quantum Deformations.* Proceedings of the Second Max Born Symposium (Wrocław, Poland, 1992). 1993 ISBN 0-7923-2251-7
53. A. Mohammed-Djafari and G. Demoment (eds.): *Maximum Entropy and Bayesian Methods.* Proceedings of the 12th International Workshop (Paris, France, 1992). ISBN 0-7923-2280-0
54. M. Riesz: *Clifford Numbers and Spinors* with Riesz′ Private Lectures to E. Folke Bolinder and a Historical Review by Pertti Lounesto. E.F. Bolinder and P. Lounesto (eds.). 1993 ISBN 0-7923-2299-1
55. R. Delanghe and F. Brackx (eds.): *Clifford Algebras and Its Applications in Mathematical Phisycs.* 1993 ISBN 0-7923-2347-5
56. J.R. Fanchi: *Parametrized Relativistic Quantum Theory.* 1993 ISBN 0-7923-2376-9

KLUWER ACADEMIC PUBLISHERS – DORDRECHT / BOSTON / LONDON

9 780792 323761

GPSR Compliance

The European Union's (EU) General Product Safety Regulation (GPSR) is a set of rules that requires consumer products to be safe and our obligations to ensure this.

If you have any concerns about our products, you can contact us on ProductSafety@springernature.com

In case Publisher is established outside the EU, the EU authorized representative is:

Springer Nature Customer Service Center GmbH
Europaplatz 3
69115 Heidelberg, Germany

Batch number: 10809439

Printed by Printforce, the Netherlands